Troubleshooting and Repairing Color Television Systems

Troubleshooting and Repairing Color Television Systems

Robert L. Goodman

McGraw-Hill

New York San Francisco Washington, D.C. Auckland Bogotá
Caracas Lisbon London Madrid Mexico City Milan
Montreal New Delhi San Juan Singapore
Sydney Tokyo Toronto

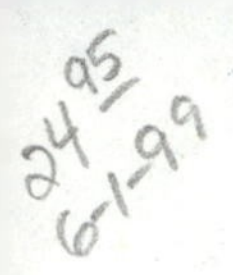

Library of Congress Cataloging-in-Publication Data

Goodman, Robert L.
Troubleshooting and repairing color television systems / Robert L. Goodman
p. cm.
Includes index.
ISBN 0-07-024571-1 (hc).—ISBN 0-07-24569-X (pbk.)
1. Color television—Repairing. I. Title.
TK6670.G5847 1997
621.388'87—dc21 96-53482
CIP

McGraw-Hill

A Division of The **McGraw-Hill** *Companies*

1 2 3 4 5 6 7 8 9 0 1FGR/1FGR 9 0 2 1 0 9 8 7

ISBN 0-07-024571-1 (HC) 0-07-024569-X (PBK)

The sponsoring editor for this book was Scott Grillo, the editing supervisor was Ruth W. Mannino, and the production supervisor was Pamela A. Pelton. It was set in Century Schoolbook by Joanne Morbit of McGraw-Hill's Hightstown desktop composition unit.

Printed and bound by Quebecor/Fairfield.

McGraw-Hill books are available at special quantity discounts to use as premiums and sales promotions, or for use in corporate training programs. For more information, please write to the Director of Special Sales, McGraw-Hill, 11 West 19th Street, New York, NY 10011. Or contact your local bookstore.

This book is printed on recycled, acid-free paper containing a minimum of 50% recycled, de-inked fiber.

Contents

Preface

Electronics products are continually becoming more complex and complicated to understand and service. This color television troubleshooting book will help the electronic technician to understand and service these new hi-tech consumer products. These chapters will show you some advanced servicing procedures with selected state-of-the-art electronic test equipment.

The first chapter has details on new test equipment used for digital color television troubleshooting. The second chapter covers color television history and basic color television circuit operations. It also includes an explanation of the new color television transmission video techniques, including NTSC and HDTV (high definition television), plus digital video concepts.

Chapter 3 covers conventional television power supplies. It includes information on power supply circuit operation, and troubleshooting tips and problems that have been known to crop up. Chapter 4 details the operation of television switch mode power supplies. Start-up and shut-down and other problems found with these power systems are discussed.

In Chap. 5 various types of color television horizontal and vertical sweep circuits are investigated, as well as convergence, sandcastle, and pincusion circuit operation and adjustments. Here it is shown how to use the Sencore TVA92 TV Video Analyzer for television sweep and HV troubleshooting.

Chroma, video, and "S" video circuit operation are discussed in Chap. 6. Chapter 6 also includes more details on using the Sencore TVA92 instrument for chroma circuit diagnosis. It concludes with solving and repairing color picture tube problems with the Sencore CR70 CRT tube checker and beam builder. Information on electronic turners and i.f. systems makes up Chap. 7.

Chapter 8 reviews analog color television signals and sync and AGC circuit operation. In Chap. 9 the "wonderful world" of digital video concepts is illuminated. How-to information on using the Sencore DSA309 Digital Analyzer and on using graphical multimeter/oscilloscope test instruments is found here.

Large-screen projection color television receivers are the subject of Chap. 10. Special features of these big-screen television sets are covered, along with troubleshooting information.

Detailed television audio and stereo information plus service tips can be found in Chap. 11. MTS signals, surround sound, audio performance tests, and stereo separation checks complete this chapter.

Television remote control operation and troubleshooting are the main topics of Chap. 12. This includes various television remote control schemes, remote control logic systems, and microprocessor troubleshooting, IR-generated remote control signals, and various remote control digital encoder functions and modes.

Acknowledgments

In this modern age of complicated digital and high-definition color television devices, you need to understand these systems and know about the latest sophisticated test instruments. This book will help you to upgrade into the latest color television servicing and troubleshooting technical knowledge.

A book of this caliber could not have been completed without the cooperation of many color television manufacturers and test equipment companies. I received a full measure of assistance from many of them.

Thanks are due to Wayne Brett, Zenith Electronics; Scott J. Stevens, Thomson consumer electronics; Beverly A. Summers, Fluke Instruments; Heather L. Wyse and Monique Hayward, Tektronix, Inc.; and George Gonos, Larry Schabel, and Don Multerer, Sencore, Inc. A special thanks goes to Frankie Rundell, who was of considerable help in developing this book.

Bob Goodman

Troubleshooting and Repairing Color Television Systems

1
CHAPTER

Color television system overview and test equipment requirements

This chapter begins with information on how the color television signal is made up and was developed and approved by the Federal Communications Commission (FCC). Next, some information on color television standards and specifications is given. Then a brief overview is presented of a simple color television transmitter and its operation. This is then followed by a block diagram overview of a typical color receiver and its various circuit operations.

The last portion of this chapter will be devoted to information about and specifications of some test equipment that you will need for fast and effective color television troubleshooting. You also will find other types of test equipment being used in other chapters of this book. In addition, the chapter discusses some digital multimeters (DMM), oscilloscopes, and graphical multimeters that are now new on the market.

The chapter concludes with information on transient-level overvoltages, safety precautions with power supply systems, and equipment ratings by the International Electrotechnical Commission (IEC).

Color television signal

The signal from a transmitter is an electrical form of energy that enters free space in some form of electromagnetic waves. No one at this time has an explanation of what makes up an electromagnetic form of energy. We do know that electromagnetic waves travel at the speed of light and that for these transmitted waves to carry intelligence, they must be varied or modulated in some way.

Color television transmits and receives visual images in full color. The U.S. color television system must be compatible with black-and-white television standards.

Compatibility occurs when the system produces programs in *color* on color television sets and in *black and white* on monochrome receivers. And of course, color television sets will receive black-and-white pictures when they are being transmitted.

The color television signal contains two main components, *luminance* (black-and-white or brightness) information and *chrominance* (color) information, which is added to the luminance signal in color receiver circuits to produce full-color pictures.

The transmitted color signal contains all the information required to reproduce a full-color scene accurately. The U.S. standard frequency channel width for a transmitted color television picture is 6 MHz.

The color signal contains not only picture detail information but also equalizing pulses, horizontal sync pulses, vertical sync pulses, 3.58-MHz color burst pulses, blanking pulses, vertical-interval test signals (VITS), and vertical-interval reference (VIR) test pulses. The horizontal blanking pulse is used to turn off the electron scan beam at the end of each scan line. Vertical blanking pulses are used to blank out the top and bottom of the picture so that the viewer will not see any of the transmitted pulses used for picture control and testing.

The color (chroma) is phase- and amplitude-encoded relative to the 3.58-MHz color burst and is superimposed on the black-and-white signal level. This video information is used to control the three electron beams (R, G, and B) that scan from left to right with 525 horizontal lines to a frame.

Signal standards

The signals from the color television transmitter are reproduced on the screen of the receiver to closely match those of the original scene. The black-and-white and color signal has a frequency-modulated (FM) sound carrier and an amplitude-modulated (AM) video carrier with a channel bandwidth of 6 MHz.

The portion of the black-and-white signal that carries the video picture information is called the *carrier amplitude,* which is the brightness or darkness of the original picture information modulation. Now, a portion of the color video signal that carries picture information is made up of a composite of color information and amplitude variations. A unique feature is the 3.58-MHz color sync burst, which is located right in back of the horizontal-sync pulse. This color sync pulse is often referred to as "sitting on the back porch" of the horizontal sync pulse. Every horizontal scan line of video color information contains the picture information, horizontal blanking, sync pulses, and a color burst of at least 8 cycles.

The picture information in a color television signal is then obtained from the red, green, and blue video signals that the camera generates as it scans the scene to be transmitted. The luminance (Y) and chrominance (C) signals derived from the basic red, green, and blue video color signals have all the picture information required for transmission. They are then combined into a single signal by algebraic addition. The final color product contains a chrominance signal that provides the color variations for the picture, and the luminance signal provides the variations in intensity or the brightness of the colors.

There are three different color television broadcast standards in the world, and they are not compatible with one another. In the United States, the National Television System Committee (NTSC) standard was devised by the National Television System Committee in 1953 and approved by the FCC. The NTSC system is now used in the United States, Canada, Japan, and other countries. Europe has the phase-alternating line (PAL) system, and the *séquential à mémoire* (SECAM) standard is used in France and Russia. Some time around the year 2000, all television video signal transmissions will be changing over to a compressed digital video format.

Color transmitters

Figure 1-1 is a simple block diagram of a color television transmitter. The red, green, and blue camera signals are fed into the top left of the drawing as inputs ER, EG, and EB. The EY, EI, and EQ signals are obtained from the matrix section with a phase-inverter network. Note the delay blocks in the EY and EQ signal paths.

The EI and EQ signals are fed into the two double-balanced modulators. The subcarrier signal for the EI modulator is developed from the 3.58-MHz burst reference via a 57-degree phase-shifting network. This adjusts the phase of the EI axis correctly.

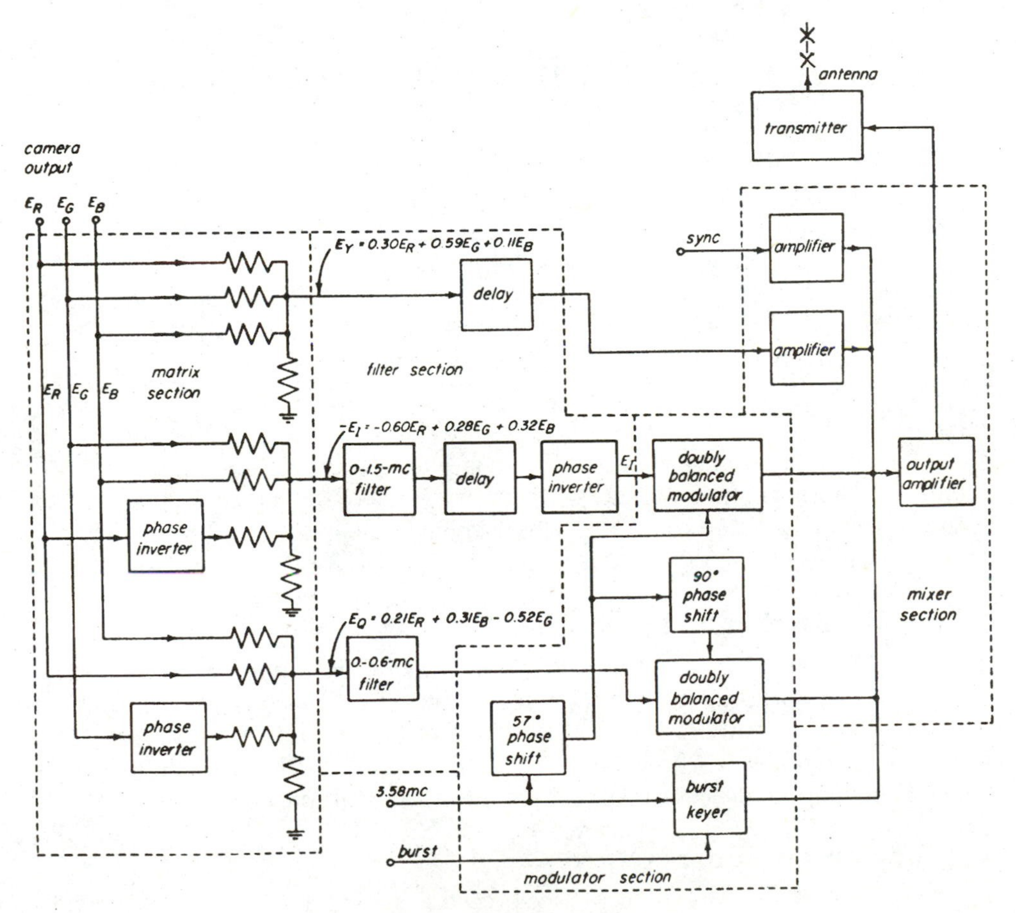

1-1 Simplified diagram of a color television transmitter.

An additional 90-degree phase-shifting network adjusts the phase of the subcarrier signal applied to the EQ modulator. The outputs of the modulators are added together and set to the mixer section, where the EY and chrominance signals combine to form the composite signal. Sync and burst signals are added in the mixer section. Burst signals, consisting of at least 8 cycles for the reference signal, are obtained from the burst keyer stage. The color bursts are timed to follow the horizontal sync pulse.

The complete color signal now goes to a buffer stage, a vestigial filter, and then a final power amplifier stage to bring the transmitter up to its designated radiofrequency (rf) power output. A feedline from this power amplifier stage now goes up the television tower to a gain antenna.

Vestigial-sideband transmission is a type of amplitude modulation (AM) in which one of the side bands has almost been eliminated. The carrier wave and the other sideband are not affected. Vestigial-sideband transmission differs from single-sideband transmission in that the carrier is not suppressed.

In television signal transmitting, vestigial-sideband transmission creates very efficient channel usage. This type of system also may be used for other types of radio transmitters.

Receiver block diagram and operation

Head end or tuner

The color television signal comes into the receiver "head end" or tuner via an antenna, cable system, or digital satellite system (DSS) dish. In Fig. 1-2, you will note that the electronic tuner has an rf amplifier, an oscillator, and mixer stages that convert the signal to a 45.75-MHz signal that is fed to the video intermediate-frequency (i.f.) stages or strip. Most modern receivers are remote channel controlled and have automatic frequency control (AFC) to lock in the station.

i.f. stages and video detector

After amplification and processing in the i.f. stages, the video and audio signals are found at the output of their detectors. The audio has now been detected in the 4.5-MHz audio i.f. section, and stereo audio is fed to the right and left audio power amplifiers and then onto the speakers. Some television sets also have MTS/SAP and DBX decoders.

Chroma amplifier and processing circuits

From the video detector, this signal goes to some video processing stages. You will find a comb filter, used to produce sharper pictures, and a delay line to compensate for the delay introduced by the narrow-band bandpass filter stages in the chrominance amplifiers. Chrominance (chroma) and video signals, along with vertical and horizontal sync signals, are also obtained from the video stages.

Chroma and luminance stages

The composite video signal from the detector is fed to the first chroma amplifier as well as to the luma (black-and-white) amplifier in the luma/chroma block. The 3.58-MHz oscillator and subcarrier generator are used to extract the chrominance

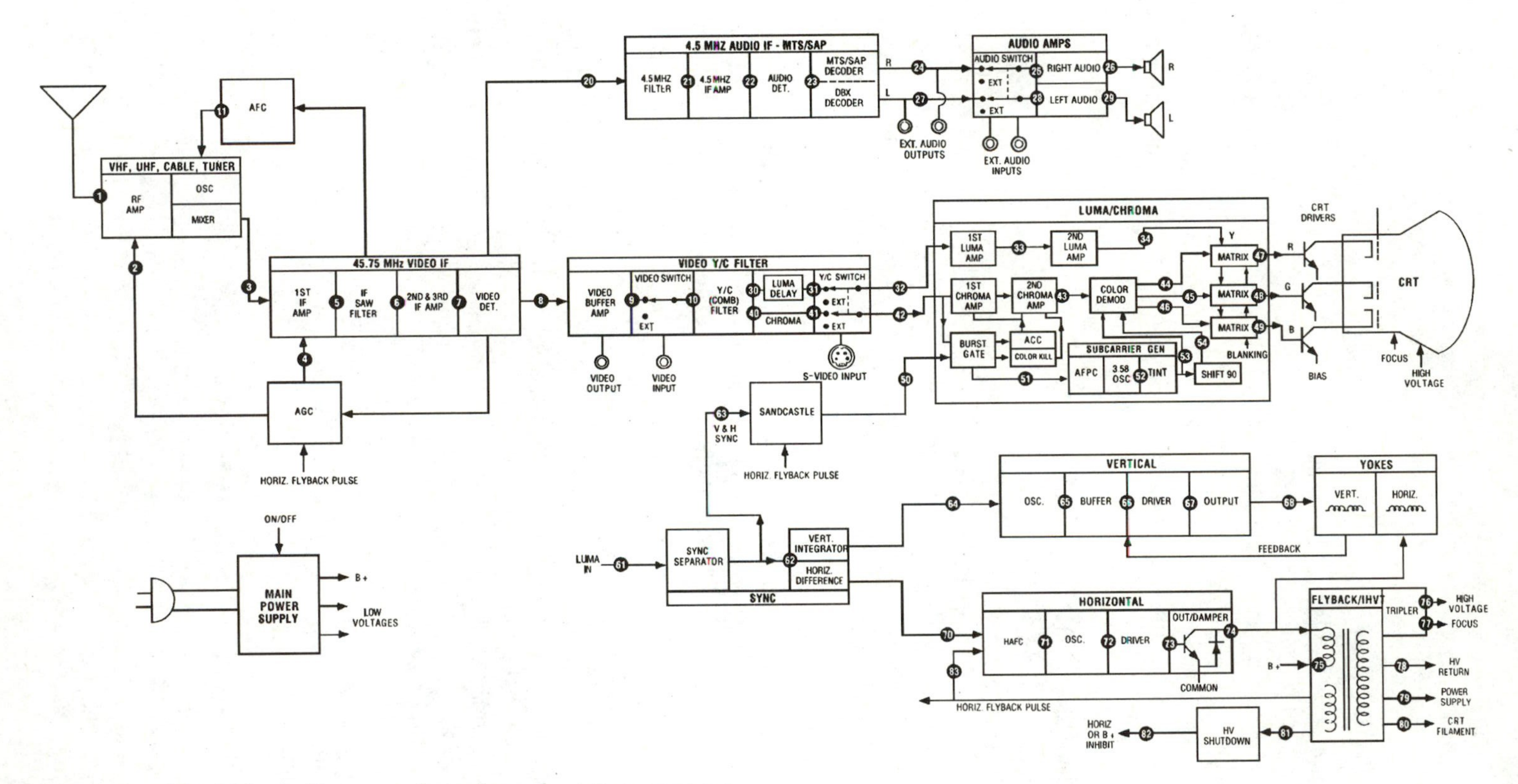

1-2 Simplified block diagram of a color television receiver.

signals. This is accomplished in the color demodulator stage, and these signals go to the matrix circuits and then onto the cathodes of the color cathode-ray tube (CRT). This block also contains blanking circuits, a burst gate, and color killer stages.

Color killer stage

The color killer circuit has a couple of basic functions. Its performance during black-and-white picture transmission is to keep high-frequency signals or noise from being amplified by the chroma amplifiers. This keeps the color snow or confetti from being seen on the picture. It also keeps the viewer from seeing color rainbows around fine details and edges of a black-and-white picture. This stage is also used to kill the color signal during weak signal conditions. Thus the killer circuit must know the difference between the color burst signal and noise.

Sandcastle circuit

In most modern color television sets you will find a sandcastle integrated circuit (IC). The sandcastle is a special signal used by design engineers to inject three mixed signals that go into one IC pin. These three signals are mixed outside the IC. The IC separates these three signals and uses them for various internal functions. The three signals are the flyback pulses, a delayed horizontal sync pulse, and a vertical pulse.

After separation inside the IC, the flyback pulse provides horizontal blanking for the output signals, while the delayed sync pulse separates the color burst from the "back porch" of the horizontal sync, and the vertical pulse provides vertical blanking. If you look at the pulse on an oscilloscope whose image is generated from this chip, the image looks like a sandcastle—hence the name of this circuit. If one of the input pulses is missing or the IC goes bad, the screen is blanked out so that you have no picture symptom to go by.

Sync circuits

The basic function of the sync circuits is to separate the horizontal and vertical sync pulses. These separated pulses are then fed to the horizontal and vertical sweep stages in order to control and lock-in the color picture. The circuit needs good noise immunity in order to maintain good, stable vertical and horizontal sync lock-in.

Some color sets have the sync and automatic gain control (AGC) circuits combined. Normally, the AGC circuit develops a bias in proportion to the sync pulse peak-to-peak level, which is then used as a direct current (dc) voltage to control receiver gain. Keyed AGC circuits are used because they provide better noise immunity.

Vertical (sweep) deflection section

The vertical sweep oscillator stage receives a sync pulse from the vertical integrator stage. This pulse keeps the vertical oscillator running at the vertical scanning rate. Some sets have digital countdown and divider circuits to perform this task. The oscillator feeds the buffer and driver stages. The output from the vertical power amplifier stage is then applied to the vertical winding of the deflection yoke located around the neck of the CRT. A pulse from the vertical output stage is used for CRT blanking. Some pulses also may be used for convergence corrections.

Horizontal (sweep) deflection stages

Older-model color sets have a horizontal circuit that consists of a sawtooth generator that drives the horizontal sweep and high-voltage (HV) transformer. This circuit is controlled by an AFC system that compares the frequency of the oscillator with the sync pulse coming from the station transmitter and then produces a correction dc voltage for an oscillator frequency drift. Deflection current for the horizontal yoke coils along with CRT high-voltage and focus voltage, and other pulses are generated by the sweep flyback transformer. This stage needs very good voltage regulation control in order to produce a good color picture.

Modern sets use a digital countdown-divider system to generate and control pulses to drive the horizontal sweep stage. The horizontal output stage in these sets is usually of a pulse-width design that not only sweeps the electron beam across the picture tube but also develops other dc voltages to operate other circuits in the color chassis. This eliminates the heavy weight and cost of a power transformer and also improves the efficiency of the alternating current (ac) power the set uses. A safety high-voltage shutdown circuit is also used.

Power supply section

As with any electronic device, the power supply is the heart of the device and makes all the other systems operate. If the television set is dead or not working properly, look at and check out the power supply section first. This could be a simple thing, such as an off/on switch that is defective, a blown fuse, or a tripped circuit breaker. And do not overlook the possibility that the ac power cord is not plugged into the wall socket or that the ac circuit breaker to the wall socket has tripped off.

This should now give you an overview of how all these blocks and circuits in a color television work together and what could go wrong.

Test equipment and safety information

This last portion of the chapter is devoted to information about and specifications of some of the test equipment needed for fast and effective color television troubleshooting. Also discussed is other types of test equipment used in other chapters of this book. In addition, some digital multimeters (DMM), oscilloscopes, and graphical multimeters that are now new on the market will be discussed.

Digital multimeter specifications

The growth in the deployment of electronic equipment throughout commercial and industrial installations has effectively increased harmonic distortion in power systems. In turn, harmonic distortion has reinforced the need to keep a critical eye on the overall health of the power system within a facility.

One result of increased harmonics is that overall power quality continues to drop. In turn, equipment can malfunction and fail, and "low power factor" can lead to higher power costs. This chain of events raises the issue of measurement tools. How appropriate and how effective are today's measurement tools at addressing problems of distortion and overall power quality?

Average-responding instruments

For years ac meters have used a technique called *average-responding rms indicating* to display the root-mean-squared (rms) value, or effective heating value, of the current being measured. The rms is the mathematical calculation on an ac waveform that returns a value equivalent to the dc level that would have the same heating value as the ac current being measured. The average-responding method provides a practical and economical way to measure the rms of a sine wave without using expensive and delicate laboratory equipment. Basically, the average value of the rectified signal is measured and then multiplied by a constant that results in the rms value, as shown in Fig. 1-3.

However, there is one important qualifier: The waveform being measured must be a perfect sine wave. Unfortunately, this cannot be guaranteed today. On the contrary, distortion in today's typical power systems guarantees errors in voltage and current readings as high as 20 to 50 percent when measured with an average-responding meter.

Signal distortion is common in heavy industry, commercial installations, and other facilities with electronically controlled equipment. Every facility is different, however. Figure 1-4 shows the current waveform at the line cord of a typical monitor for a personal computer (PC) in an office. Instead of a perfectly shaped sine wave, the monitor is drawing power from the line in an inconsistent, or *nonlinear,* fashion. The result is distortion that causes harmonics in the power system. As

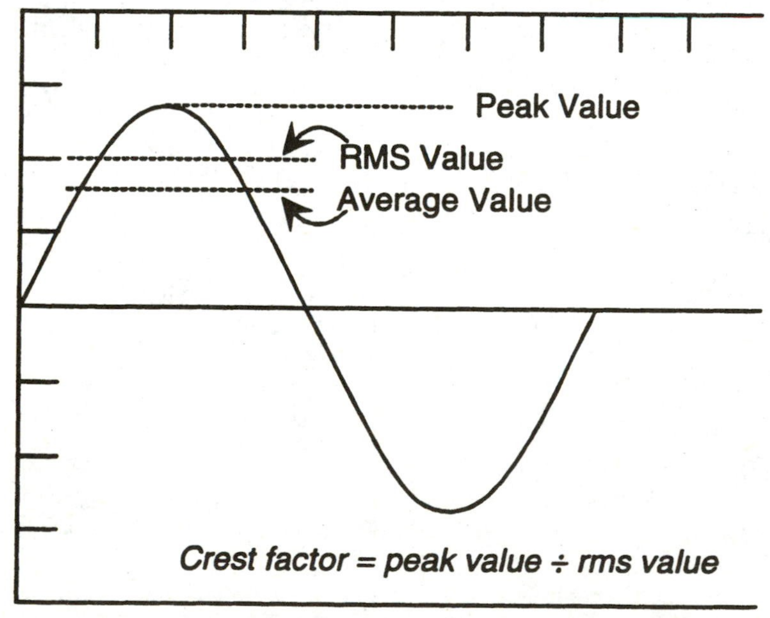

1-3 Average-responding method of measuring ac voltage.

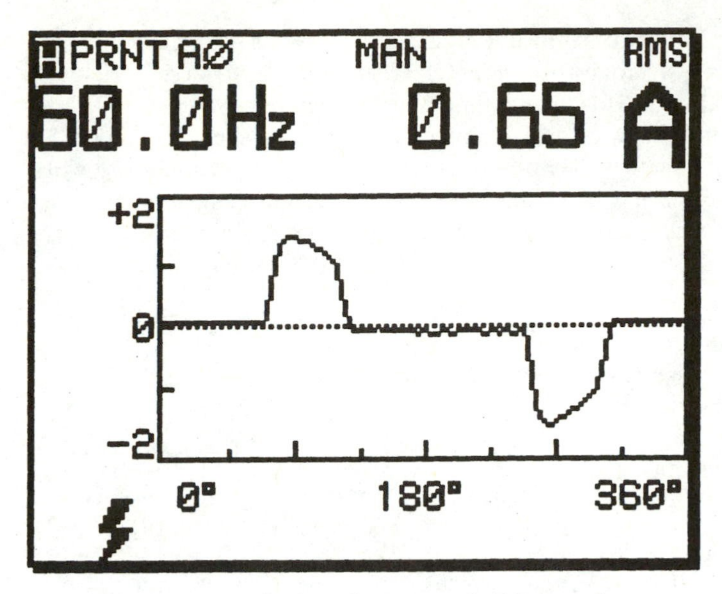

1-4 Current waveform of a typical PC monitor captured by a power harmonics analyzer. (*Courtesy of Fluke Corporation.*)

shown in Fig. 1-4, a reading by a Fluke Model 41 meter indicates a true-rms current of 0.65 A. The same current, measured with an average-responding meter, was 0.47 A—nearly a 28 percent error!

If the magnitude of the nonlinear loads is minor compared with the magnitude of the traditional linear loads (loads that draw power in a linear fashion relative to the voltage waveform), the problems are minimal. However, as more and more nonlinear loads are added (more PCs, fax machines, copy machines, lighting fixtures with electronic ballasts, etc.), the nonlinear loads begin to predominate. The results are varied, depending on the installation and the loads. Typical symptoms include overheating transformers and distribution panels, as well as breakers that trip even when an average-responding meter indicates that the current is within the acceptable range.

Harmonic distortion is certainly not new; it has been a fact of life in power systems ever since the introduction of solid-state components to control and regulate power. Today, however, with the balance between traditional electrical equipment and electronic equipment shifting toward the latter, the full consequences of harmonics are starting to come to light.

Plant engineers typically use a "walking tour" of a plant to get a first visual reading of impending problems. On the walking tour of the facility, a plant engineer would notice an early indication of harmonics: heat. An electrical panel and a transformer are hot to the touch. Another panel is emitting a humming sound. An experienced plant engineer knows that these are clues to the presence of harmonics. At this point, the engineer takes readings of voltage, current, and frequency with a hand-held digital multimeter (DMM). However, the readings obtained will be incorrect if an average-responding meter is used. Let us see why.

True rms value versus rms from average-responding instruments. As explained earlier, average-responding instruments give correct readings for pure sine-wave signals only. However, *true-rms instruments*—such as the Fluke Model 76 digital multimeter—give correct readings for signals of all shapes. An average-responding instrument will produce unpredictable readings—in the worst case, as much as 50 percent low. Waveforms containing harmonics cannot be measured accurately by average-responding instruments, which still make up the vast majority of low-cost electrical measurement tools (Table 1-1 and Fig. 1-5).

Table 1-1. Comparing the performance of average-responding and true-rms meters

		Response to		
Type of meter	**Measuring circuit**	**Sine wave**	**Square wave**	**Distorted wave**
Average-responding	Multiplies rectified average by 1.1	Correct	10% high	Up to 50% low
True rms	rms-calculating converter calculates heating value	Correct	Correct	Correct

The performance of any ac meter is limited by its bandwidth and crest factor ratings. *Bandwidth* refers to the range of frequencies within which the meter can make accurate measurements. Distorted waveforms contain multiple frequencies that are higher than the fundamental frequency (typically 50, 60, or 400 Hz in ac power systems). With a bandwidth of at least 1 kHz, any commercial-quality meter can handle the range of frequencies found in most commercial and industrial power systems. However, if the application is in audio or communications systems, consider a meter with a higher bandwidth.

Crest factor is the ratio of the peak value of a signal to its rms value. Refer again to Fig. 1-3. The crest factor of a sine wave is 1.414; therefore, in a system with pure sine-wave voltages and currents, any meter with a crest factor rating above 1.414 would be up to the task. However, systems with harmonics have higher-peaking signals, which increase the crest factor of those signals. This is why you should look for a meter with a crest factor rating of 3 or higher at full scale.

When evaluating true-rms instruments, be sure to consider other specifications. In power system environments, you will need an instrument that is designed to

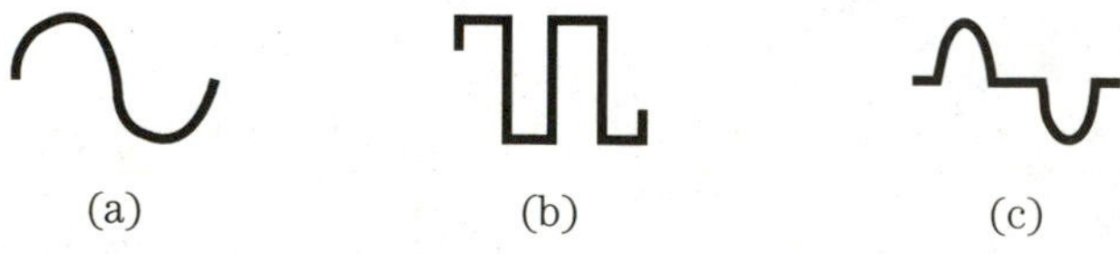

1-5 (*a*) Sine wave; (*b*) square wave; (*c*) distorted wave.

1-6 Digital multimeter. (*Courtesy of Fluke Corporation.*)

handle at least 600-V ac and dc signals. Crest factor specifications should be at least 3 at full scale. And be sure to demand an instrument that measures continuity, capacitance, and frequency. All these capabilities should be available in a meter in the $200 price range. The Fluke Model 76 shown in Fig. 1-6 is a digital multimeter that meets all these qualifications.

Fluke Model 76 DMM

A new hand-held digital multimeter from Fluke, the Model 76 Digital Multimeter (DMM), is a true-rms meter designed to IEC-1010 category III standards and priced to set a new standard in medium-priced multimeters. At $199 (U.S. list price), the Fluke Model 76 is the first multimeter in its price category to meet the requirements of IEC-1010-1 for use in overvoltage category III locations. As a category III instrument, the Model 76 is designed to withstand up to 600 V ac or dc continuously between any terminal and earth ground, with impulse-withstand protection up to 6000 V.

To ensure accuracy in the presence of electromagnetic interference, the Model 76 meets the generic standard for the CE-mark electromagnetic compatibility (EMC) requirements. The Model 76 is certified by or has pending certification from the following agency approvals: Underwriters Laboratories (UL), Canadian Standards Association (CSA), CE, and TÜV.

At less than $200, the Model 76 offers guaranteed true-rms response. This means that the Model 76, like higher-priced multimeters, guarantees accurate response to the growing number of distorted (nonsinusoidal) signals found in today's power and electronic environments. (By contrast, average-responding multimeters are accurate only when measuring undistorted sine waves.) The Model 76 has a rated crest factor of 3 at full scale. Crest factor indicates a unit's ability to accurately measure severely distorted signals, such as those with a high harmonic content.

The combination of category III certification and true-rms response makes the Model 76 ideal for use in a variety of environments, including field service, electrical facilities maintenance, and electronic original equipment manufacturer (OEM) design.

The Model 76 is a 3½-digit, 4000-count multimeter that offers many of the capabilities of premium DMMs. Measurement modes include true-rms ac voltage and current, dc voltage and current, resistance, frequency, capacitance, continuity, and diode test. With a basic dc voltage accuracy of 0.3 percent and an ac voltage accuracy of 1.5 percent, the Model 76 is an effective all-around troubleshooting tool, especially in environments where harmonics are a problem.

Other features include autoranging, Smoothing, and automatic Touch Hold mode. With autoranging, the user simply chooses the measurement function while letting the meter select the range with the greatest accuracy and resolution. (The user also can override autoranging and set the range manually.) In the Touch Hold mode, the meter automatically captures and holds a reading in memory, allowing the user to concentrate.

Fluke's mission is to be the leader in compact, professional electronic test tools. Fluke's products are used by technicians and engineers in service, installation, maintenance, and manufacturing test and quality functions in a variety of industries throughout the world.

Two new Fluke CombiScope oscilloscopes

An even wider range of user needs is covered by new entry-level and high-bandwidth models just introduced by the Fluke Corporation. Fluke makes its range of CombiScope oscilloscopes, a combined analog and digital storage oscilloscope, even more comprehensive by adding two new models. The new entry-level Model PM 3370, shown in Fig. 1-7, offers a cost-effective solution for two-channel applications up to 60 MHz. Also new is the PM 3390A (Fig. 1-8), a competitively priced two-channel model but with a full 200-MHz bandwidth.

With the addition of these two models, Fluke's CombiScope oscilloscope family now includes seven instruments, covering virtually all applications and budget requirements.

Key to all these instruments is Fluke's proven CombiScope oscilloscope concept: an autoranging data set optimizer (DSO) combined with a state-of-the-art analog scope for no-compromise troubleshooting performance. The CombiScope instruments address today's real-world needs, for example, in electrical and electronic service and maintenance situations where a mix of analog and digital signals exists. These instruments have the full digital storage power to capture elusive short-duration spikes and glitches, plus the effectively infinite display resolution, fast updating speed, and variable intensity of an analog scope for precise viewing of video signals and other complex waveforms.

PM 3370A: Performance and economy.

Offering an unmatched combination of performance and economy, the new PM 3370A, with a U.S. list price of $3370, has simultaneous 100 megasamples per second sampling on both channels with equivalent time sampling at up to 10 gigasamples per second on repetitive signals. This model has a 4K per channel acquisition memory, with optional 16K per channel, and the ease of operation of every Fluke CombiScope oscilloscope. For example, the unique autoranging timebase and attenuators provide the fastest and easiest way to make the scope settings match the signals applied and keep the signals on screen, even if these signals change. Probing various test points no longer requires changing instrument settings because the scope does all that automatically.

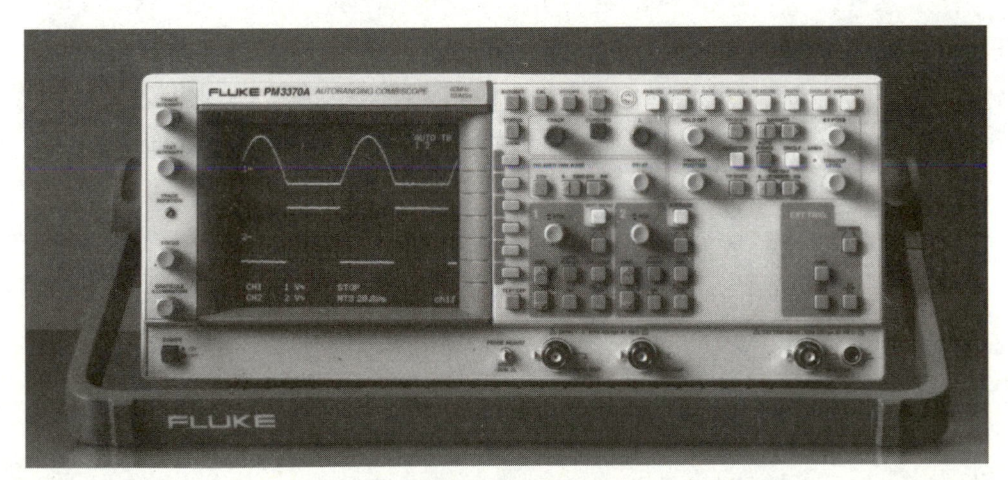

1-7 CombiScope. (*Courtesy of Fluke Corporation.*)

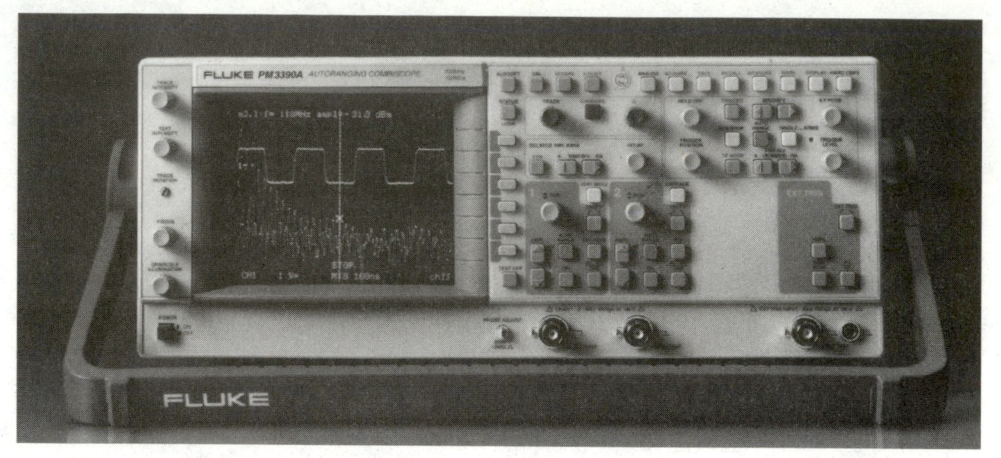

1-8 Full 200-MHz bandwidth oscilloscope. (*Courtesy of Fluke Corporation.*)

PM 3390A: Equivalent time sampling at up to 25 gigasamples per second. With its higher 200-MHz bandwidth, the PM 3390A, with a U.S. list price of $4790, increases equivalent time sampling performance to 25 gigasamples per second. This model has two channels plus an external trigger view (acting as a third channel), a 4K per channel acquisition memory (with an optional 16K per channel), and a full autoranging timebase for fast display of any signal.

Like all the Fluke CombiScope oscilloscopes, both the PM 3370A and PM 3390A have a standard RS-232 interface that supports full remote control and enables direct connection to a PC, printer, or modem. Also available is a GPIB/IEEE-488.2 interface for system use, which provides full remote control capability and trace dump facility to a printer or plotter.

The range of seven Fluke CombiScope oscilloscopes now gives users a choice of 60-, 100-, or 200-MHz bandwidths; 100 or 200 megasamples per second sampling rates; acquisition memories up to 32K; and either 2-, 2+2-, or full 4-channel input capability.

Extending the functionality of these versatile instruments is the optional FlukeView PC software package, which offers extensive and easy-to-use capabilities for the management and processing of waveforms.

Fluke Model PM3380A CombiScope FlukeView software. Users of the Fluke PM3380A (shown in Fig. 1-9) CombiScope oscilloscope can now apply the power of the PC to waveform documentation, archiving, and analysis. FlukeView software integrates the CombiScope oscilloscope into Microsoft's Windows environment easily and professionally. FlukeView software enhances the versatility of Fluke's autoranging CombiScope oscilloscopes, which combine both analog and digital storage functions in one unit.

The FlukeView CombiScope software user interface matches today's PC standards. For example, the toolbar offers instant mouse access to many of the program's commands and features and copy/paste functions via the Windows clipboard. Ease of use could not be more simple. The function of each button in the toolbar is revealed when the user simply places the mouse pointer on the button.

FlukeView software for CombiScope oscilloscopes is priced at $295 U.S. list and is available immediately. A demo disk is available to allow customers to evaluate the program prior to purchase, and a new informational brochure is also available.

FlukeView software for CombiScope oscilloscopes has three main functions:

1. *Documentation.* FlukeView receives waveforms as a screen "snapshot," complete with cursers and measurement information, or as waveform data with actual DSO sample values. Both formats can be copied easily to the Windows clipboard for inclusion in other Windows applications such as word processors and spreadsheets (Fig. 1-10). The waveform also can be sent directly to the PC's printer for an immediate hard copy of the screen.
2. *Archiving: Waveform annotation, storage, and retrieval.* FlukeView CombiScope software is extremely efficient for waveform storage and retrieval. A typical example is the creation of personalized waveform libraries. Text annotations, such as measurement conditions and instrument setups, can be added to the waveforms and stored to create complete records of measurement-related data. Stored waveforms can be used for reference and comparison purposes on the PC. Screen snapshots can be saved on disk in several Windows-compatible formats (.PCX, .BMP, and .HPL) (Fig. 1-11). Waveform data can be saved as ASCII, bitmap .BMP, and .PCX. To complete the archive function, oscilloscope settings also can be stored on disk and later downloaded to the CombiScope oscilloscope for reproduction of earlier test setups.
3. *Analysis: Obtaining valuable extra data.* The spectrum function of the FlukeView CombiScope software enables users to see the harmonic components of a waveform. This is a valuable tool for checking signal purity and tracking sources of interference. The spectrum can be displayed with linear or logarithmic scaling and with absolute or relative values. More

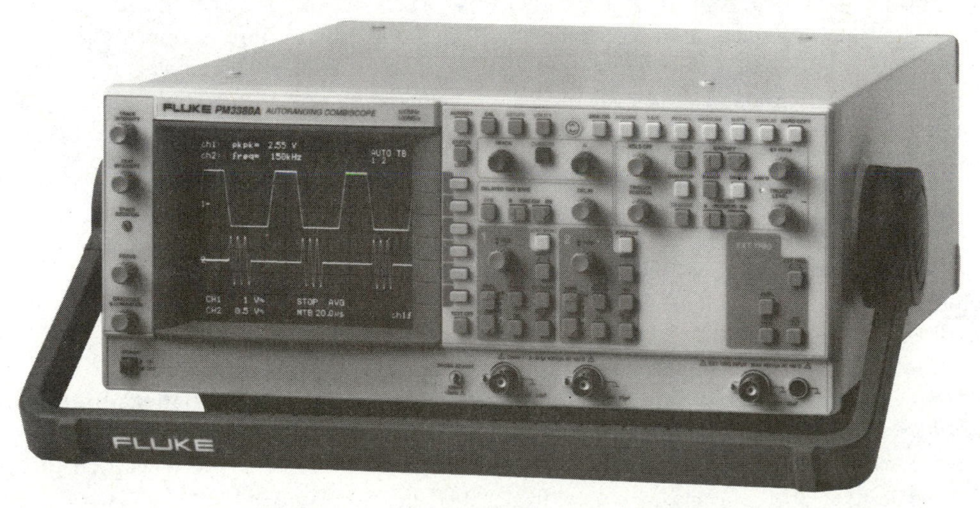

1-9 Autoranging CombiScope. (*Courtesy of Fluke Corporation.*)

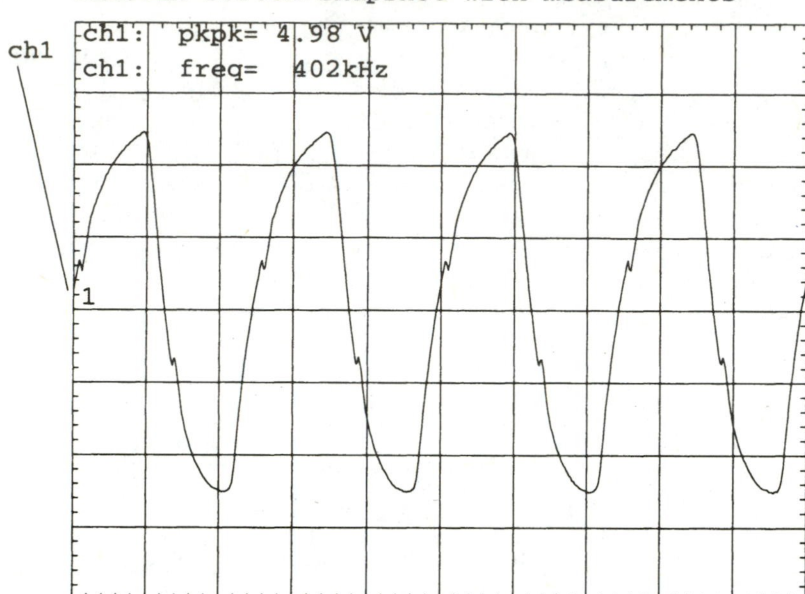

1-10 Waveform screen snapshot. (*Courtesy of Fluke Corporation.*)

detailed analysis, such as statistics, can be done by importing waveform sample point values to a spreadsheet program. Waveforms can be saved in ASCII format to facilitate this or can be copied as sample values to the Windows clipboard and pasted into the spreadsheet (Fig. 1-12).

FlukeView CombiScope software is suitable for use with any of the Fluke PM 3380A and PM 3390A Series CombiScopes. These instruments offer a choice of 100- or 200-MHz analog and digital bandwidths, with sampling rates of up to 200 megasamples per second with a choice of 2-, 2+2-, or full 4-channel operation. The CombiScope oscilloscope gives users the best of both worlds: the power and measurement capabilities of digital storage operation and the real-time signal display of an analog oscilloscope that allows accurate visualization of complex or fast-changing waveforms.

Overview of graphical multimeters

The increasing sophistication of electronic and electromechanical systems, including everything from computer-controlled industrial equipment to consumer audio/visual products, has created the need for compact, easy-to-use troubleshooting tools that can perform a variety of functions.

In response, Fluke Corporation, a world leader in test and measurement equipment, has developed the 860 Series Graphical Multimeter (GMM) test tools. These are the first high-performance test tools to combine an advanced multimeter with graphical capabilities, including waveform display, trend plotting, in-circuit component testing, and logic activity testing, for faster, more effective troubleshooting, maintenance, installation, and calibration. Designed with a simple-to-use rotary knob and intuitive user interface (patent pending), the 860 GMM test tools have the familiar feel of a DMM yet provide a much broader range of capabilities in a single tool.

Graphical multimeters for new test and measurement challenges

Fluke has a long and successful history of identifying emerging trends and responding to customer needs. Since Fluke introduced its first instrument in 1949, the company has used a simple strategy as part of its product-development process—to go directly to customers for answers. Listening to the everyday needs of customers, watching them work, and seeing them use their tools has provided Fluke with the kind of invaluable information that cannot be learned in a research and development (R&D) laboratory.

In developing the 860 Series GMM test tools, Fluke spoke with more than 2000 customers by way of extensive telephone interviews, focus groups, usability tests, and on-site research. The critical data Fluke gathered allowed the company to create a multipurpose tool that significantly enhances users' troubleshooting capabilities. The result was a revolutionary new product, the 860 Series GMM test tools, which are redefining multipurpose service tools. Following are several key issues Fluke considered in its development of the 860 Series GMMs.

More complex signals. The proliferation of electronic and electromechanical equipment has increased the number of complex signals a technician must interpret. The more common use of switching power supplies and the development of

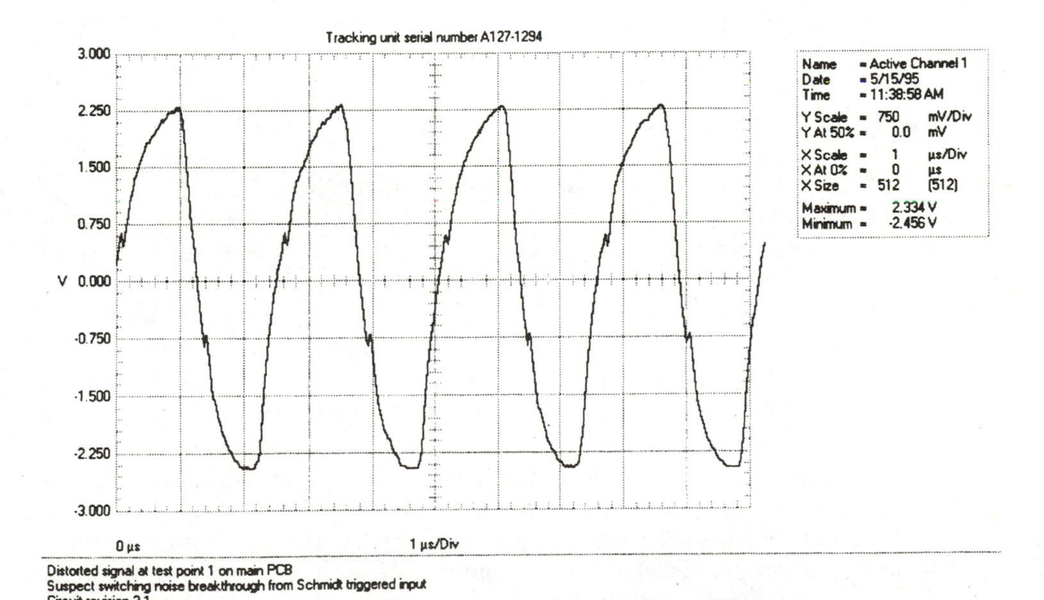

1-11 Waveform saved as a bit map on Microsoft Word for Windows. (*Courtesy of Fluke Corporation.*)

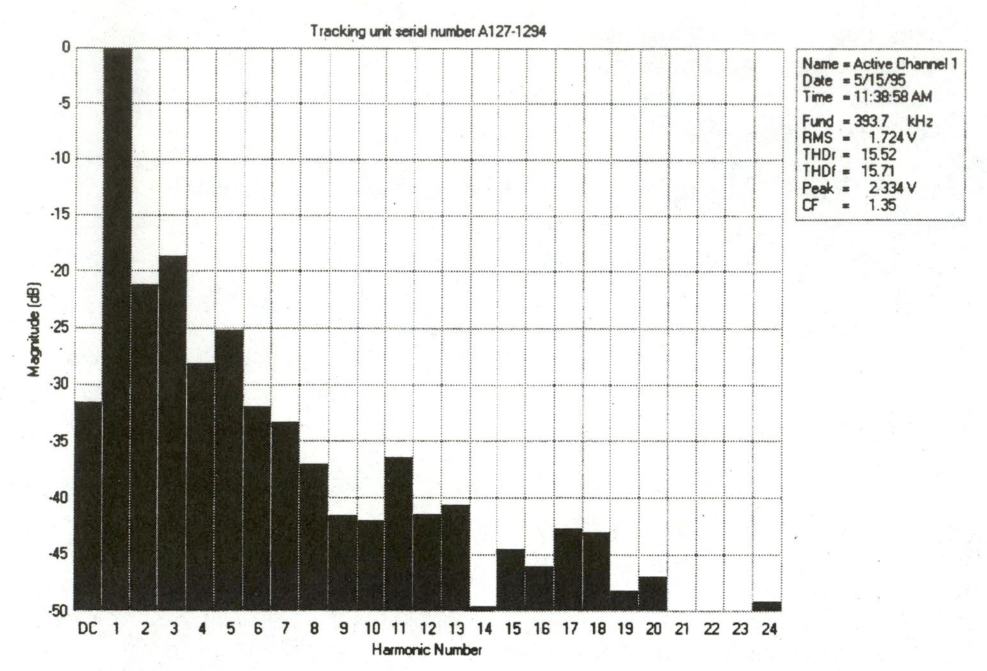

1-12 Diagram showing the frequency spectrum of the waveform drawing in Fig. 1-11. (*Courtesy of Fluke Corporation.*)

products that integrate audio, video, electronic, and mechanical technologies have created complicated, unfamiliar systems for many technicians. In the past, technicians were fairly specialized; today, a single technician must work with many diverse technologies. To troubleshoot these complex signals effectively, technicians need more than a digital reading. They need graphical information, including waveform displays, graphs of measurement trends over time, and in-circuit component test patterns, to identify the problem. In addition, technicians require versatile tools that can troubleshoot an array of electrical, electronic, and mechanical systems.

Less training and documentation on specific systems. Technicians are receiving less specific training, in part due to the rapid proliferation of new makes and models of equipment they are working on. New models are often updated with revolutionary new technology every 6 months to a year, and service training and documentation does not keep pace with these rapid new model introductions. In addition, there are fewer technicians receiving military training, which had provided experience in repairing and troubleshooting a wide range of equipment.

Another reason for less training and documentation is the increased number of proprietary components being used in new products. Often, manufacturers do not release complete documentation on new systems to avoid giving away proprietary technology. Designers are also relying on fewer standard parts in favor of more complex devices to create more elaborate electronics systems. Many technicians are not familiar with custom devices and cannot get detailed information about the internal workings of ICs. Multipurpose tools that offer a broad range of graphical pattern-recognition capabilities will help technicians troubleshoot successfully with less application-specific documentation or training.

From analog to digital to graphical instrumentation

Fluke's reputation as a leader has been the result of many years of technological innovations and superior customer service and support. Following is a list of milestones in Fluke's multimeter development that shows how the company has responded to the needs of customers for innovative tools that provide good value, improve performance, and simplify the job of test and measurement.

1949. Fluke debated its first product, the Model 101 VAW meter, a benchtop analog meter that was one of the *first power meters* on the market. In addition to powerful electronics that enabled the instrument to operate at much higher frequencies than the competition, Model 101 also offered users one very practical new feature: a rugged industrial case designed to withstand harsh environments. (Many instruments on the market were housed in mahogany cases, appropriate only for the pristine environment of the laboratory.) Fluke's Model 101 established Fluke as a significant new player in the growing field of test and measurement.

1955. Fluke introduced the Model 800A, the industry's *first differential voltmeter,* and created a new category of instruments. For the first time, laboratory accuracy was available in a rugged, compact unit designed to be usable by even nontechnical people with no loss of accuracy—something unheard of at the time. The Model 800A was used in the most demanding industrial and aerospace applications.

1969. Fluke introduced its *first digital multimeter,* Model 8100A. This high-resolution, high-accuracy meter was battery operated, offering customers the first general-purpose, truly portable multimeter. The technical innovation and competitive low price of Model 8100A helped establish a new market for affordable, compact electronic test tools.

1977. Proprietary microelectronics paved the way for the introduction of Fluke's *first hand-held DMM.* The compact size and low price of Model 8020A literally put a DMM into the hands of those professionals even casually interested in electronics. The Model 8020A also featured breakthrough liquid-crystal display (LCD) technology that provided a large, easy-to-read digital display, faster updates, and a longer battery life.

1983. Fluke's 70 Series was the first family of autoranging multimeters to combine *analog and digital displays,* establishing a new class of versatile multimeters that gave users the dynamic quality of an analog meter with the accuracy and durable performance of Fluke's hand-held DMMs. Equipped with digital microprocessors, the 70 Series multimeters were the first low-cost instruments on the market with such advanced circuitry. The 70 Series was also the first family of test instruments produced on an automated production line, which substantially reduced prices and carved out an enormous market for the now-classic 70 Series.

1991. Fluke's ScopeMeter test tools revolutionized field service by combining a full-featured *dual-channel 50-MHz digital storage oscilloscope and digital multimeter capabilities* in one rugged, portable package.

1994. Bridging the gap between multimeters and ScopeMeter test tools, Fluke created an entirely new category of troubleshooting tools. Fluke's 860 Series graphical multimeters are the first hand-held tools to combine the *industry's most advanced multimeter with analog, digital and graphical data displays.* The new GMM high-performance multimeter capabilities, combined with the graphical power of waveform display to 1 MHz, trend graphs, in-circuit component testing, and logic activity testing, provide customers with a powerful multifunction service tool. Yet the simple rotary switch and familiar interface (patent pending) of the GMMs make it easy for DMM users to operate this new class of highly integrated instruments.

Fluke's business focus is on new and growing markets for compact, professional electronic test tools. Fluke products are used in service, manufacturing, test, and quality assurance functions in a variety of industries throughout the world. Fluke was founded in 1948, has over 2600 employees worldwide, and is headquartered in Everett, Washington. Fluke is ISO-9001 registered for all its North American facilities.

Fluke high-performance troubleshooting equipment

Fluke Corporation, a leading manufacturer of compact, professional electronic test tools, has announced the introduction of its revolutionary new 860 Series Graphical Multimeter (GMM) test tools. Graphical Multimeters are a new category of test instruments that combine the industry's most advanced multimeter capabilities with the visual power of waveform display, in-circuit component testing, trend plotting, and logic activity detection together in one easy-to-use, hand-held instrument. The Model 867 is shown in Fig. 1-13.

The Fluke 860 Series is a family of three GMM test tools that are the first high-accuracy, high-performance multimeters combined with analog, digital, and graphical displays. The 860 Series selectable display modes allow users to view information in the form best suited for their application for faster, more effective troubleshooting. And for the first time, this graphical power is accessed through an easy-to-use rotary switch that many multimeter users are already familiar with. This simplified user interface virtually eliminates long learning curves.

"The 860 Series GMM test tools are the next step in the evolution of multimeters," said Kim Boyer, senior product planner for Fluke's Service Tools Division. "With so many complex, sophisticated electronic and electromechanical systems to maintain, today's technicians need multipurpose tools that are easy to operate and can diagnose problems on the spot. The 860 Series GMM instruments give users the troubleshooting capabilities they need most often in a single, rugged, compact tool. And the GMM test tools are genuinely easy to use for anyone familiar with a multimeter," Boyer said.

Fluke's new 860 Series GMM test tools are multipurpose, multifunctional tools designed for a wide variety of applications, including troubleshooting, maintenance, installation, and calibration of industrial, medical, and production equipment; repair of computers and office machines; and repair and maintenance of telecommunications and home entertainment systems.

860 Series graphical multimeters.

Meter mode. This mode offers the highest accuracy (0.025 percent dc) and most comprehensive set of functions available in a hand-held meter. The GMM test

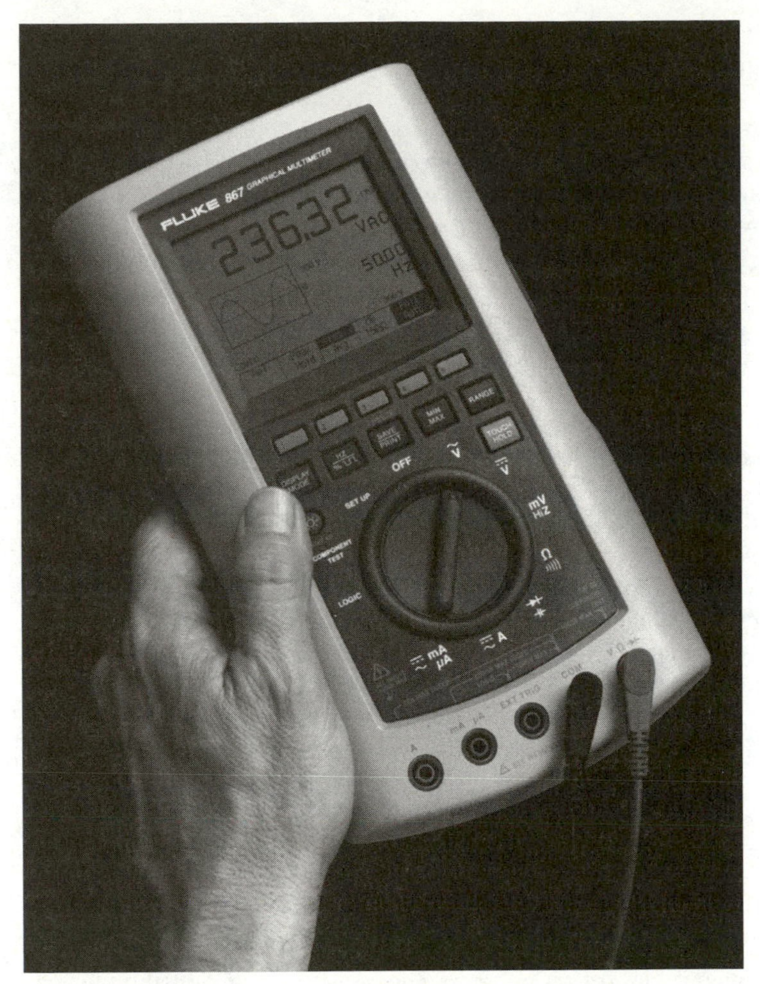

1-13 Graphical multimeter. (*Courtesy of Fluke Corporation.*)

tools' meter functions can be used to troubleshoot industrial equipment, consumer electronics, and precision components; calibrate industrial equipment; and perform quality control or production tests on components. The 860 Series GMMs offer high-precision, 32,000-count (4.5-digit) resolution, a dual digital display that gives users additional information about the parameter being measured, and an analog NeedleGraph display. Meter mode enables users to directly measure current, resistance, conductance, capacitance, frequency, duty cycle, pulse width, period, decibels, and ac and dc volts. The GMM test tools also offer a new AutoDiode feature (patent pending) and optional temperature and pressure measurements and are the first hand-held test tools to offer true IEC-1010 Class III 1000-V protection in all functions.

Waveform display. This feature provides a clear picture of noise, waveform distortion, intermittent failures, and glitches. The waveform display capability (up to 1-MHz signal bandwidth) enhances the numeric display by providing more qualitative information about a signal. The AutoGraph display setup automatically scales voltage, time base, triggering, and position for fast and easy test setups. Manual setup, or external triggering, is also available.

TrendGraph mode. This mode plots high-resolution meter readings for up to 30 h, in intervals from 1 s to 15 min, to detect such anomalies as power sags and surges and droops. The TrendGraph feature saves users time by collecting and graphing information automatically so that they do not have to "baby-sit" a test setup.

In-circuit component test. This feature allows technicians to view component signatures in-circuit, without having to remove and handle sensitive and expensive components. The in-circuit component test feature enables users to troubleshoot problems by comparing component signatures of known, functioning circuits with those of defective circuits. Users also can check components safely without having to power up the entire circuit.

Logic Activity mode. This mode offers a simple way to isolate digital failures. The Logic Activity feature indicates logic transitions or state changes to 10 MHz and shows if a circuit is active or stuck high or low. It also shows the frequency of activity and the average dc voltage.

While the 860 Series GMM test tools offer a host of advanced troubleshooting capabilities, users will appreciate the easy-to-operate rotary knob and intuitive user interface (patent pending) that has the familiar look and feel of a DMM. Functions are selected by simply turning the knob, so there is no need to memorize long menus or complicated test setups, and users can see at a glance which function the GMM test tool is in.

The fully integrated 860 Series GMMs also save time and money by using only two inexpensive meter leads for all measurements. Users do not have to buy scope leads or current clamps (for measurements up to 10 A continuous or 20 A for 30 s) or change test leads when changing display modes.

Using the RS-232 interface, technicians can upload known waveforms or component signatures from a computer for on-screen comparison or download information from the GMM instrument to a serial printer or computer. This allows information to be transferred to reports for documentation or analysis or logged directly into a data file.

For increased flexibility, the 860 Series instruments operate on a variety of power sources, including an external battery eliminator, alkaline batteries, or a rechargeable NiCad battery pack.

Available models.

Model 863, priced at $795, offers meter and graphical capabilities, including meter mode, waveform display, and TrendGraph mode. Basic dc accuracy is 0.04 percent. *Model 865,* priced at $995, offers comprehensive meter and graphical capabilities, including meter mode, waveform display, TrendGraph mode, in-circuit component testing, and Logic Activity mode. Basic dc accuracy is 0.04 percent. Model 865 also provides internal battery charging, line-voltage adapter/battery charger, and LCD backlight. *Model 867,* priced at $1295, offers comprehensive meter and graphical capabilities in addition to improved dc accuracy (0.025 percent). Model 867 also provides an optically isolated RS-232 cable and companion software for immediate interfacing with a printer or PC, rechargeable NiCad battery pack, internal battery charging, line-voltage adapter/battery charger, LCD backlight, and both a basic and deluxe set of meter test leads.

For more information on the new Fluke 860 Series GMMs, designed and manufactured in the United States, contact Fluke Corporation, P.O. Box 9090, Everett, WA 98206.

Safety standards for DMMs.

The increased occurrence and levels of transient overvoltages on today's power systems have once again called attention to requirements for safety in troubleshooting, maintenance, and repair. Safety concerns have become paramount as occupational-safety risks are recognized and businesses take action to safeguard their employees. The safe use of test instruments is one critical matter. Fortunately, formal definitions now exist to guide users on the selection of appropriately rated test instruments for a specific overvoltage environment.

High-voltage and high-current environments are perhaps the greatest concern. Engineers, technicians, and electricians working on power systems must seriously consider the overvoltage environment in which they are working. *Overvoltage environment* refers to an installation environment in which a high transient voltage could occur on a system being maintained. The level of the transient, in thousands of volts, defines the specific overvoltage category for the environment. Typically, the farther the system is from the main distribution lines coming from a utility, the lower is the potential transient level as a result of inherent damping in the system.

The transient alone may or may not have sufficient energy to be life-threatening, but the fact that the transient typically rides on top of a power source, such as 440 V at 200 A, could have devastating follow-through effects. Thanks to the efforts of safety agencies in the United States and Europe, definitions of installation environments and the risks inherent in each have been formalized in writing.

Transient level overvoltages.

The International Electrotechnical Commission (IEC) develops international general standards for safety of electrical equipment for measurement, control, and laboratory use. IEC in 1988 voted to replace an older standard, IEC-348, with a more stringent standard, IEC-1010-1. IEC-1010-1, including amendment 1, is used as the basis for the following standards:

- U.S. standard: ANSI/ISA-S82.01-94
- Canadian standard: CAN C22.2 No. 1010.1-92
- European standard: EN61010-1:1993

Here are some of the common questions about IEC-1010-1, especially as it applies to electrical and electronic test equipment.

Q: How are IEC-1010-1 and IEC-348 different?

A: There are many differences between these standards. One of the main differences concerns spacing requirements. The IEC-1010-1 standard goes a step further than IEC-348 in the spacing requirements for a given maximum input-voltage rating. The standard refers to these spacings as "creepage" distance (along surfaces) and "clearance" distance (shortest distance through the air). Larger clearance distances enable the meter to withstand the higher overvoltage transients that may be found on the system being measured.

Q: What is an overvoltage category, and what is the significance of overvoltage category III?

A: IEC-1010-1 specifies overvoltage category ratings, which relate to the probability of a voltage occurring (at some location) that is significantly higher than the voltage expected. For example, an unexpected voltage transient would be considered an overvoltage. Overvoltage categories range from I to IV, in increasing order of overvoltage level expected. They are not necessarily related to the nominal voltage level of the system.

Overvoltage category IV is typically the overhead or underground utility service to an installation and is beyond the scope of IEC-1010-1. Overvoltage category III typically refers to mains voltage feeder or branch circuit lines that are separated from the utility service by at least a single level of transformer insolation. Categories I and II are successively more isolated and distant from the utility service than categories III and IV. Instruments designed to meet overvoltage category III are able to withstand overvoltage transients better than instruments designed to overvoltage category II.

Anyone working on mains voltage feeder or branch circuit lines up to 600 V should use instruments rated for at least overvoltage category III.

Q: How does IEC differ from standards such as those published by UL and CSA?

A: Underwriters Laboratories (UL), the Canadian Standards Association (CSA), and TÜV (a German standards organization) are approval/listing agencies. IEC is an international standards writing organization. Local governments may use the IEC standards to write their own national standards, complete with enhancements influenced by local needs (e.g., ANSI/ISA-S82.01-94 based on IEC-1010-1 in the United States). The approval/listing agencies are independent testing laboratories that test products against national standards or their own standards, such as UL3111 based on IEC-1010-1.

Q: What is meant by "UL listed"?

A: A manufacturer may state that a product is designed to meet the requirements of IEC-1010-1, UL 3111, or both. However, for a product to be UL listed, CSA certified, or TÜV certified, the manufacturer must employ the services of the approval/listing agency to test the product's conformity to the standard. Only on successful completion of this independent testing and receipt of authorization from the approval/listing agency may the manufacturer display the mark of the agency on the product.

A manufacturer may claim that its product is "designed to IEC-1010-1 standards," but this does not mean the instrument has passed successfully the independent evaluation and testing of an approval/listing agency such as UL. The corollary is also true: A product with a UL listing, for example, may or may not meet the requirements of IEC-1010-1, depending on whether UL tested the product to UL 1244 (the old standard) or UL 3111 (standard based on IEC-1010-1).

Q: How do these safety standards apply to the Fluke Model 76 DMM?

A: Fluke designed the Model 76 DMM to meet the requirements of IEC-1010-1 overvoltage category III. To meet the standards, Fluke increased the internal spacing of components and added new input protection components. By meeting the new standards, the Fluke Model 76 exceeds the requirements of both the old IEC-348 and UL 1244 standards.

Q: How can I compare two products if one was rated according to IEC-348 and the other according to IEC-1010-1?

A: Because IEC-1010-1 takes into account the higher-voltage transients that occur in some locations, overvoltage protection in a new meter rated for 600 V could be better than that of an older meter rated for 1000 V under IEC-348. Here is an example. A meter rated for an input voltage of 1000 V may have been designed to meet the IEC-348 safety standard. The clearance distance for 1000 V under IEC-348 is 8 mm. On the other hand, a meter rated for an input voltage of 600 V and developed in 1995, such as the Fluke Model 76, has been designed to meet overvoltage category III under IEC-1010-1. Category III specifies a clearance distance of 11.5 mm for a 600-V rating. Thus the protection from overvoltage transients is actually better in the 600-V Fluke Model 76 than in the earlier 1000-V meter because the increased distance will withstand a higher-voltage transient before breaking down.

Tektronix real-time oscilloscope

Three Tektronix Series TDS 300 oscilloscopes are shown in Fig. 1-14. The TDS 300 Series oscilloscopes are the first low-cost scopes with the power to acquire

1-14 Digital real-time oscilloscopes. (*Courtesy of Tektronix Corporation.*)

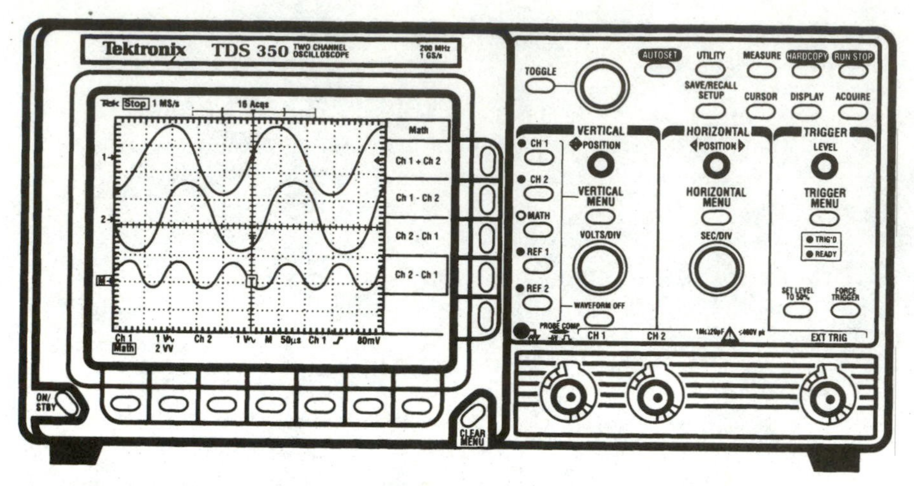

1-15 Real-time oscilloscope characteristics. (*Courtesy of Tektronix Corporation.*)

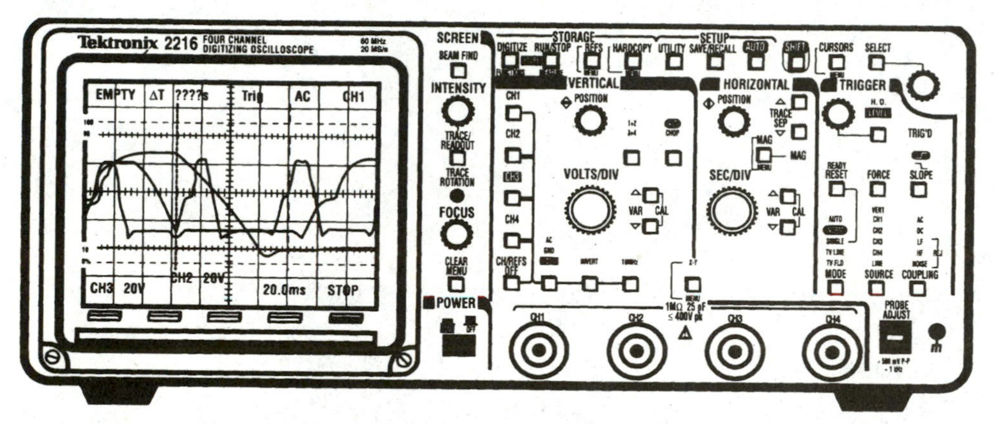

1-16 Digital-analog oscilloscope characteristics. (*Courtesy of Tektronix Corporation.*)

signals over their entire bandwidth with minimal aliasing distortion—even when measuring high-speed transient and single-shot events. Their intuitive interface combines dedicated knobs with menu-driven digital functions. Twenty-one automatic measurements, information-enhancing display modes, and both edge and basic video triggering come standard. The front panel of the TDS-300 scope is shown in Fig. 1-15.

The four-channel Model 2216 scope is a low-priced digital plus analog oscilloscope with readouts, on-screen cursers, X-Y store and nonstore operation, and parallel printer port. The 2216 scope features a familiar, convenient front panel layout and automatic setup for fast, accurate measurements. Communications interface options allow remote operation for automated data collection. See Fig. 1-16 for control locations.

1-17 Portable television signal monitor and signal generator. (*Courtesy of Tektronix Corporation.*)

Video test instruments

The Tektronix 91 Series, shown in Fig. 1-17, is a portable television signal monitor that is used for setup, maintenance, and operation of various television systems. The television technician can quickly verify correct connections of the system and components and adjust equipment timing and levels for best television signal quality.

The Tektronix TSG90 Pathfinder NTSC signal generator, also shown in Fig. 1-17, has 16 video test signals and many other features. This unit can be used for verifying microwave links, field servicing, or system installation. Its powerful combination of test signals, identification capabilities, and audio tones makes it a "must-have unit" for every television engineer's tool case or work bench. In addition, this "little baby" fits right in the palm of your hand. .

2

CHAPTER

Analog color television preview

This chapter begins with an overview of analog television principles and operations. You will see how a color television camera transmits an optical image and how the scanning process is accomplished. The chapter then looks at scanning rate, fields, and bandwidth requirements. The two color difference signals and how color separations are obtained are then explained. An explanation is also given of how black-and-white television and color television are made compatible.

Any discussion of digital television which is covered in Chap. 9, would be incomplete without a review of analog television principles. Although I will not pretend to give you a complete discussion of analog television principles, I have found that it is usually useful to remember some of these underlying principles. This is true partly because the new world of digital video has cut across the lines of different disciplines and is bringing together the worlds of traditional television (broadcasters and video production engineers), telephone, and computers. If you are well versed in the field of analog television, then you may want to skip to other chapters while the rest of us do a quick review of this subject.

To produce a television signal, some sort of transducer is necessary to change an optical image into an electrical signal. We will now examine the television camera, since it is the most typical transducer.

In a color television signal, the light from the image is split into the three primary colors, red, green and blue, as it enters the camera. Any color in the image can be reproduced by a combination of these three colors. Figure 2-1 shows a simplified camera system.

The incoming light from the scene is split into the three component colors by the dichroic mirrors and sent to the corresponding imaging device. Each imaging device converts the image into an electrical signal.

In order to transmit an optical image electrically, the image is broken up into thin horizontal stripes by the scanning process. Television images being produced

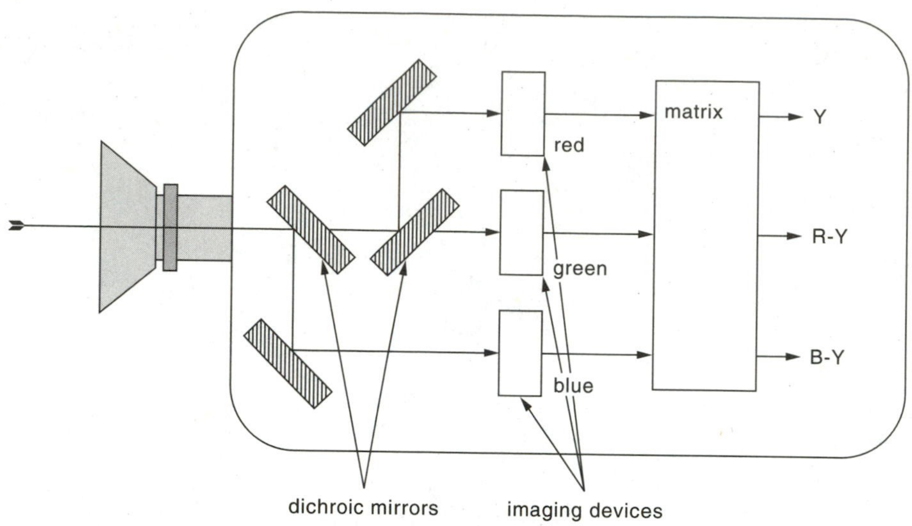

2-1 Color television camera system. (*Courtesy of Sencore, Inc.*)

for broadcast are currently divided into 525 or 625 horizontal scan lines or, as they are more commonly called, *lines*. The United States, in the early days of television, standardized on dividing the images into 525 lines. In Europe, 625 lines was chosen; thus two incompatible systems were developed and have remained so to this day.

Each total image, made up of 525 (or 625) lines, is called a *frame*. The total image is scanned 30 times each second (25 times in the European system). This repeated scanning allows the image to appear to portray motion in objects being shown.

A technique called *interlace* is used to reduce the flicker effect that would be caused by showing as few as 30 frames each second. Figure 2-2 shows the principle of interlaced scanning.

The image is scanned beginning in the upper left-hand corner, scanning across to the right side of the image. The scanning is then turned off and returned rapidly to the left-hand side of the image, to a position that is one line lower than the previous scan. Thus an unscanned line is left between two successive scans. Every other line in the image is scanned, a process that takes one-half the frame time, or $\frac{1}{60}$ s.

The scanning then recommences at the top and covers the remaining image areas that had been left unscanned before. This fills in the missing image stripes, and after an additional $\frac{1}{60}$ s, the second field is scanned and the frame is complete.

The first 20 lines of each field are not used to convey picture information but are part of the vertical blanking interval. The *vertical blanking interval* (VBI) is the time allotted for retrace, when the electron beam is returned to the top of the screen after the previous field has ended and the beam has reached the bottom of the screen. These first lines therefore do not have picture information. Some of these lines are used for specific purposes, such as test lines, vertical-interval reference (VIR), ghost canceling signals, and so forth. Line 10 is commonly designated as the point at which switching between two synchronized video sources occurs.

As the lines are scanned, the variations in light are converted into varying voltages. Thus, during the time period of the scan line, 63.5 μs, there is a continuously varying voltage produced by the scanning device, in this case a camera. During the retrace period, the output is blanked, or set to 0 V. Thus, at the output of these three imaging devices, signals are produced whose voltages represent the magnitudes of the red, green, and blue light reaching the corresponding imaging device.

Although our elementary camera does not show the circuitry that produces the synchronizing pulses, it is necessary to insert horizontal sync pulses that signal the receiving equipment that it is time to do horizontal retrace, which returns the scanning beam to the left edge of the screen. After the image is scanned, vertical sync pulses must be inserted to trigger the vertical retrace, which returns the beam to the top of the image. The time period between the end of one scan line and the beginning of the next is known as the *horizontal blanking interval* (HBI). As mentioned earlier, the period between the end (bottom) of one field and the beginning (top) of the next field is called the *vertical blanking interval* (VBI), or *vertical interval.*

At the receiver, the same three (RGB) varying voltage signals are developed and used to modulate the electron beams in the television's cathode-ray tube (CRT). By sweeping the beam across the screen in sync with the original imaging device, one is able to reproduce the image that was originally encoded into an electronic signal.

The bandwidth necessary to convey reasonable picture detail of a monochrome (black-and-white) image and the accompanying audio is about 6 MHz. Thus directly transmitting the three signals of an RGB color signal would occupy the equivalent bandwidth of two or three normal (NTSC) television channels. This, of course, is unacceptable in a real-world situation. However, most of the high-frequency information, the fine details of an image, is contained in the luminance, or brightness, portion of the image. The color information can be bandwidth-limited without loss of detail in the picture.

High frequencies in the luminance component of the image are due to abrupt changes in brightness level. Thus the edge of an object that is overlapping another

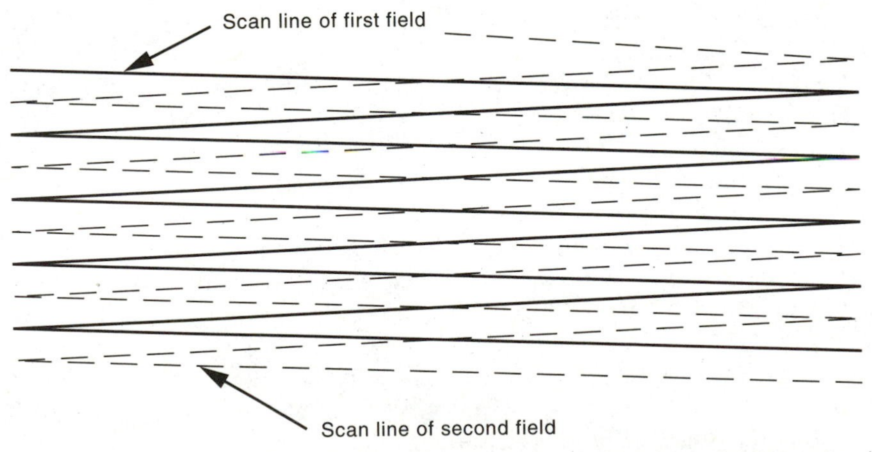

2-2 Principle of interlaced television scanning. (*Courtesy of Sencore, Inc.*)

object of difference brightness will produce an abrupt change in voltage level, or a high-frequency component. An abrupt change in color between two parts of an image is almost always accompanied by an abrupt change in brightness level as well. Thus, if the edge of the object is well defined by a clean high-frequency edge in luminance, the color information can be allowed to vary more slowly without any loss of apparent picture quality.

Therefore, the system that was developed in the early days of television combined the red, green, and blue signals in a matrix to produce a luminance signal, called the *Y signal,* and two color difference signals. The Y signal is produced by adding the RGB signals in the proportions of 0.299ER+0.587EG+0.114EB.

The two color difference signals are produced by subtracting the luminance from the red signal (called the *R-Y color difference signal*) and subtracting the luminance from the blue signal (called *B-Y*). All three original RGB signals can be redeveloped at the receiver from these three signals, and thus they convey all the original information.

Separating the color information from the luminance or brightness information has important consequences for transmission of the signal. Since the color difference signals carry less high-frequency information, as explained above, they can be bandwidth-limited and transmitted in a narrower bandwidth than the luminance signal. This characteristic of the luminance-chrominance bandwidth relationship is also exploited in digital television as we will see in Chap. 9.

An additional consideration in the days of conversion from monochrome to color television was the need for a compatible signal for monochrome television sets and the ability to view monochrome signals on a color set. The Y signal is a monochrome signal and can be viewed directly on a monochrome monitor or television set.

The currently popular analog videotape recording formats are Sony's Betacam SP and Panasonic's MII. These systems record all three component signals, keeping them separate throughout the recording process. If component routers and production switchers are used, the highest possible *analog* video quality is possible.

Let us examine a waveform representation of the various signals involved. Figure 2-3 shows the familiar color bars as we would see them on a waveform monitor as RGB and as Y, R-Y, and B-Y signals.

The waveforms in Fig. 2-3 only represent the active picture portion of the horizontal line; the synchronizing portion of the line is not shown. During sync time, the signals are held to black level (or the zero color level for R-Y and B-Y). The synchronizing signals may be distributed separately on a fourth wire or can be combined with one of the signal lines, often the green or luminance channel.

Most other videotape recording formats and over-the-air broadcasts encode the three components into one composite signal. One such composite signal is called *NTSC,* named for the committee that established the standard in the United States, the National Television Systems Committee. By using the largest part of a 6-MHz bandwidth for the Y signal while modulating the two color difference signals onto a subcarrier, all (or most) of the original RGB signal, along with accompanying audio, can be transmitted in one 6-MHz bandwidth.

It is worth noting, however, that encoding the information in this way is accompanied by some significant sacrifices, most notably the presence of NTSC artifacts

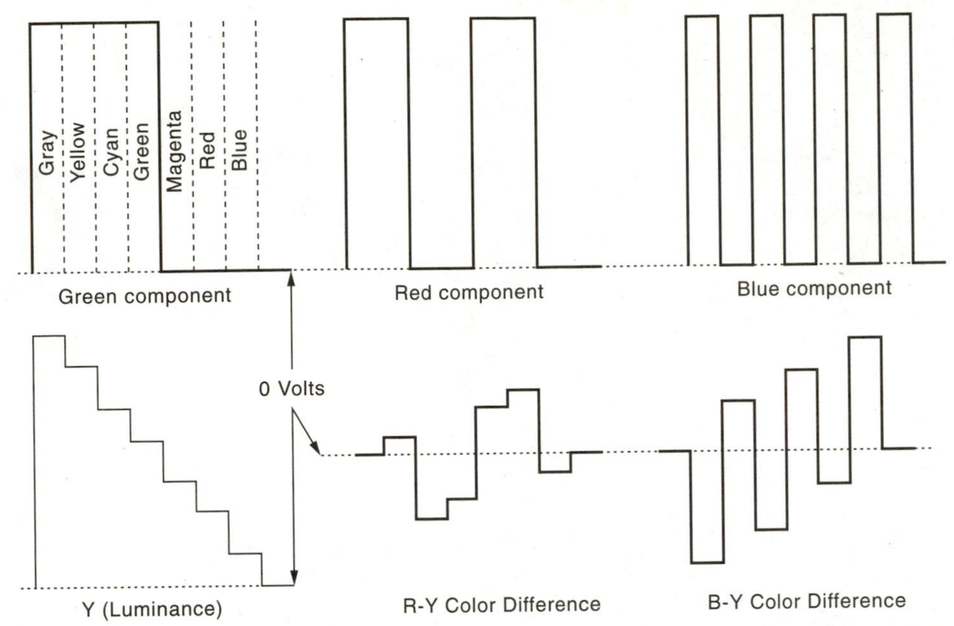

2-3 Color television test bars as seen on a waveform monitor or receiver. (*Courtesy of Sencore, Inc.*)

(dot crawl, color bleeding, etc.). It is thus preferable to minimize the use of composite video wherever possible.

Two other major composite video encoding systems exist in the world. They are based on a 625-line, 50 field per second scan rate. The phase alternating line (PAL) system, used in much of Europe and other countries around the world, has a structure similar to NTSC, in that all three components are transmitted in one 6- to 8-MHz channel. The methods of modulating the chrominance subcarrier and the subcarrier frequency are different. The third system, the *séquential à mémoire* (SECAM), has some additional differences from the PAL system and is used primarily in France, French territories, former French colonies, and Russia and the former Soviet Union.

This limited discussion of analog television is not intended to present an exhaustive discussion of its nature. Its purpose is rather to define common terms and explain basic principles that will allow persons not familiar with the technology to understand the principles of digital video that will be covered in detail in Chap. 9. Refer to Chap. 8 for more in-depth analog television information and video troubleshooting procedures.

3
CHAPTER

Conventional power supplies

This chapter will review some of the older model conventional color television power supplies. Power supply circuit operations will be explained along with troubleshooting tips and procedures.

The use of an isolation variable ac line adjuster and ac line monitor to safely troubleshoot "hot" ac chassis television sets will be explained. The chapter concludes with some power service repair information and case history problems and solutions.

Power supply circuitry

A television power supply circuit is used to convert alternating current (ac) or unregulated direct current (dc) to a smooth, regulated direct current. The two main types of television ac-to-dc power supplies are linear-regulated and switching-regulated circuitry. Refer to Chap. 4 for more information on switch-mode- and pulse-width-regulated television power supply operation and troubleshooting.

A linear power supply may have either a series or shunt voltage regulator. The series-pass regulator with a *pass transistor* is the most commonly used circuit configuration.

The rectifiers, usually a *bridge,* provide unregulated dc to a filter circuit that contains capacitors and/or choke coils. A transformer provides ac power line isolation. Output from the diode rectifiers is a steady dc with an alternating voltage superimposed. The filter network section smoothes out the ac voltage (ripple) and routes it to the pass transistor, which is in series with the load. Regulation is performed by a variation of current through the series-pass transistors that results from changes in the line voltage or circuit current loading.

The Tektronix DMM155 digital meter shown in Fig. 3-1 can be used to check ac and dc voltages in power supply circuits and monitor the dc circuit regulation.

3-1 Digital meter for power supply checks. (*Courtesy of Tektronix, Inc.*)

Adjustable isolation transformer

An isolation transformer that is adjustable not only helps you to perform faster repairs but also lets you do this more safely. It eliminates downtime of your test equipment and customers' sets that are damaged because they were not properly isolated from the chassis under test. Damaged sets and equipment must be repaired or replaced, and this comes right out of your profit. Let us now review why ac line isolation is needed when repairing modern color television sets.

Half-wave rectifier chassis

One side of the ac line is connected directly to the chassis in a half-wave "hot" ground power supply.

Bridge rectifier chassis

There is approximately 67 V of potential between the television chassis and the earth ground in modern television set bridge rectifier power supplies (Fig. 3-2*b*).

Switching power supply

Similar to a bridge rectifier chassis, the switching transistor converts the bridge's dc output into a square wave that is in turn filtered for a dc output. Switching power supplies are now found in many newer-model color television and VCR chassis.

There is no problem with connecting grounded test equipment to the half-wave rectifier chassis circuit shown in Fig. 3-2*a*. One side of the ac line is connected to the chassis. Many technicians make the proper connection by measuring the voltage

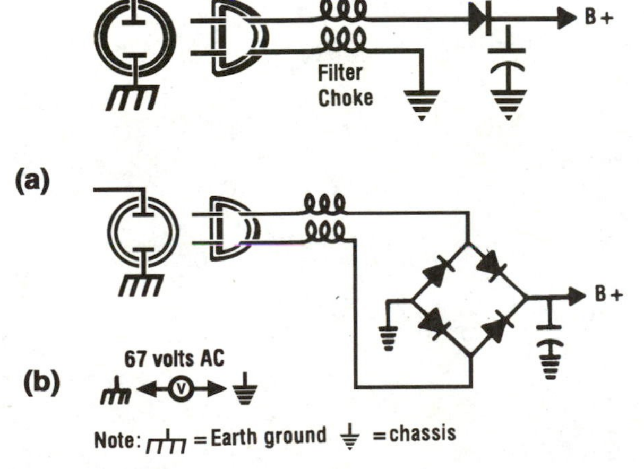

3-2 (*a*) Half-wave rectifier chassis circuit. (*b*) Bridge and switching power supply circuits. (*Courtesy of Sencore, Inc.*)

between the chassis and the grounded test equipment. A high ac voltage reading shows that the "hot" side of the line is feeding the chassis. If so, they just reverse the ac plug.

Unfortunately, this approach does not work with the bridge and switching power supply circuits shown in Fig. 3-2*b*. In these power supplies, the metal chassis is at half the ac line, no matter how the ac plug is connected. Thus you can cause costly damage to your test equipment and equipment under test. Connecting your scope or any other grounded test instrument to the television receiver or VCR chassis shorts out half the power supply, as illustrated in Fig. 3-3. Damage results from excessive current flowing through the power supply's input choke and rectifier diodes. Chances are the high current will "melt" your common test lead or even the scope probe.

A solution would be to break the ground path that causes the short circuit. However, breaking the test instrument's third wire safety ground by cutting the ac plug ground pin or using a three-wire adapter plug leads to two other problems. First, the instrument's third-wire ground return is often necessary to properly shield sensitive circuits such as microprocessors and high-impedance amps to prevent erroneous readings. Defeating the shield causes you to spend lots of "no pay" time chasing electrons up the wrong path. The second problem with defeating the third-wire ground is that a serious shock hazard is formed between the instrument's metal case and any metal object connected to earth ground. Accidentally placing your body between these two ac points may cause permanent downtime.

Both these problems can be eliminated if you make it a "golden rule" to *always* plug every television, VCR, or any other electronic ac-powered equipment into an isolation transformer before servicing it.

Powerite variable ac supply

You will find that the Sencore PR57 adjustable isolation ac power supply can be a tremendous time saver for your service bench. Most television technicians will agree that problems in regulator or shutdown safety circuits are the most time-

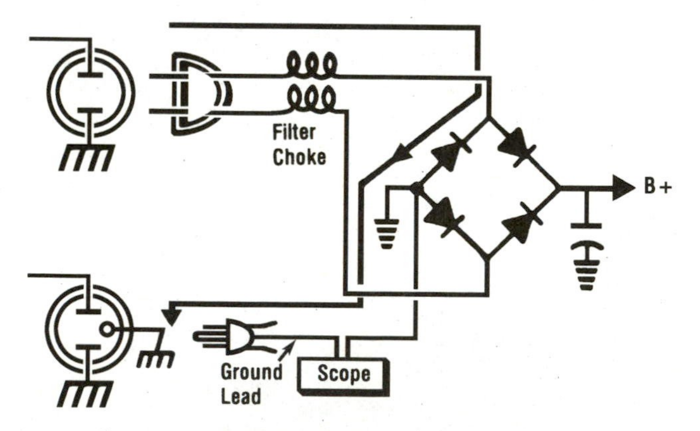

3-3 Shorting out half the power supply by connecting a ground lead to a bridge rectifier chassis without using an isolation transformer. (*Courtesy of Sencore, Inc.*)

consuming problems to troubleshoot. The set manufacturers' service literature recommends that a variable ac line voltage to about 90 V ac will enable you to service shutdown problems. The reason for this is that by lowering the ac line voltage to about 90 V ac keeps the chassis from going into the shutdown mode, thus letting you make waveform and voltage measurements to quickly find the circuit or component causing the shutdown condition.

Parts saver hints

You probably have replaced the horizontal output transistor, the driver transistor, and the power supply regulator—and then powered up the chassis, only to have all the parts "zapped out" or "go up in smoke" because you could not check the current drawn or control the input ac line voltage. Replacing horizontal output transistors can be an expensive procedure, because you can only charge the customer for the first one, since the next ones are "on the house" so to speak. Thus the ones you blow up come out of your profit. If the chassis fails after a few days or a few weeks of operation, it becomes even more expensive to repair because you now have the extra parts expense plus a costly and time-consuming call-back expense. It only takes one of the horizontal output transistor's tuned capacitors to change value and change the timing of the horizontal flyback pulse, as illustrated in Fig. 3-4. A shortened flyback time means increased horizontal trace time, increased duty cycle, and an increase in the current drawn. More current for a longer period of time means that more heat is generated than the horizontal output transistor is designed to dissipate. The result is thermal runaway and eventual part failure. A variable ac power control can keep you from destroying parts and be helpful in locating the problem at the same time. The Sencore Powerite lets you monitor the current to 4 A and wattage to 470 W at the push of a button. Always remember to never "fire up" a repaired solid-state set at full line voltage. Always start the line voltage out at 90 V ac and you will cut down on parts loss.

Safety leakage test

Another good use for the Powerite unit is the safety leakage test, which will help protect you, the test instruments, and very important, your customer. In many shops the technicians do not perform the safety leakage test because it takes too much time.

One reason given for not making the test is that the usual procedure is very complicated and time-consuming. It requires a good ground, a resistor-capacitor combination "made up" or located with lots of other shop parts, and the chassis disconnected from the bench power isolation. The Powerite safety leakage test eliminates this time-consuming procedure. You need only plug the chassis under test into the Powerite and use the supplied safety probe to touch and test all exposed (around the cabinet) metal parts. Any leakage in excess of 500 μA is a "hot" chassis problem that should be corrected before the set is returned to the customer.

Just takes one!

Most shops realize that they should be performing a leakage test on every chassis that leaves the shop. Some shops may not think they can justify the time for testing

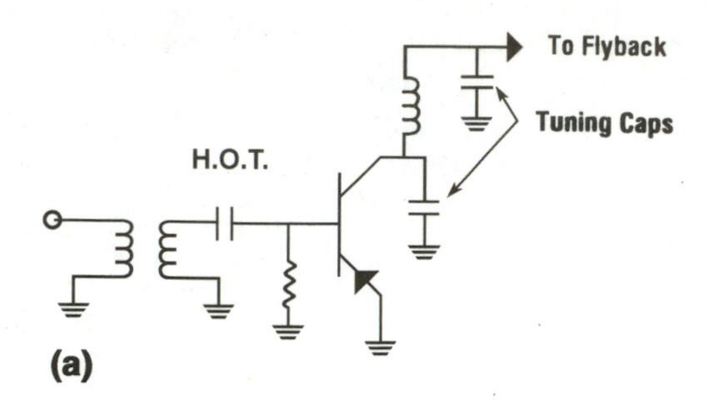

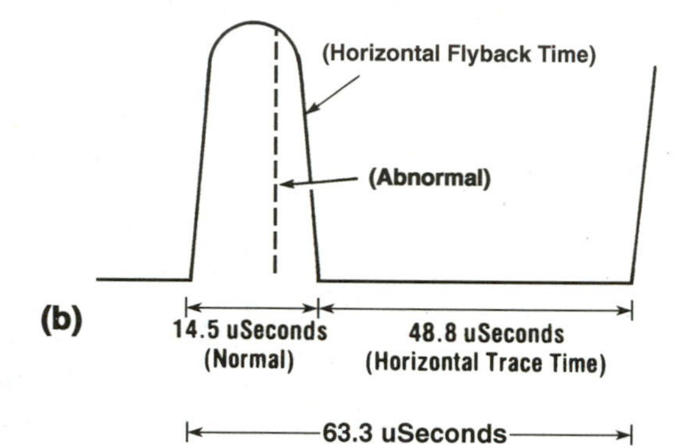

3-4 Tuning capacitors (*a*) determine the horizontal flyback pulse and (*b*) allow the horizontal output transistor to overheat. (*Courtesy of Sencore, Inc.*)

every set when only 1 of 100 sets will show up as a shock hazard. However, this is the good reason why you need to make a safety check of every chassis serviced in the shop. It only takes one to damage your health so that you cannot work or to damage your test equipment or your customer's television or VCR. Then you realize that it costs you more *not* to make the leakage test on every repaired chassis.

Let us now review the ways that a variable isolation transformer such as the Sencore Powerite can work for you on the service bench:

1. Isolation transformer: lets you be more efficient and is a good safety factor.
2. Variable ac source: lets you troubleshoot problems quicker.
3. Line voltage (ac) monitor: lets you solve shutdown problems quickly.
4. Power monitor: keeps you from smoking parts.
5. Safety leakage test: protects you and your business.

Scoping the electronic regulators

Good dc voltage regulation is required for the sensitive solid-state electronic devices found in all consumer equipment of today. In most electronic devices, Zener diodes, Zener-controlled transistors, or ICs smooth out the dc variations and hold the output voltage constant under various circuit loads. Zener regulators react to changes so quickly that they are often referred to as *electronic filters.* With these electronic regulators now used in the power supply, large filter capacitors are no longer required. Some complete electronic regulators are now all packaged within one IC chip.

The electronic regulators now found in solid-state equipment are very good, but they can and do develop problems. A fault in a Zener regulator may give some odd picture symptoms in television circuits and in some cases can be intermittent. These symptoms can be small white horizontal lines that creep up or down the screen or flicks and flashes that dart intermittently across the screen.

If you suspect trouble on the regulated B+ line, use the scope to look for any hash spikes or ripple that might be present. Under normal operation and with maximum vertical amplifier scope gain, you should see only a smooth horizontal line across the scope screen. A faulty Zener diode may go into oscillation and show up as hash on the scope line, or an intermittent breakdown within the regulator transistor can put spikes on the B+ line. As shown in Fig. 3-5, a technician is using a wide-band scope to check out a power supply problem. These types of problems will not show up on a DMM.

A problem was found in a late-model Zenith color chassis that indicated considerable ripple on the regulated B+ line, as shown in the Fig. 3-6 scope waveform. The ripple caused the picture to weave, and there was an intermittent loss of color. The ripple was traced to a defective Zener regulator control diode.

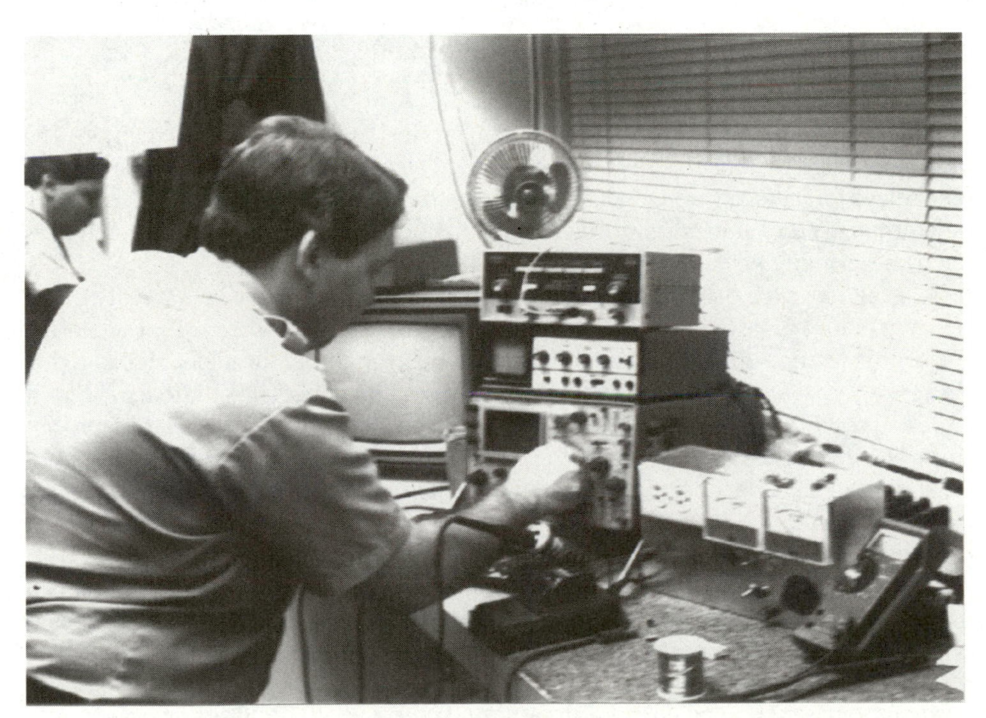

3-5 Scope for a power supply ripple check. (*Courtesy of Sencore, Inc.*)

3-6 Oscilloscope trace showing ripple in a power supply caused by a defective Zener diode. (*Courtesy of Sencore, Inc.*)

Color television power supply

As you look at the power supply schematic in Fig. 3-7, note that many of the transistors, Zeners, and other parts used to regulate the line voltage in earlier-model color television sets are no longer used. This is so because of the voltage-regulating transformer (VRT) now being used. The transformer suppresses transient pulses from the incoming power line and regulates the secondary voltage.

The VRT has a loosely coupled primary winding to secondary winding. The secondary winding is tuned to the 60-Hz line frequency by capacitor C248. Regulation of the transformer output is accomplished by this tuned circuit. The VRT is a *ferroresonant power transformer* and is also known as a *constant-voltage-supply.* It is a tuned transformer that acts as a filter. The transformer is heavy and it hums, but it is very stable and efficient. The voltage in the secondary winding will increase until the core of T205 is saturated. Once the core is saturated, the voltage will remain relatively constant. At this point, the resonating circuit is clipping the peaks off the ac voltage, and the output waveform resembles a square wave.

The circuit breaker F201 is for overcurrent protection. Under a continuous short-circuit condition, F201 will trip out, providing protection, and it can be reset. This is possible because of the loose coupling of the primary and secondary windings, which limits short-circuit current to approximately twice the normal operating current. Because of the parallel tuned circuit and saturated core, the VRT will supply all voltages required with reasonable regulation for a powerline variation from 95 to 140 V.

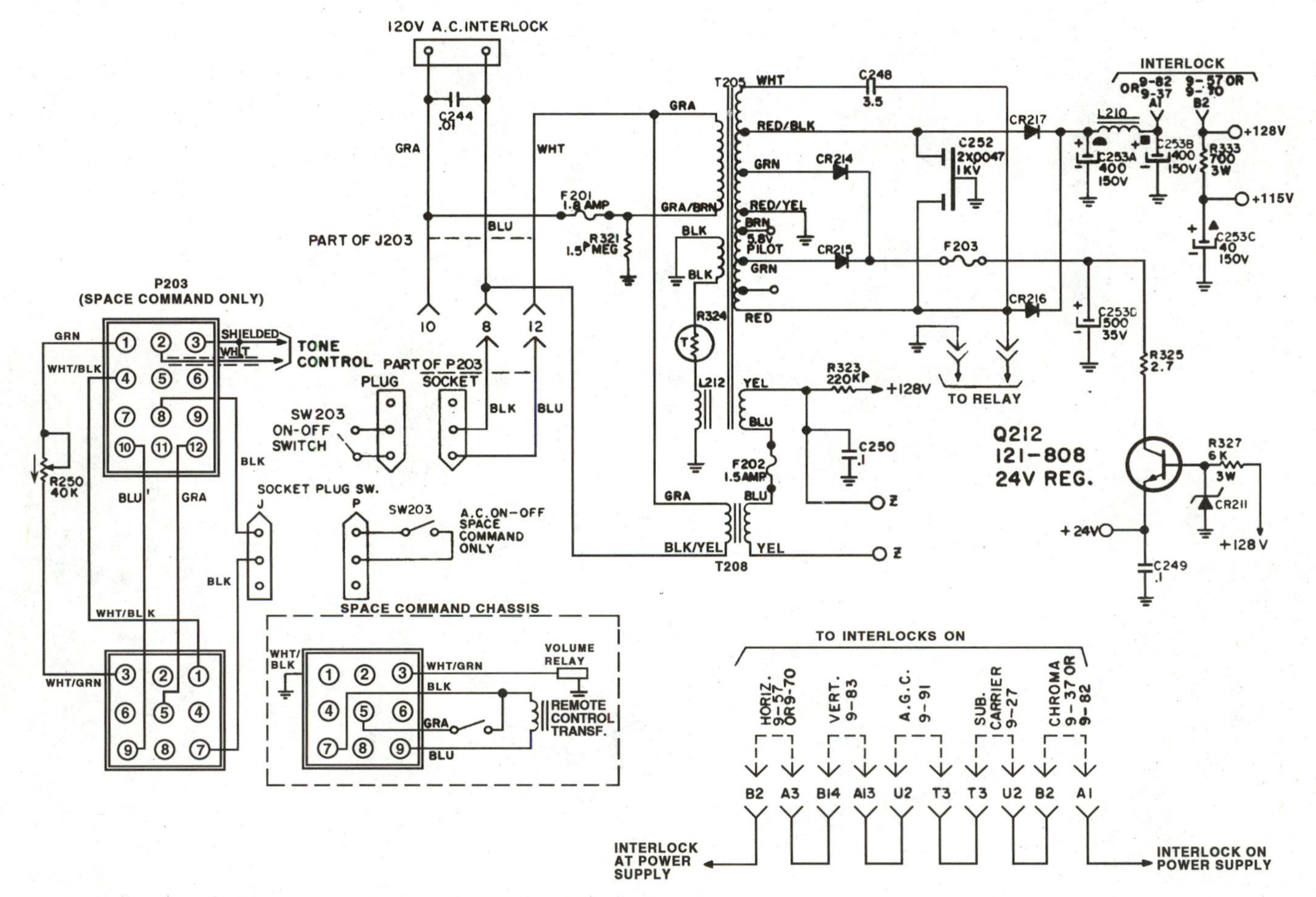

3-7 Voltage-regulating transformer (VRT) used in a television power supply circuit. (*Courtesy of Zenith Electronics Corporation.*)

In order to provide a picture almost immediately when the set is turned on, a standby transformer T208 is used. This transformer is connected in parallel with the on/off switch. When this switch is open or in the off position, the standby transformer is in series with the primary winding of the VRT. Because of the large difference in impedance between the two primaries, most of the voltage will be developed across the primary of the standby transformer. The standby transformer now steps down the voltage to approximately 5.15 V in the secondary winding. Since the secondary is in series with the VRT heater winding, the voltage is further decreased to 4.8 V. The phase relation is such that the current in the VRT filament winding is opposing the VRT primary winding. This phase relation will keep the voltages generated during the standby mode in the other secondary windings in the VRT to a minimum.

With the on/off switch in the on mode, the primary winding of the standby transformer is shorted. No voltage is generated in the standby transformer, and the VRT is now generating the filament voltage. The voltage on the VRT filament winding will be approximately 7.8 V rms, but since this winding is in series with the standby transformer secondary winding, the voltage on the CRT filament will be reduced to 6 V rms.

The +128 dc source is developed by full-wave rectifier diodes, capacitors C252 and C253A, and filter choke L210. Resistor R333 drops approximately 13 V, developing the +115-V supply. Capacitor C253C provides the additional filtering necessary due to the application of +115 V in the audio circuit.

Diodes CR214 and CR215, in conjunction with capacitor C253D, develop the voltage necessary to drive the 24-V regulator transistor Q212. Note that Q212 has an ac grounded emitter via capacitor C249. Zener diode CR211 in the base circuit is connected to the +128-V source and regulates the transistor output.

Servicing information and tips

Servicing this television power supply is not too difficult because the circuitry is quite simplified when compared with some previous solid-state circuits. Since the circuitry is simplified and most power supply systems show common symptoms when defective, you can isolate the defective area by just observing the symptoms.

Low B+ (+128 and +115 V) symptoms.

These symptoms involving low B+ can be isolated to the defective stage by measuring the three B+ sources: +128, +115, and +24 V. If the +128- and +115-V sources are low by approximately equal amounts, you check filter capacitors 253A, B, and C for leakage. Additionally, if choke L210 should become resistive, then low B+ will be the result. As in any power supply system, leaky rectifier diodes can cause the same problem. Since this is a voltage-regulating transformer system tuned to 60 Hz, should capacitor C248 become leaky or change value, or if transformer T205 becomes defective, the system will no longer operate in saturation, resulting in low B+ and very poor regulation.

Low +24-V symptom.
If the +24-V supply is low while the +128-V and +115-V systems are normal, check rectifier diodes CR214 and CR215 as well as filter capacitor C253D for leakage. Capacitor C149, Zener diode CR211, and transistor Q212, if leaky, can cause a low output from the 24-V supply. If Zener diode CR211 opens, the +24-V supply will have poor regulation, and the output voltage may be too high.

No B+ voltage symptom.
The most common cause of no B+ voltage is a shorted rectifier diode or shorted filter capacitor. If either condition is present, circuit breaker F201 will trip off. Also, at this point you should check out the ac input circuit and also test with a DMM for correct ac line voltage.

Picture hum symptom.
Hum in a dc power supply will result in horizontal bending in the picture that moves vertically through it. Filter capacitors are usually the most common cause of hum problems. With your scope, check the ripple voltage on the +24-V power supply. It should not be greater than 0.1 V peak to peak. Next, check the ripple voltage on the +128-V source. It should not be greater than 1.5 V peak to peak on the scope waveform.

Horizontal bars and vertical weaving in the picture can be caused by defective elements in the +24-V regulator circuit, such as transistors and diodes. Ripple in any of the power sources also may cause audio hum and buzz.

Symptoms of critical adjustment of controls or improper operation for some of the television chassis circuits often can be traced to the +24-V supply section.

Conventional television power supply operation

The Zenith television power supply shown in Fig. 3-8 utilizes a two-section regulated dc supply system. Separate regulation is provided for the +24-V supply and for the +125-V supply. Additional power is also furnished for the pilot lamps, automatic degaussing, and to heat the picture tube filaments when the receiver is not working and is in the standby mode. Since the chassis is solid state, the picture will come on quickly along with the audio.

The degaussing circuit

The automatic degaussing circuit is in series across a portion of the secondary winding of the power transformer. The degaussing coil L212 is mounted around the front circumference of the CRT bell. Thermistor R324 allows ac to flow through the coil, accomplishing demagnetization of the picture tube shadow mask. In a very short period of time the current flowing through the coil and R324 causes the thermistor to heat up, which causes a corresponding increase in its resistance. This continues, and quite rapidly the current through the degaussing coil is reduced to some very small value. This complete cycle takes place within a second or two. Degaussing

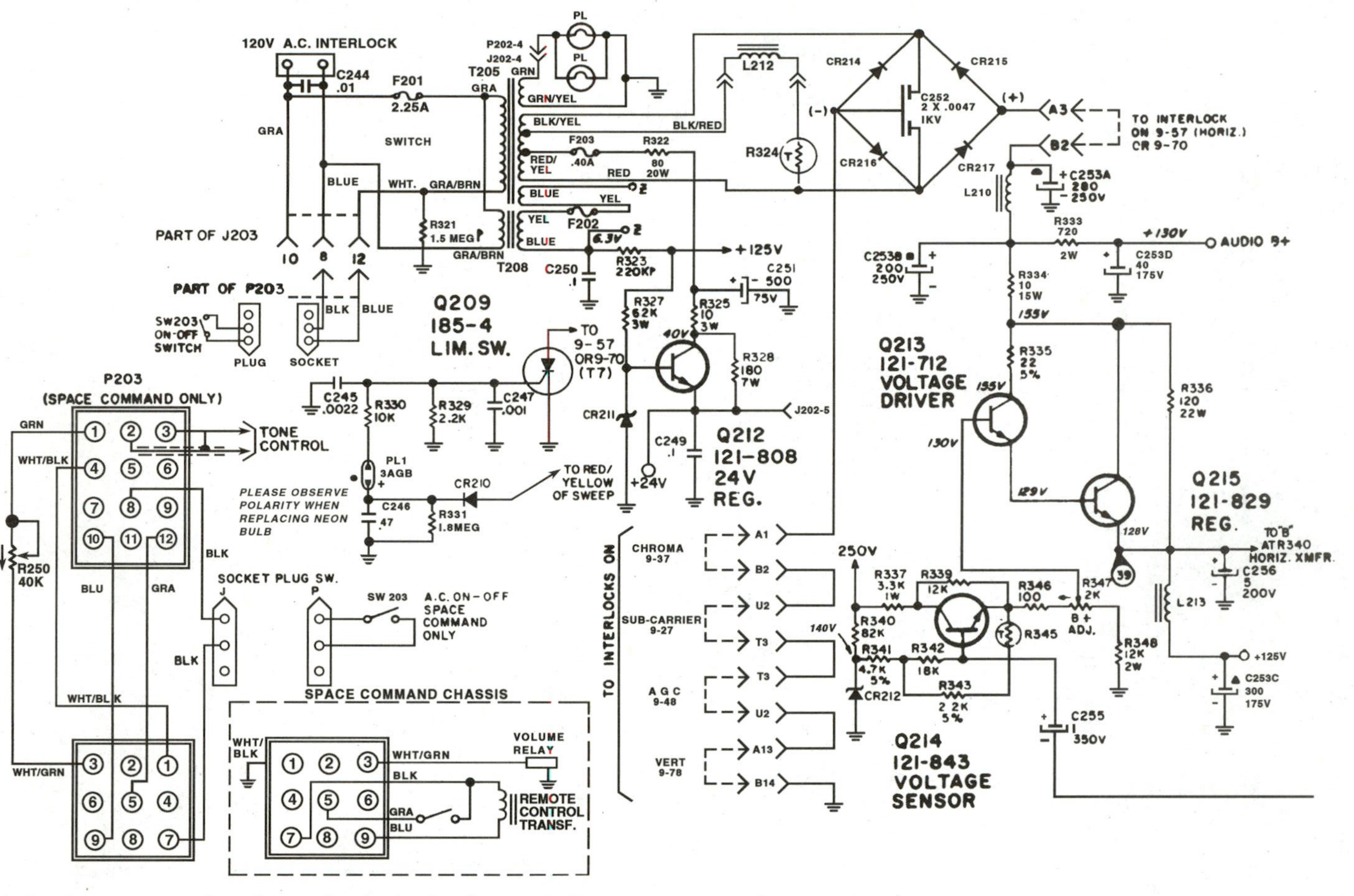

3-8 Conventional color television bridge rectifier power supply circuit. (*Courtesy of Zenith Electronics Corporation.*)

will not occur again until the television receiver has been shut off and the thermistor has cooled down sufficiently to allow current flow through the degaussing coils.

Bridge rectifier circuit

This linear dc supply uses a conventional power transformer. The power supply uses a full-wave bridge consisting of diodes CR214, CR215, CR216, and CR217 to provide approximately 160 V that is used after filtering for audio B+ and 130 V used by Q215 to provide the regulated +125-V supply. By means of a tap on the power transformer, CR214 and CR216 provide a dc voltage that is controlled by Q212, resulting in the development of a +24-V source. The ground return for the rectifiers in the power supply and the output of the full-wave bridge rectifier are interlocked via the plug-in module pins, thus disabling the power supply voltage when modules are pulled for chassis service or replacement.

+24-V circuitry

The +24-V supply is regulated by transistor Q212, which functions as a series regulator in conjunction with Zener diode CR211. Bias for Q212 is set by CR211, which is biased into the Zener mode by R327. Capacitor C251 provides some prefiltering for the regulator transistor.

Regulator transistor Q212 serves as a series regulator, and virtually all the current drawn by the +24-V loads flows through the device as collector current. Any change in current serves to rebias the transistor, thereby maintaining a steady current through the device and a stable voltage at the emitter. Resistor R328 connected across the regulator transistor serves to reduce the dissipation of transistor Q212.

+125-V circuitry

The +125-V regulator circuitry is somewhat more elaborate than that in earlier-model sets but essentially similar to the +24-V regulator just discussed. The current demands of the +125-V loads require greater current gain. One way of increasing current gain while maintaining regulation is the use of a dc-coupled amplifier to drive the regulator transistor.

The purpose of Q213 and Q214 and associated components is to provide a constant voltage to the base of regulator transistor Q215 regardless of the line voltage variations or changing load requirements. The transistor is basically regulating in the same way as the +24-V regulator. The operating point of the regulator is adjusted by R347, the B+ adjustment control.

The source of the voltage to forward-bias Zener diode CR212 in the +125-V regulator circuit comes from the horizontal sweep circuit. The horizontal sweep circuit supply comes from the +125-V regulator. The horizontal circuit cannot operate until the regulator passes current, and the regulator cannot operate until the horizontal circuit functions. For this reason, R336 is placed between the emitter and collector of the regulator transistor, allowing initial operation of the horizontal circuit to provide the required +250 V to operate the +125-V regulator. This resistor also lowers the dissipation of the regulator transistor because it carries some of the current drawn by the +125-V loads.

Pincushion circuitry

The side pincushion circuit is part of the power supply. A 60-Hz sawtooth waveform is taken from the green and red leads of the vertical output transformer. Diode CR208 is placed in series with R303, R301, R300, and the emitter of the pincushion amplifier transistor Q211. This is used to remove a large spike from the waveform and, with C238, serves to shape the waveform into a parabola that is amplified and coupled via capacitor C255 to the base of Q214, the voltage sensor transistor. This 60-Hz parabola is amplified by Q214, Q213, and Q215. This effectively modulates the B+ fed to the horizontal sweep transformer. Since this 60-Hz signal is synchronized to the vertical circuitry and is injected into the power supply, it has the effect of changing the width with the vertical scan. This width change occurs during vertical scan time to compensate for pincushion distortion by pushing out the sides of the picture's middle section. Since this 60-Hz ripple is unwanted in other circuits, the audio and the +125-V supply must be decoupled. The audio is decoupled by resistor R333 and capacitor C253D. The +125-V output is decoupled by C256, L213, and C253C.

Limit switch operation

The power supply shown in Fig. 3-8 incorporates a limit switch, which alters operation of the receiver in the event the high voltage increases beyond acceptable limits. The horizontal sweep transformer supplies pulses through diode CR210 to capacitor C246. Capacitor C246 achieves a charge that depends on the amplitude of the pulses. The value of the charge on the capacitor is not sufficient to cause the neon lamp to fire under normal operating conditions. Should the high voltage exceed limits, the pulses increase in amplitude and along with the charge on C246 reach a value sufficient to cause the neon bulb to fire. The charge on C246 then supplies current to keep the neon bulb lit between pulses from the sweep transformer. Should an increase in high voltage cause neon bulb PL1 and system clock reference (SCR) Q209 to fire, the horizontal sweep system is disabled, since the collector circuit of the oscillator transistor is effectively grounded through the limit switch. Thus high voltage is no longer available.

Should the SCR be fired accidentally by an arc, the receiver will have to be turned "off" for approximately 10 s before normal operation can again be achieved. This is the length of time it takes for the various capacitors in the receiver to discharge sufficiently to cause the limit circuit to return or reset to its normal operation.

Trouble symptoms and solutions

Let us now look at some troubles that have "cropped up" in conventional power supplies and use the symptoms to help isolate the defective components.

Main power supply fuse "blows"

If the main power supply fuse keeps "blowing," you should check the load on the +125-V distribution system for a short circuit. This is easily accomplished by measuring the resistance from the +125-V test point to ground. It should be in excess of

4000 Ω. If the resistance is less than 1000 Ω, you should check the horizontal, vertical, and audio output transistors for shorts, as well as the damper diode, retrace capacitors, regulator transistors, and filter capacitors.

Small raster on the screen

Leaky capacitors and increased resistance in the regulating network will cause reduced B+ voltage, resulting in a reduction in the raster size on the picture tube.

Audio, video, and vertical sweep problems

Failure of the +24-V supply circuit will cause a loss of audio, video, and vertical sweep deflection. If these are the symptoms, you should check out the +24-V regulator circuit and any of its load sources.

Hum problems

Hum in either the +125-V source or +24-V source is most likely to be caused by defective filter capacitors. The symptoms will be buzz in the audio and hum in the video stages if the +24-V section is defective.

Critical adjustment of controls

Symptoms of critical adjustment of controls or improper operation of many of the chassis circuits often can be traced to faulty operation of the +24-V power supply section.

Limit switch notes

Firing of the limit switch to prevent excessive high voltage may be caused by the horizontal circuits, arcing in the CRT or spark gaps, and changing television channels. To reduce the possibility of premature firing, a 10,000-Ω resistor at 0.5 W was added in series with the gate of limit switch SCR Q209, as well as a 0.01 microfarad disk capacitor with a 500-V rating, from the anode to the cathode of the SCR.

Massive failure of components in the +125-V source may be caused by the horizontal circuitry. For example, if the damper diode should become defective, causing an increase in current through the +125-V dc source, many of the components in the power supply can easily be changed in value. Therefore, it would be a smart troubleshooting procedure to check out the horizontal sweep circuitry before repairs of the power supply are completed.

4
CHAPTER

Switch-mode power supplies

This chapter presents the basics of switch-mode television power supply systems and how they operate. The chapter also reviews various troubleshooting techniques for repairing these circuits. In addition, some ways to dynamically test power transistors, switching transformers, and other components are discussed. These power supplies are sometimes pulse-width circuits, scan-derived systems, "choppers," "switchers," and so forth.

Switch-mode power supplies (SMPSs) are now found in most modern color television receivers and at times can be troublesome to many technicians. A first look at a modern SMPS schematic reveals strange circuit figures and parts. However, once you understand their basic operation, common troubleshooting techniques used in conjunction with proper test equipment will quickly find the problem.

Basic switch-mode circuit operation

The basics of a switching power supply include a source of dc voltage, a high-frequency transformer, a power transistor, and a pulse generator. Figure 4-1 shows the basic switching circuit. Circuit operation is very simple. The transistor is switched on and off by the drive signal closing and opening the path for dc current flow. This produces a changing magnetic field in the transformer primary winding. The changing magnetic field induces voltage in the secondary winding, where it is rectified and filtered into ac voltage.

The power transistor operates as a switch; it is either on (saturated) or off (cut off). The amount of energy delivered to the secondary is determined by the "on time" of the transistor. The output voltage can be varied by changing the frequency or duty cycle of the transistor drive signal. These two techniques of regulating the output are called *pulse-width-modulated regulation* (varying duty cycle) or *pulse-rate-modulated regulation* (varying frequency of pulses). Figures 4-2 and 4-3 show how each form of regulation changes the total "on time" of the switcher transistor.

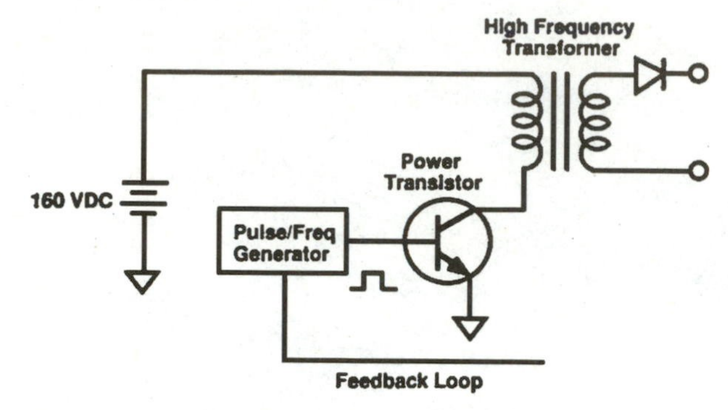

4-1 Basic switched-mode power supply (SMPS), including a dc voltage source, high-frequency switching transformer, power transistor, and pulse generator. (*Courtesy of Sencore, Inc.*)

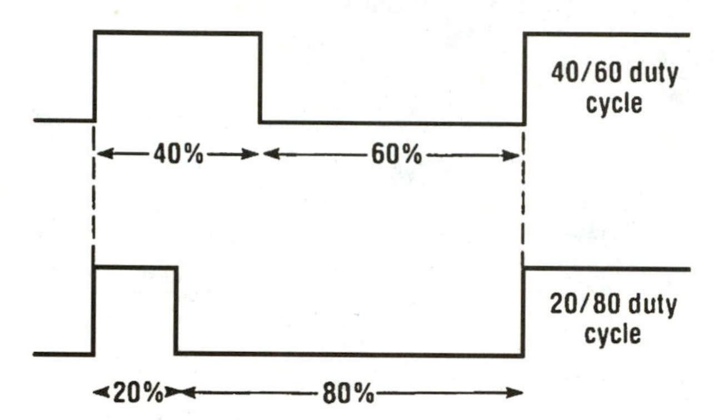

4-2 Pulse-width regulator varying the switching transistor's conduction time by varying the width of the pulses. (*Courtesy of Sencore, Inc.*)

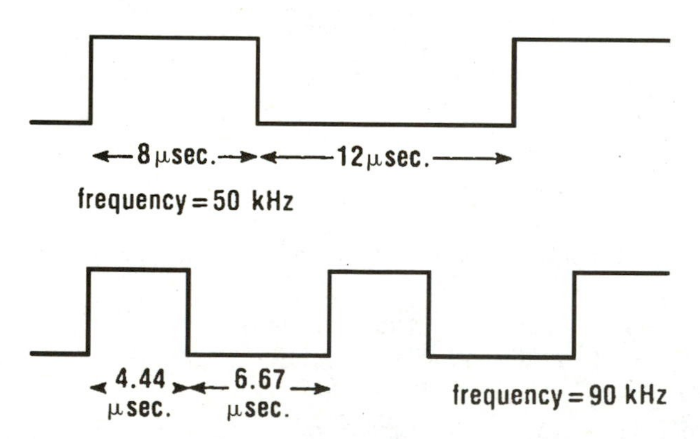

4-3 Pulse-rate regulator varying the switching transistor's conduction time by varying the frequency of the pulses. (*Courtesy of Sencore, Inc.*)

One of the chief benefits of SMPSs is their ability to closely and efficiently regulate the power delivered to the load. Regulation in an SMPS is obtained by comparing a dc voltage from the output with a reference voltage. The comparison provides feedback to the pulse generator, which alters the drive signal to compensate for the output voltage change. This comparison/compensation occurs continually and provides a closely regulated output. Note that only enough voltage as necessary is provided; thus no excess power needs to be dissipated, as in conventional shunts or voltage-regulator systems.

The block diagram in Fig. 4-4 shows a typical television SMPS. The basic blocks include the transformer, power transistor, pulse generator, and feedback voltage-regulator blocks. The transformer input is unregulated B+ supplied by rectified and filtered ac line voltage. Optoisolators may be used to isolate the transformer secondary circuits from the ac line.

Overcurrent protection is also included in most modern SMPSs. This circuit usually samples the voltage drop across a resistor in series with the switching transistor. If the current rises abnormally, the voltage will exceed a reference level and shut down the pulse generator. This circuit provides protection to components in the event a problem occurs.

Most modern receivers have a standby power supply that runs all the time the television is plugged in. This supply is needed to power the microprocessor, memory circuit, remote control receiver, and SMPS's startup system circuitry. In some newer-model receivers, the standby supply is also an SMPS.

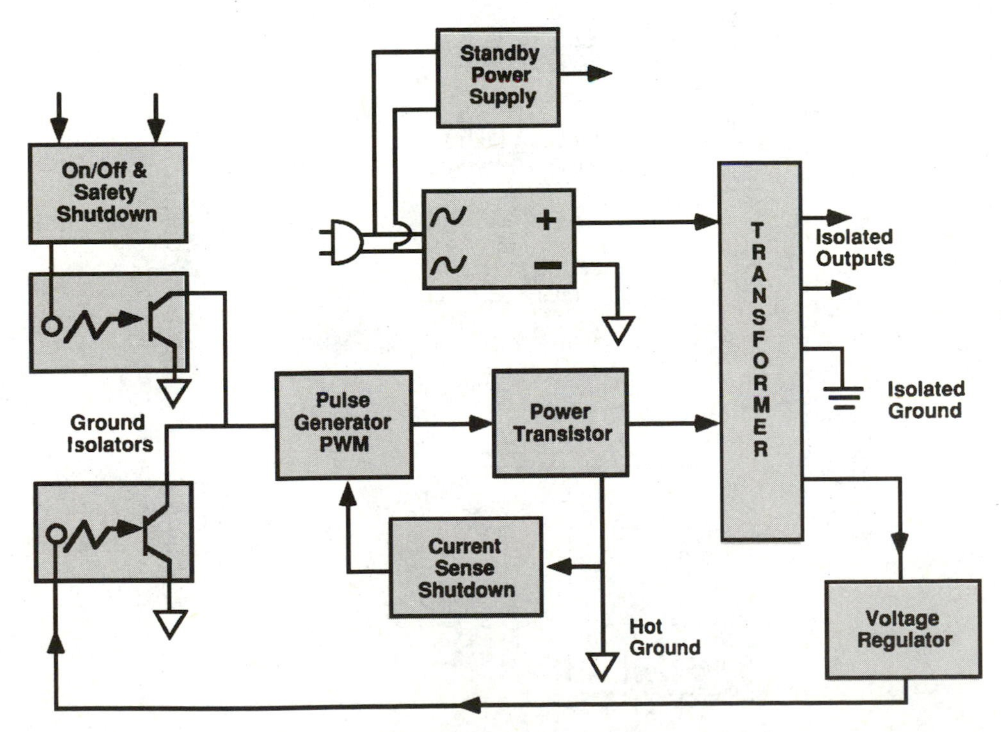

4-4 Block diagram of a typical SMPS. (*Courtesy of Sencore, Inc.*)

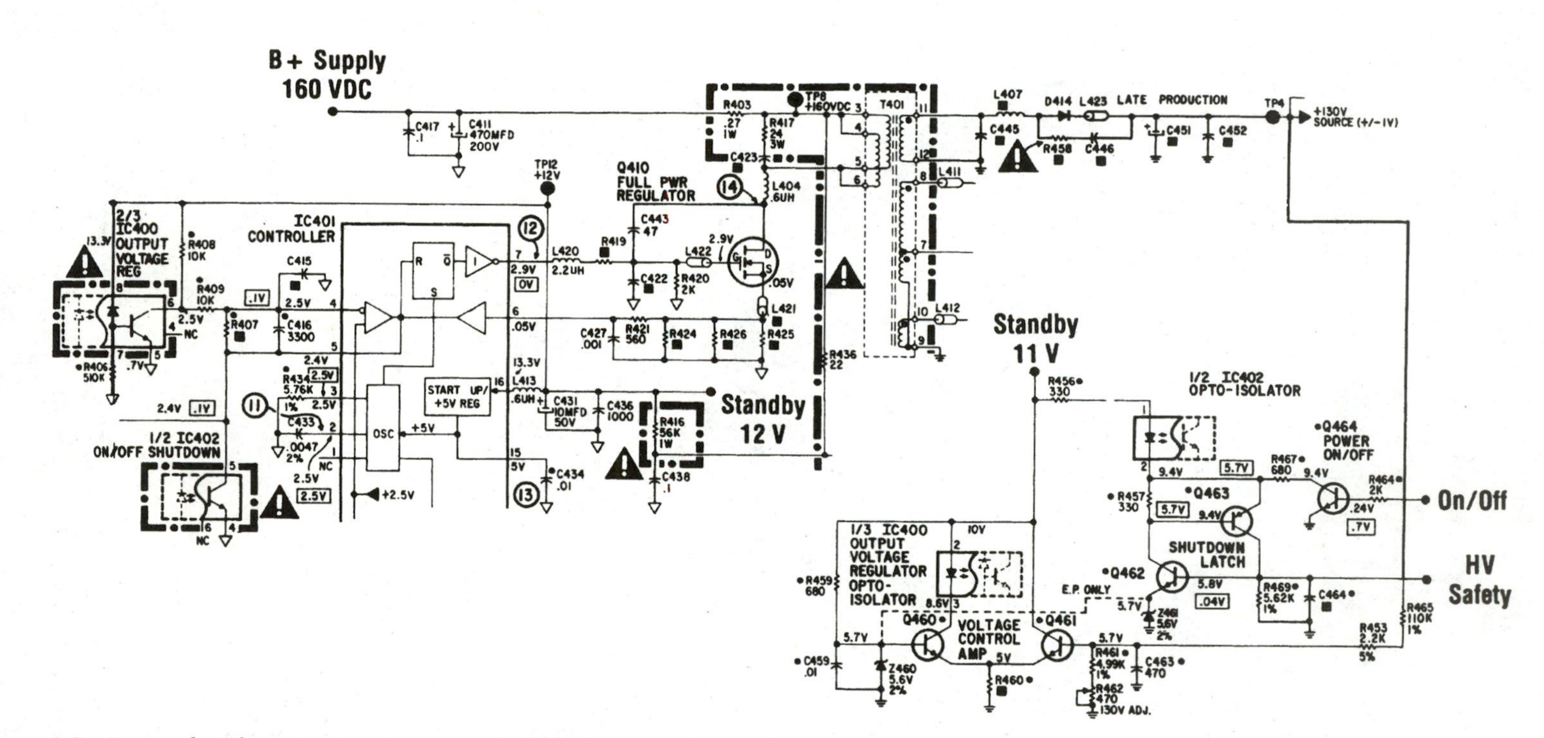

4-5 Example of an SMPS that can be troubleshot efficiently by using the proper test equipment techniques. (*Courtesy of Sencore, Inc.*)

Now that we have a good basic understanding of how a modern television SMPS works, let us transfer the basic blocks to a real troubleshooting process you can deal with.

Isolating the defect

Please refer to the block diagram in Fig. 4-4. You can take virtually any SMPS, no matter how complicated it first appears, and fit it into blocks. Let us now look at a typical color television power supply schematic, as shown in Fig. 4-5. In this SMPS, the standby switched-mode circuitry has been omitted for simplification.

The switching transistor output provides the best starting point when troubleshooting an SMPS. Note the B+ voltage and analyze the waveform amplitude, frequency, duty cycle, noise, and ripple. The Sencore SC3100 Waveform Analyzer (Fig. 4-6) is ideal for this because it provides all the voltage and waveform analyzing tests automatically with one test probe. Connect the SC3100 to the terminals on the transistor that connects to the primary of the switching transformer (normally the collector or drain), and connect the ground to the switching power supply's primary or "hot" ground. Set the SC3100 volts/DIV switch to 100 and the timebase frequency to 10 kHz. Now be ready to observe events when you power up the television chassis. These results will provide valuable clues to the television set's problems.

4-6 Using the Sencore SC3100 Auto-Tracker oscilloscope. (*Courtesy of Sencore, Inc.*)

Remember to always use an ac line isolation transformer for the chassis under test. Select the dc volts button on the SC3100, and read the dc volts to the switching transistor. With the set turned off, you should see about 150 to 160 V. If the voltage is low or missing, check the unregulated supply and the switching transformer primary, or test for a shorted switching transistor. Also, check the B+ supply for excessive ac 120-Hz ripple, which may indicate a main filter capacitor problem.

If the unregulated B+ output is good, you are ready to test the switching power supply. Turn the set on, and check for switching pulses, comparing the wave shape to that shown in the service data. In many cases, some ringing on the waveform is normal. Test the frequency of the switching pulses on the SC3100 by pushing the Freq button and reading the frequency automatically on the CRT display. Confirm that the switching frequency is normal by comparing it with the frequency shown in the television set's service data.

At this time, connect the second channel input of the scope to the SMPS's regulated B+ output or to the horizontal output transistor's collector. Also attach the second probe ground lead to the secondary ground. If the B+ voltage is correct, test the other supply voltages or circuits to isolate the receiver's defect.

If no pulses are observed at the switching transistor output, the SMPS is dead, not starting, or is going into immediate overvoltage/current shutdown. Move the test probe to the input, usually the base or gate element of the switching transistor. Reduce the voltage/DIV control setting on the oscilloscope, and test for an input drive signal. This will determine if the problem is an open transistor or a lack of drive from the pulse generator. If no drive pulses are present, the pulse generator is not working.

To isolate problems with the pulse generator, first test the standby power supply voltages with your scope. If these voltages are low or have excessive ripple, they may cause "no start" problems. If the standby power supply voltages are good, check the power on/off control circuitry. Press the television on switch to ensure that the microprocessor is providing the proper turn-on signal. The turn-on signal should turn other circuits on, providing the input to the pulse generator. A defective voltage regulator circuit also can prevent the SMPS from starting. Check for improper voltages in this circuit with a DMM or scope, and then test any suspected components.

Another common symptom observed at the switching transistor's output is momentary pulses that quickly stop. This is an indication that the pulse generator is being started and then shut down. The most likely cause is the high-voltage shutdown control circuits. This can be due to high B+ voltages supplied to the horizontal circuits or problems in the horizontal circuits. This type of problem can be especially frustrating because nothing stays on long enough to test.

To isolate this problem, defeat the horizontal output stage by opening the regulated B+ path to the collector of the horizontal output transistor. Simulate the horizontal circuit load by connecting a high-wattage 500- to 1000-Ω resistor between the B+ point you opened and ground. Turn the set on, and test the regulation of the SMPS's output by varying the ac output line voltage. If the B+ output does not regulate properly, you need to isolate problems in the voltage regulation feedback circuits.

If you have good regulation, test the high-voltage shutdown circuits by substituting the high-voltage sense input using a dc power supply. Increase the dc voltage above the reference level, and test for normal shutdown. If regulation and shutdown are normal, a problem exists in the horizontal output timing or shutdown reference circuits.

The final waveform symptom that is sometimes observed at the switching transistor's output is fast, short-duration spikes. This waveform may be accompanied by a squealing sound or a hotter than normal switching transistor. These symptoms are indications that the SMPS is quickly going into current limiting during each drive pulse. This current limiting may be caused by a defective transformer or a short in the secondary circuit.

Scoping the SMPS circuits

The last step to quickly troubleshoot these circuits is to locate the defective part or parts. Replace them, and watch the set start to operate properly. However, proving suspect parts good or bad can be difficult. Swapping specially ordered parts, such as switching transformers and special power transistors, just cannot be cost- or time-effective. Randomly replacing parts also can lead to mistakes or parameter changes that compound the original problems.

Components such as optoisolators or IC pulse generators are best tested in-circuit by thoroughly analyzing input/output voltages and waveforms. Use a wide-band oscilloscope to check out these waveforms. Verifying proper working voltages, checking waveforms, and tracing defective signals isolate defects to these components in the least amount of time.

Switching transistor checks

Occasionally, you may suspect that a switching transistor is breaking down at the circuit operating voltages. You can check for breakdown by using the leakage test of the Sencore LC102 Auto-Z instrument. Remove the transistor and connect a 1000-Ω resistor in series, as shown in Fig. 4-7. Program the LC102 to the same or a slightly

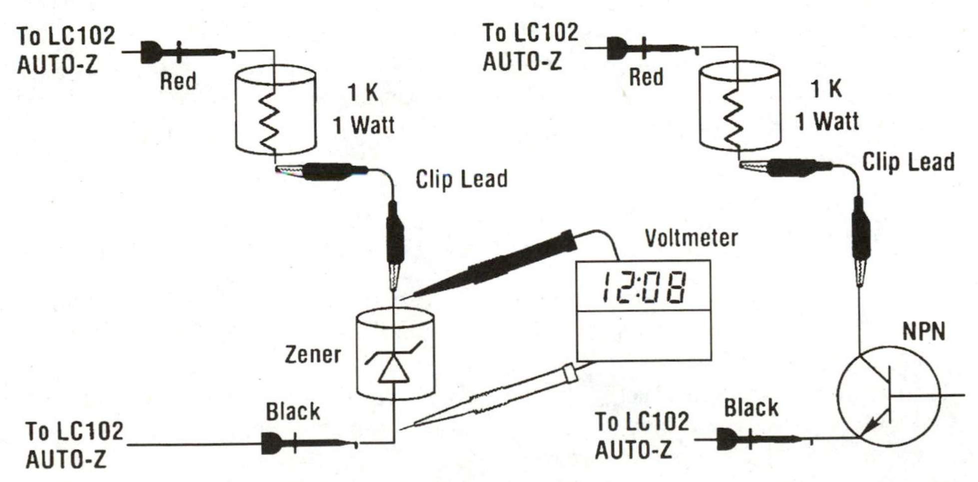

4-7 Test setup for finding the breakdown voltage of a transistor. (*Courtesy of Sencore, Inc.*)

higher voltage as that supplied by the unregulated dc power supply to the transistor. Press the capacitor leakage button, and check for leakage current. If the LC102 shows leakage current, the transistor is breaking down and should be replaced.

Zener diode checks

A similar test can be used to check the reference voltage of Zener diodes. Connect a 1000-Ω resistor in series with the Zener diode, and use the Auto-Z's leakage test to apply a reverse bias voltage as shown in Fig. 4-7. Connect a voltmeter, and measure the voltage across the Zener diode. Apply a reverse voltage with the leakage test that is slightly higher than the Zener's rated voltage. Increase the applied voltage until current is indicated on the Z-meter when the leakage button is pressed. The voltmeter will indicate the Zener voltage. This voltage should be within the Zener's rated tolerance.

Switching transformers

You can test the switching transformer for a single shorted turn using the LC102's ringer test feature. With the switching transistor removed, perform the ringing test across the primary winding of the switched-mode transformer while in circuit. If the transformer rings 10 or more times, it has no shorted turns. If it rings less than 10 times, remove the transformer from the circuit and perform the ringing test again. If any winding rings 10 or more times, the transformer is good. If the transformer is completely removed from the circuit and does not ring 10 or more times, it is defective and should be replaced.

Electrolytic capacitors

You will find that electrolytic capacitors are especially troublesome in SMPSs because of the power supply's high operating frequency. A capacitance value test alone rarely finds bad electrolytic capacitors. Thoroughly testing electrolytics requires testing the capacitor for value, leakage, ERS, and dielectric absorption. The Sencore LC102 Auto-Z dynamically tests capacitors with automatic good/bad tests to EIA established parameters. Simply program the capacitor's parameters into the LC102 and perform all four good/bad capacitor tests by pressing the individual test buttons. The value and good/bad read-out are automatically displayed, telling you the precise condition of the capacitor.

Solving startup and shutdown problems

The first step in troubleshooting startup and shutdown circuits is to understand how the circuits operate. For these checks, we will be using the Sencore VA62A Video Analyzer shown in Fig. 4-8. Since the scan-derived power runs most of the circuits in a modern television set (except the horizontal output stage), we will start with a look at the scan-derived system: the regulated dc source, the output transistor, and the scan-derived voltage source.

4-8
Sencore VA62A Universal Video Analyzer. (*Courtesy of Sencore, Inc.*)

Regulated dc source

The regulator is driven by a full half-wave power supply that provides between 140 and 170 V dc. The regulator can be as simple as a pass-transistor and a Zener reference. In many cases, however, the regulator is an SCR driven by a separate oscillator that is controlled by scan-derived power. Use a DMM to prove that the voltage supplied to the regulator is OK, as well as to monitor the regulator output.

Horizontal output transistor

The SCR regulator provides a constant voltage to the horizontal output transistor, which in turn keeps the 25- to 35-kV high voltage stable for any changes in the ac line voltage and different flyback loading. Note that the special drive signals from the VA62 let you inject up to the horizontal output transistor to prove that the horizontal stages are working.

Scan-derived voltage sources

The flyback supplies the high voltage to operate the CRT, the keying pulses required in other stages, plus the scan-derived dc voltages used by all other circuits in the receiver. A typical flyback transformer, for example, will have a low-voltage winding to power the i.f., video, color, and sound stages and a higher-voltage winding to power the vertical and video output stages. And always keep in mind that the horizontal oscillator is also powered from the flyback output stage. The VA62A provides a special current-limited dc power source, allowing you to substitute in for the horizontal oscillator power and get the set operating. Thus you have with this circuit a basic loop situation:

- There are no scan-derived voltages unless the horizontal output is operating.
- The horizontal output is driven by the horizontal oscillator.
- The horizontal oscillator is powered by a scan-derived voltage.

In such a case, the "loop" must be complete before the television will start up and operate.

Start circuits: Trickle and kick start

There are two types of horizontal oscillator starters, *trickle start* and *kick start* (Fig. 4-9). The starter used depends on whether the chassis is connected to the ac line directly or is electrically isolated from the "hot" ground by the flyback transformer.

Trickle start mode

This starting circuit is less complicated because there is only one ground. A large-value resistor is simply connected between the unregulated B+ line and the horizontal oscillator (see Fig. 4-9). The resistor will not supply enough current to operate the receiver but will allow the oscillator to operate. The flyback output supplies power to the horizontal oscillator as soon as the horizontal oscillator has started up, closing the power loop.

Kick start mode

The kick start mode supplies a small amount of voltage to the horizontal oscillator for just a few seconds after the receiver is turned on, as shown in Fig. 4-10. This small power supply has its own ac line isolation transformer to prevent electrical connection between the chassis and the ac line. The primary of the transformer is connected between the unregulated B+ and a large electrolytic capacitor returned to the "hot" ground. The transformer supplies an output only during the time the capacitor is charging. If the horizontal oscillator has not started by the time the capacitor is fully charged, the power supply loop is not completed, and the receiver will not operate.

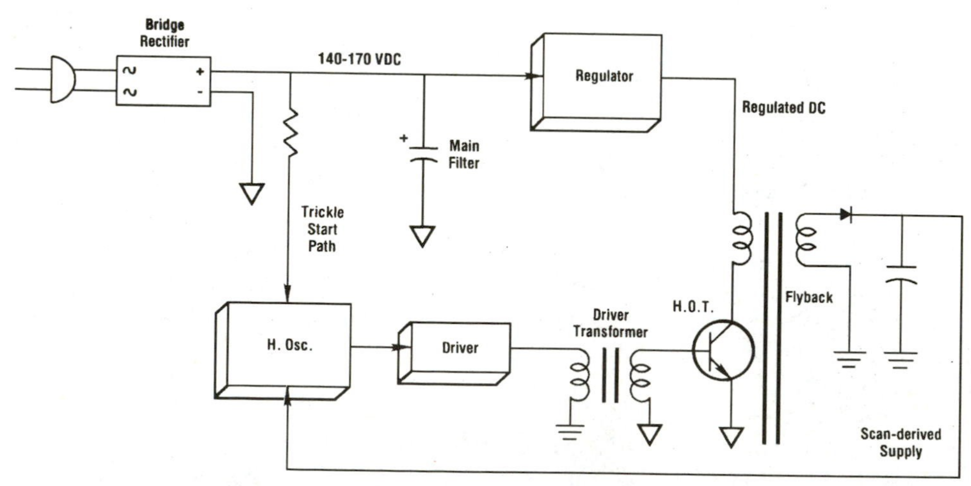

4-9 Scan-derived power system, including the regulated dc source, the horizontal output transistor, and the scan-derived supply. (*Courtesy of Sencore, Inc.*)

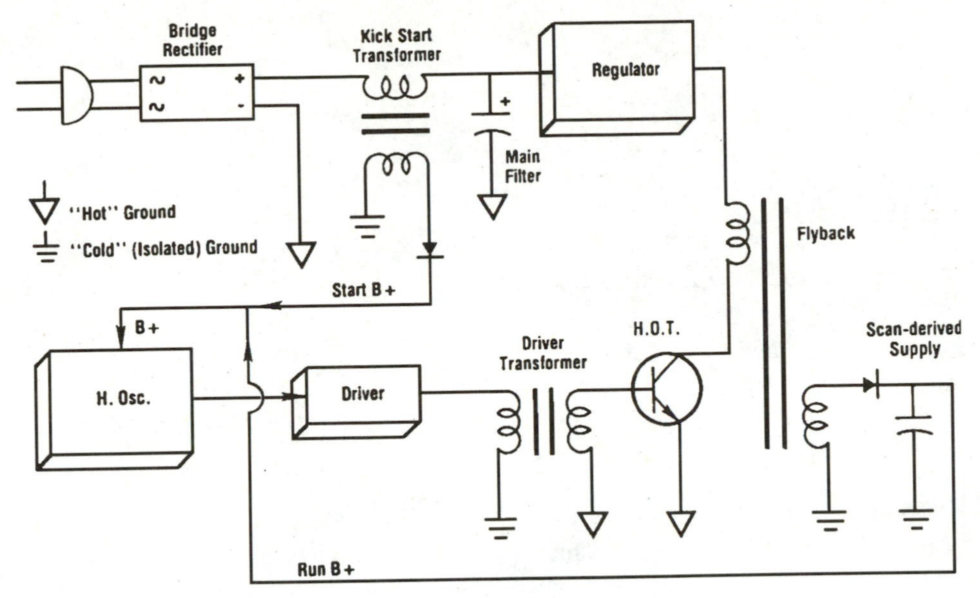

4-10 Block diagram of a scan-derived power supply. (*Courtesy of Sencore, Inc.*)

Each type of starter circuit gives a different shutdown symptom, which provides some valuable troubleshooting clues. Note that the kick start system only allows the horizontal oscillator to start once each time the receiver is turned on. You may hear a short audio rushing sound or the crackle of high voltage when the receiver is first turned on, but there will be complete silence after that. The trickle start circuit allows the set to restart each time the shutdown circuit is activated. This produces a "put-put-put" sound as the audio and high-voltage circuits are alternately turned on and off.

Shutdown circuits

The shutdown circuits can prevent the television receiver from operating. Unsafe conditions that produce X-radiation or stress components in the horizontal output stage will activate the shutdown circuits. Three basic shutdown circuits are used, as shown in Fig. 4-11. Some receivers use only one of these circuits, while others have all three.

- Horizontal oscillator shutdown
- Output power supply shutdown
- Horizontal frequency-shift shutdown

Horizontal oscillator shutdown

This is the most common type of shutdown circuit. It simply disables the signal from the horizontal oscillator to the output transistor.

1. The signal may be killed in the oscillator itself.
2. The oscillator output may be shorted to ground.

This shutdown circuit is activated by a pulse from the flyback transformer. The pulse amplitude tells when everything is operating normally. When the pulse is too large, the receiver is developing excessively high voltage. A level detector, triggered by this flyback pulse, turns on an SCR or transistor in the oscillator (or driver stage) to kill the oscillator signal.

Output power supply shutdown

This power supply circuit senses the amount of current in the emitter of the horizontal output transistor. It interrupts the regulated dc (by stopping the separate regulator oscillator used in these chassis) if the voltage across a small-value resistor in the emitter circuit is too high. The receiver shuts down when the regulated supply is interrupted.

Horizontal frequency-shift shutdown

RCA has used the frequency-shift scheme for shutdown protection in some of its earlier-model color television chasses. However, this scheme does not result in complete shutdown, but rather pulls the horizontal oscillator off frequency to reduce the high voltage to a safe level. This circuit is triggered from any of three input samples:

1. The 180-V dc supply
2. The CRT screen grid and focus voltages
3. A pulse from a separate winding on the flyback transformer

Note that the horizontal hold control has no effect if and when this shutdown circuit is activated.

Problem isolation

Let us now isolate the problem to the power supply or shutdown loop. You can use the receiver's symptoms when it is first turned on to decide whether to trou-

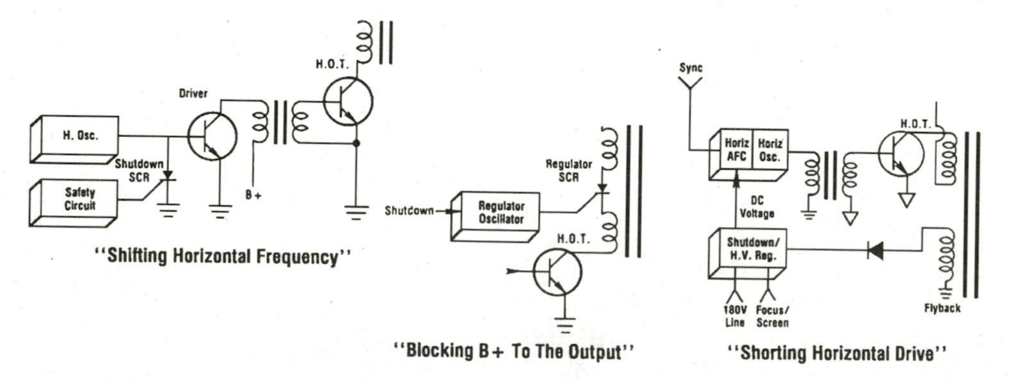

4-11 Safety circuits triggering one or more shutdown circuits to stop the horizontal outputs. (*Courtesy of Sencore, Inc.*)

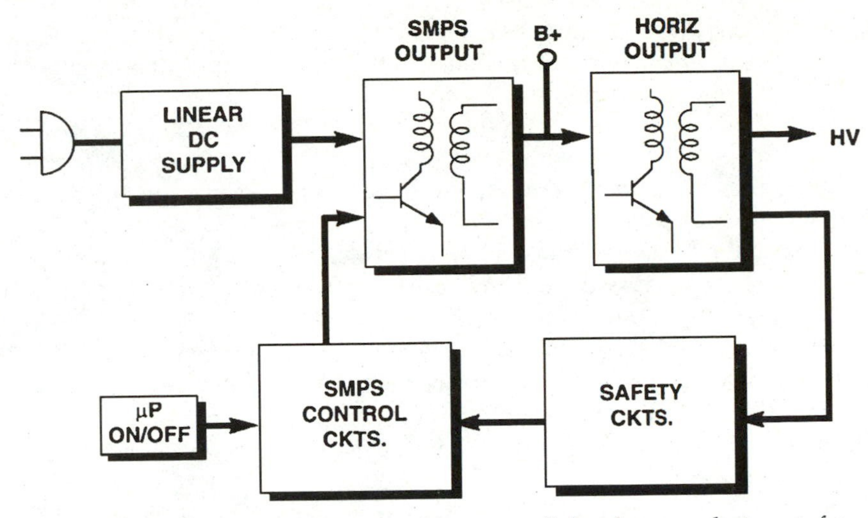

4-12 Problems in the horizontal output, flyback secondaries, safety shutdown, or control circuits preventing normal power supply operation. (*Courtesy of Sencore, Inc.*)

bleshoot the power supply or the shutdown loop. Generally, a power supply problem results in a totally dead receiver. A shutdown problem, on the other hand, allows the set to operate for a split second before the shutdown circuits take over and kill the operation. For this check, just turn the volume up halfway before turning the receiver on, and listen for a rushing sound or a "put-down" symptom. Either one indicates a shutdown condition.

If you have a Sencore VA62A, its digital meter, drive test signals, and dc power substitution supply will give you the basic tools needed to solve startup and shutdown troubles. And with the SC3200 Auto Tracker, with its speed and accuracy, you will have an unbeatable troubleshooting combination.

Horizontal output load test

After many years of television troubleshooting, it is quite evident that "power-up" problems involve testing more circuits than just the power supply. Note in Fig. 4-12 the relationship between the SMPS, horizontal output/flyback, safety circuits, and control circuits. When the television will not "power up" properly or parts are instantly damaged, there is no procedure to make measurements to help you isolate the problem.

Through trial and error it was found that the horizontal output stage was the key to detecting power supply loading and "power-up" problems. The horizontal output stage is responsible for taking current (power) from the B+ supply and transferring it to most of the circuitry via the flyback transformer. Thus the timing or resonant action of the horizontal output stage determines the amplitude of the flyback pulse used to develop normal high voltage.

To meet these needs, Sencore's TVA92's horizontal output load test was developed to fill these power supply service voids. This test checks the horizontal output circuit and associated loads with no power applied to the receiver. Let us look at how this test works and how you can use it to troubleshoot B+ power supply loading, horizontal output stage problems, and shutdown problems.

The Sencore TVA92's horizontal output load test enables you to simulate the normal operation of the television set's horizontal output stage with no ac power applied to the set. Simulating the operation of the horizontal output stage requires the following:

1. B+ voltage to the flyback primary
2. A transistor switch that switches flyback primary current to ground at a 15,750-kHz rate with 30-μs "on time"

The TVA92A's horizontal output load test fulfills these requirements with a 15-V B+ substitute supply and a power transistor switched at the proper rate and time, simulating the action of the horizontal output stage. During the loading test, ac is produced in the television's flyback and yoke, closely matching the full-power operation of the horizontal output stage.

The Sencore TVA92A's horizontal output load test requires three simple connections to the chassis. The B+ test lead is connected to the television's B+ terminal of the flyback transformer primary. The other two connections are made to the emitter and collector of the horizontal output transistor. These test lead connections are shown in Fig. 4-13.

Good/bad test read-outs

The horizontal output load test checks the operation of the horizontal output stage and monitors the current load demand of the television's B+ supply. Separate milliamp and microsecond read-outs are provided to detect abnormal conditions, with good/bad indicators for each.

The TVA92A uses 15 V as a substitute for the television's B+ supply, which is approximately ⅒ the normal B+ voltage found in most sets. At this reduced B+ level, the television's horizontal output stage operates similarly as it would if the full B+ voltage were applied but with approximately ⅒ the current and voltage amplitudes.

The TVA92A measures and displays the current supplied by the horizontal output load test's 15-V supply. Read-outs between 5 and 80 mA represent a normal range of current for a wide variety of horizontal output stages and are displayed as

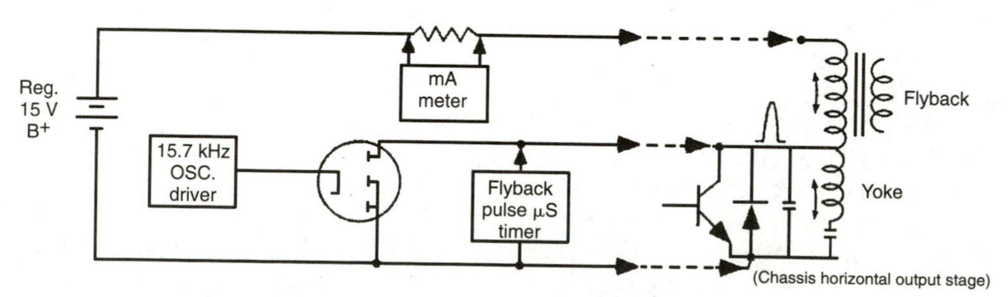

4-13 Sencore horizontal output load test simulating the normal action of the horizontal output stage without applying ac power to the television set. (*Courtesy of Sencore, Inc.*)

"good" by the TVA92A. Current levels less than 5 mA indicate improper test lead connections or an opening in the horizontal output circuits. Readings greater than 80 mA indicate heavy current load demands from the horizontal output, flyback, or other B+ circuits. Readings outside the 5- to 80-mA range are shown as "bad" by the TVA92A Video Analyzer.

The milliamp read-out of the horizontal output load test may be used to forecast the television's B+ power supply current at full operating voltages. To approximate the load current, simply multiply the milliamp read-out of the horizontal output load test by 10. In most receivers, the television's actual B+ current at full voltage will be slightly higher due to additional CRT current. Readings higher than 80 mA relate to excessive B+ power supply current demands of 1 A or more at full B+ operating voltages.

During the horizontal output load test, a flyback pulse is produced at the collector of the horizontal output transistor. The TVA92A measures and displays the microsecond duration of this flyback pulse. The duration or "on time" of the flyback pulse depends on the operation of the horizontal output circuits of the television, primarily the flyback, retrace timing capacitors, yoke, and yoke capacitor.

The pulse-time read-out indicates if proper timing and resonant action are occurring in the flyback and yoke circuits. Readings between 11.3 and 15.9 μs is the normal range. Readings above or below this range are considered "bad" and indicate improper timing, flyback defects, or a loading problem.

When both the milliamp and microsecond read-outs indicate "good," the horizontal output stage and flyback secondaries are operating within a normal range. In the majority of cases, this indicates a 100 percent problem-free horizontal output stage.

A "bad" reading in one or both of the horizontal output load test parameters indicates a problem in the horizontal output circuit, flyback, or flyback secondaries. You can use the "bad" read-out to help isolate the problem.

Troubleshooting B+ shorts

The horizontal output load test milliamp read-out can be used to determine the severity of the loading problem. There are three common types of B+ supply shorts or leakage problems:

1. Short (low-resistance dc current path to ground)
2. Dc leakage (higher-resistance dc current path to ground)
3. Ac short or leakage (added ac current or power demand due to a shortened turn in the yoke/flyback or a defect in the flyback secondary circuits)

The maximum current by the TVA92A's horizontal output load test is 250 mA. A readout near 250 mA during the horizontal output load test indicates a low-resistance short on the B+ power supply. A likely cause of a B+ short is a *shorted horizontal output transistor.* The horizontal output transistor and/or damper, if good, will not draw excessive current, affecting the horizontal output load test results. If the milliamp read-out changes to "good" after removing the horizontal output transistor, you have confirmed that the horizontal output transistor is shorted. If the milliamp read-out remains "bad" after removing the horizontal output transistor, continue to open possible dc shorted paths to isolate the problem.

Loading problems consisting of higher-resistance shorts produce read-outs on the horizontal output load test ranging from 80 to 200 mA. The first step in isolating a loading problem is to determine if the added load current is caused by dc loading or ac loading. Refer to ac and dc leakage path drawings in Fig. 4-14. You can determine this by disconnecting the yellow test lead (collector). This stops the switching action in the horizontal output stage by removing any ac in the flyback and yoke.

The remaining current indicated by the horizontal output load test display is dc current to the horizontal output stage, driver, and other B+ powered circuits. If the current is higher than 5 mA, suspect a dc short or leakage path on the B+ power supply circuits. To isolate dc short or leakage paths, you should open the circuit paths while comparing milliamp read-outs with your earlier readings.

If the current read-out is greater than 80 mA during the horizontal output load test but less than 5 mA when the yellow (collector) lead is removed, the high current is a result of a large ac load in the horizontal output stage. The current demand may be caused by shorted turns in the flyback/yoke or by a short or leakage on any of the secondary circuits of the flyback.

To isolate an ac loading problem, use the Ext PPV & DCV input of the TVA92A to measure dc voltages and peak-to-peak flyback pulse amplitudes on the secondary windings of the flyback. The voltage and current levels should be approximately $\frac{1}{10}$ the normal values indicated on the schematic. Waveforms or dc voltages that are missing or much lower indicate a possible problem associated with the flyback winding. Open the current path by unsoldering a flyback lead, scan diode, etc. Repeat the hori-

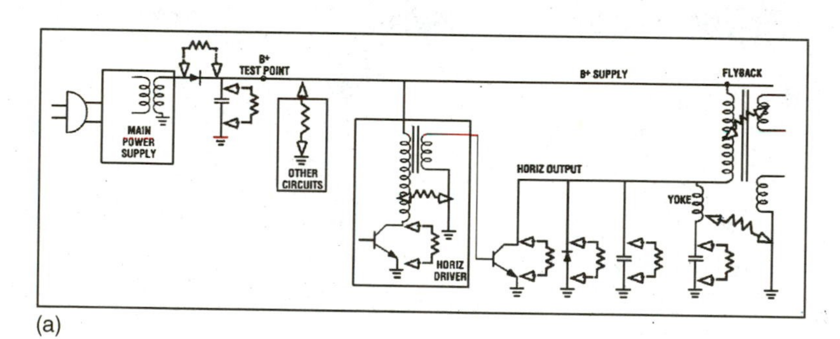

(a)

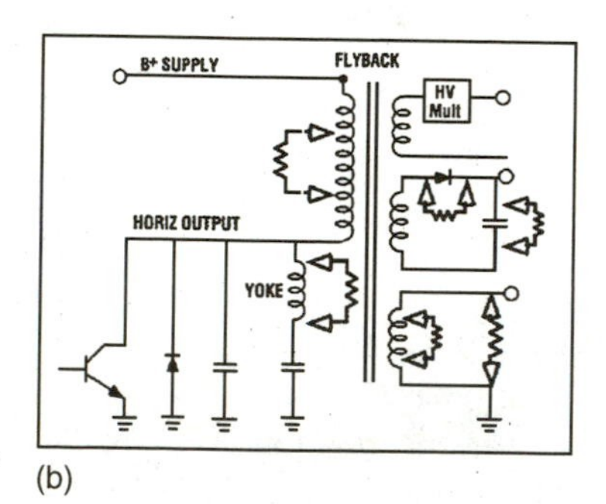

(b)

4-14 Some possible short or leakage paths that can load down the B+ power supply: (*a*) dc leakage paths; (*b*) ac leakage paths. (*Courtesy of Sencore, Inc.*)

zontal load test and compare the milliamp read-out with the previous value. A large decrease in the milliamp read-out confirms a problem with the load on that winding.

If all the secondary voltages appear normal or equally reduced, the problem is likely caused by a shorted turn in the flyback/yoke, or a component is breaking down in the horizontal output stage.

Horizontal sweep timing problems

The horizontal output load test microsecond read-out provides an indication of the resonant timing change in the horizontal output stage or problems affecting the flyback pulse wave shape. There are types of problems in the horizontal output flyback circuits that will be evident on the μs read-out load test.

A blank μs read-out indicates that flyback pulses normally produced during the horizontal output load test are not present. This read-out can be an indication of improper test lead connections or an open circuit path from B+ through the flyback primary to ground. The open circuit may be in the flyback primary, B+ path, or emitter ground path. A blank μs read-out also may be caused by a short in the horizontal output stage. A short is indicated when this blank read-out is accompanied by a "bad" milliamp read-out.

A steady "bad" microsecond read-out indicates a value change in one of the critical timing components in the horizontal output stage. A microsecond read-out greater than 16 μs usually indicates a problem with the yoke components, so check the yoke, yoke series capacitor, and other components associated with the yoke.

Read-outs that vary by several microseconds or more indicate an abnormal flyback pulse wave shape. Varying microsecond read-outs may be caused by multiple flyback pulses or abnormal flyback pulse ringing. This symptom is typical of loading problems in the flyback secondary or horizontal output stage.

SMPS troubleshooting techniques

Switch-mode power supplies (SMPSs), also may be called *choppers, switchers,* or *pulse-width power supplies,* are now used in most ac-powered devices, such as computers, monitors, television receivers, and VCRs. You will now see that SMPSs call for a different troubleshooting approach than linear power supplies.

Conventional linear power supplies usually continue to supply an output, even if a load shorts, a filter capacitor opens, or half the bridge quits. Isolating the defective components in a linear power supply usually consists of tracking too low, too high, or lost dc voltages. The SMPS, on the other hand, requires a different troubleshooting process.

First off, replacing a burnt resistor, shorted diode, or bad transistor in an SMPS does not guarantee a fix as with linear power supplies. Just repairing the SMPS by replacing a defective or burned-out part could well mean that the part will fail again soon.

Also, when troubleshooting a linear power supply, a variac is used to slowly bring up the ac line voltage while you monitor the current. This is not the case with SMPSs. You have virtually no chance to back off the ac power once the SMPS kicks in. It draws either full current or no current at all.

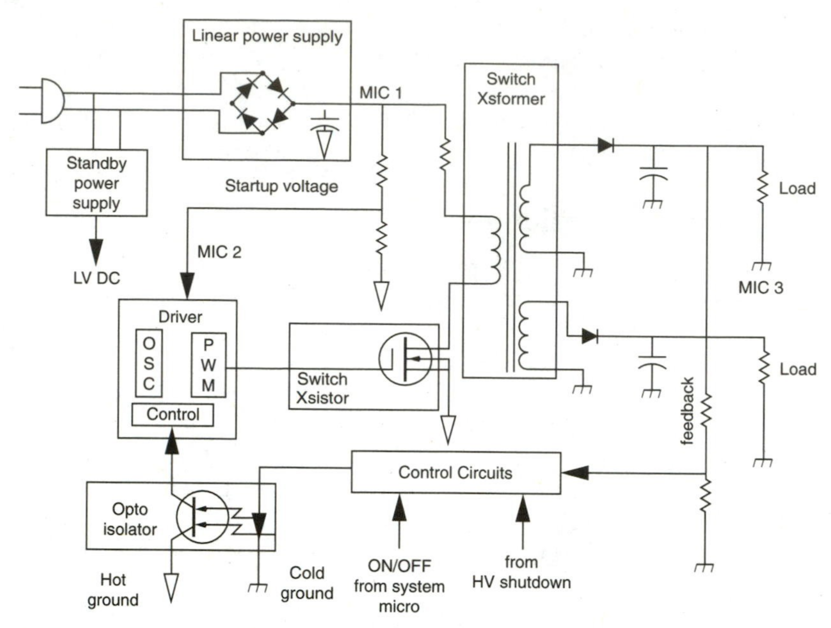

4-15 SMPS's four key circuits. (*Courtesy of Sencore, Inc.*)

Perhaps the biggest difficulty with troubleshooting an SMPS is a direct result of one of the SMPS benefits—its ability to protect itself from overvoltage or overcurrent conditions by shutting down. Most SMPS component failures or load changes cause the SMPS to completely shut down and produce a "dead chassis" symptom. This can make troubleshooting difficult and confusing. Is the shutdown caused by too much high voltage? Is the B+ being pulled down? Is there too much load current? Is there a supply component failure? Or is there a defective safety circuit?

Without a logical procedure, SMPS troubleshooting can be frustrating. However, you can break the SMPS shutdown loop and quickly isolate the defective problem area with the following six easy steps.

However, before we look at the six steps, let us briefly review the four general circuits that make up an SMPS, as shown in Fig. 4-15.

- *Circuit 1: Unregulated B+.*
  This circuit includes the linear power supply, standby supply, primary of the switching transformer, and the switching transistor.
- *Circuit 2: Startup and drive.*
  This section provides the control signal for the switching transistor. The heart of this circuit is the driver network. It can be a single-stage transistor or a current-mode controller IC.
- *Circuit 3: Secondary circuits.*
  The secondary circuits include the secondary windings of the switching transformer and the components (diodes, capacitors, etc.) that provide power to the loads. Most SMPSs have two to five loads.
- *Circuit 4: Feedback and control.*
  Most SMPS feedback loops provide four functions:

- Output voltage sampling for regulation
- High-voltage monitoring
- System control micro for power on/off
- Ground isolation through optoisolators

The six-step troubleshooting procedure

The following six steps are proven to be safe, effective methods for isolating the problem to a specific key circuit. Combining these steps with dynamic component analyzing will get even the toughest SMPS systems up and running quickly.

Keep the following things in mind when performing the SMPS troubleshooting procedures:

- Always use the correct ground reference when making a measurement. Using the wrong ground reference will result in an incorrect reading.
- "Hot" grounds (⏚) are usually found on the primary side of the switching transformer. Use this ground for all circuit 1 measurements.
- Chassis (⊥) grounds are found on the secondary side of the switching transformer. Use this ground for circuit 2, 3, and 4 measurements.
- The optoisolator input (from the control circuits) is measured with respect to chassis ground.
- The optoisolator output (to the primary-side driver or controller stage) is measured with respect to the "hot" ground.
- Be prepared to make all parameter measurements. Efficient troubleshooting depends on your ability to quickly measure all the different signals and voltages from dc tenths of a volt to 160 V, signal voltages from 2 to 400 V peak to peak, and frequencies from 40 to 150 kHz.

The flowchart in Fig. 4-16 outlines the six troubleshooting steps. Perform the steps in order. Each step funnels the problem to a specific circuit and suspect components. Any one of the troubleshooting steps may isolate the SMPS problem. Often, you may not need to do all six steps because the problem will be found in one of the first steps. Follow along, and let us go over each of these troubleshooting steps.

1. Check the standby supply.

The standby voltages to the driver and microprocessor must be correct before the SMPS can operate (not all SMPSs have standby supplies). Check for standby voltage with the chassis plugged into an isolation transformer set for an output of 117 V but with the chassis power (on/off switch) turned off.

Some chassis use a second, smaller SMPS as the standby supply. The fact that the standby supply is working eliminates many suspect components. The IC driver in circuit 2 is always a suspect in a shutdown condition and is often needlessly replaced. The standby switcher is physically smaller and has lower power-handling capabilities than the main switcher, but it is driven by the same driver IC as the main switcher. Therefore, if the standby switcher is running, the driver IC is likely OK. The shutdown condition is caused by something else that is preventing the driver IC from supplying a control signal to the main switching transistor.

If the standby supply voltage is correct but the main SMPS is still not operating, then you go on to step 2.

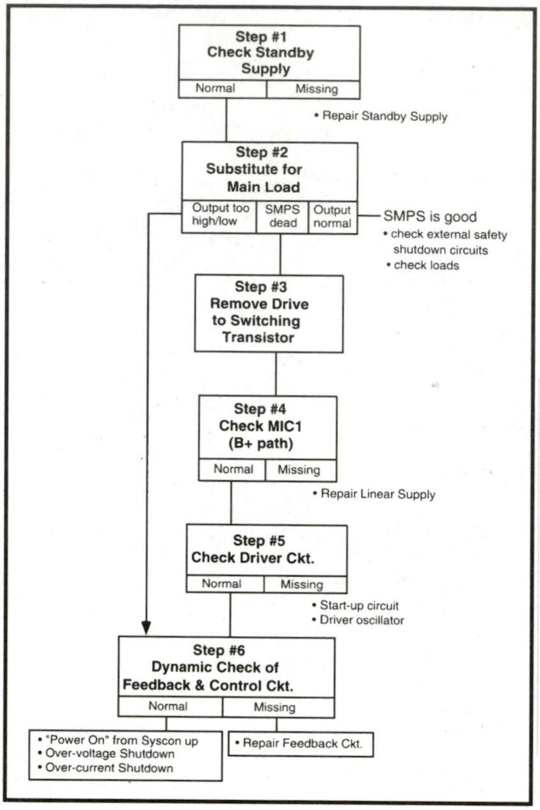

Step #1
Check Standby
Supply
Normal
Missing
• Repair Standby Supply

Step #2
Substitute for
Main Load
Output too high/low
SMPS dead
Output normal
SMPS is good
• check external safety shutdown circuits
• check loads

Step #3
Remove Drive
to Switching
Transistor

Step #4
Check MIC1
(B+ path)
Normal
Missing
• Repair Linear Supply

Step #5
Check Driver Ckt.
Normal
Missing
• Start-up circuit
• Driver oscillator

Step #6
Dynamic Check of
Feedback & Control Ckt.
Normal
Missing
• "Power On" from Syscon up
• Over-voltage Shutdown
• Over-current Shutdown
• Repair Feedback Ckt.

4-16 Steps to quickly and safely isolate SMPS problems. (*Courtesy of Sencore, Inc.*)

2. Substitute for the main load.

An important step in troubleshooting SMPS problems is to separate the SMPS output from the rest of the chassis. This helps you determine if the shutdown supply symptom is due to the SMPS supply itself or if the symptom is due to an outboard circuit or load problem.

Most SMPSs will not operate without an adequate load current. Therefore, you cannot just disconnect the loads. Instead, most service notes recommend replacing the main load with a lightbulb that has approximately the same wattage rating. The lightbulb provides current limiting and a suitable constant load for testing of the SMPS without any damage.

The main B+ load is the output of the SMPS that contains the feedback divider network. In a television receiver or computer monitor, this is the output voltage that powers the collector or the horizontal output transistor. By substituting for this load, you effectively disable the safety shutdown controls that come from the external circuits.

The size of the lightbulb you use depends on the load you are substituting. For example, if you are substituting for the load on the 130-V dc B+ supply in a television receiver or computer monitor, use a standard 60-W, 120-V ac lightbulb. If you are substituting for the 15-V B+ supply in a VCR power supply, use a 12- or 18-V bulb.

You will need to open the circuit path to remove the normal load. Make sure to break the circuit after the feedback take-off point. Removing the horizontal output transistor in a television receiver or computer monitor will break the circuit, but do not connect the bulb in place of the horizontal output transistor. The primary of the flyback is not designed to handle a continuous current. Connect the lightbulb ahead of the primary, as shown in Fig. 4-17.

After you substitute for the load, you will see one of four conditions when you turn the SMPS on:

- The bulb lights, and measured voltage is normal. This means that the SMPS is working properly. Something external to the SMPS is causing the shutdown. Possibilities include excessively high voltage, excessive current drawn by one of the loads, or a defective safety circuit.
- The bulb does not light (the SMPS will not start up).
- The bulb comes on but goes out (the SMPS starts but then goes into shutdown).
- The bulb is very bright (indicates possible regulation problems).

The last three conditions indicate that something is wrong with the SMPS. Continue with the remaining steps until you locate the problem.

3. Remove drive signal from the main switching transistor.

Open the signal path between the drive and the gate or the base of the switching transistor. This is done easily by unsoldering and lifting any one of the components in the signal path. Disconnect the input signal to the main switching transistor. This allows you to safely troubleshoot the SMPS circuits while the chassis is turned on without accidentally producing an output from the SMPS.

4. Check circuit 1.

Circuit 1 includes all the B+ paths from the output of the linear supply to the ground point of the emitter or source of the switching transistor. Begin by checking for B+ voltage at the switching transistor:

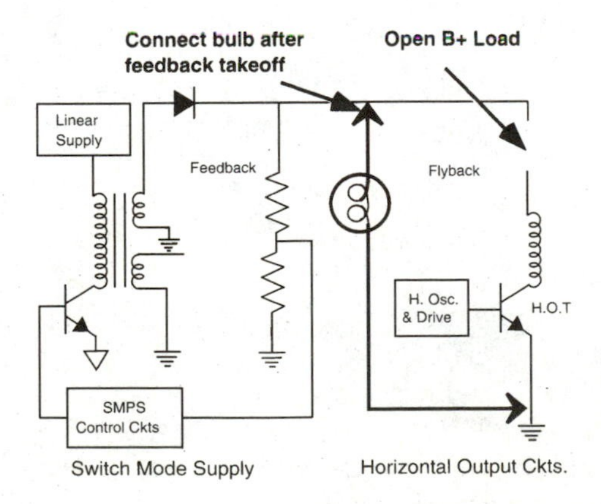

4-17 A lightbulb used to check circuit loading. (*Courtesy of Sencore, Inc.*)

1. Connect an oscilloscope to the switching transistor's drain or collector. Set it up to measure dc voltages.
2. With an adjustable ac control, set ac voltage to zero.
3. Gradually increase the ac voltage to the television set while monitoring the output current.

You should observe one of the following conditions:

- Low current, normal B+ (approximately 160 V dc) with line input voltage at 117 V ac. This means that the B+ supply is good. But there still might be a problem in circuit 1.

Check the switching transistor to make sure that it is not open. Check this with a dynamic transistor tested. Also check the resistor(s) in the emitter or source lead of the switching transistor. If you suspect they may have changed in value, replace them with the exact replacement. These resistors are precision tolerance and are critical to safe operation of an SMPS.

If the transistors and resistors are good, proceed to the next step.

- No dc and no ac current draw. There is an opening in the B+ supply. Check for fuse, safety resistor, diode, and switching-transformer primary problems.
- No (or low) dc and increasing ac. Low or missing dc voltage along with increasing ac is caused by a short in the B+ supply itself or somewhere within circuit 1. Check the switching transistors, bridge, and filter capacitor. Also check the primary winding of the switching transformer for a dc short to the core or to another winding.

5. Check the driver circuit.

First, confirm that the driver IC has startup voltage. In most SMPSs, the startup voltage is obtained from a resistor divider network off the linear unregulated B+ supply. Always check for startup voltage before checking if the oscillator is running. Connecting a probe to the oscillator test point can kick start the oscillator into action.

Second, check all the oscillator test point waveform parameters: dc, PPV, and frequency. The oscillator frequency must run at the switch-mode supply frequency. If the frequency reads high (the oscillator waveform may be noisy and contain glitches), confirm the frequency using the delta function on the waveform analyzer. If the frequency is more than 10 percent too high, the controller IC is probably defective.

> **Note:** Usually, a current-mode controller will not output a drive signal to the switcher with the SMPS disabled. Make your measurements at the oscillator test point on the controller IC.

6. Perform dynamic circuit checks.

This troubleshooting step confirms proper operation of the feedback and control circuit 4. Often failures in this circuit are caused by defective transistors that shut down the entire feedback loop. The dynamic feedback circuit check will quickly isolate any of these problems.

To check these circuits, you will need to apply an external voltage (equal to the normal main B+ output of the SMPS) and confirm that the circuits respond properly to it. Use a variable-current limited dc supply for these checks. The hookup connections for these tests are shown in Fig. 4-18.

1. Disconnect the substitute lightbulb load from the main B+ output of the SMPS.
2. Connect the test dc power supply to the B+ point, where the lightbulb was connected.
3. Connect the waveform analyzer to the control input of the driver (output of the optoisolator). Press the "dc volts" function of the scope.
4. Set the ac line voltage to 117 V ac, and turn on the television receiver.
5. Vary the dc power supply voltage from 5 V below the normal B+ voltage while monitoring the dc reading on the waveform analyzer.

If the feedback circuitry is working, you will see an increase in the dc voltage at the driver input as you increase the applied voltage above the normal B+ level. A 1-V dc change in the applied voltage typically may only show a 0.1-V dc change at the driver.

If there is no change at the driver input, check the optoisolator. (Remember to reference to the correct ground.) With a changing dc into the optoisolator, there should be a corresponding change in the output dc voltage. Continue checking the circuit 4 feedback loop until the defect is located. This includes the "power-on" command from the system control microprocessor and the output from the safety circuits, such as overvoltage and overcurrent shutdown. Be sure to check the electrolytic capacitors for open, leakage, and effective signal radiated (ESR).

SMPS component analyzing tips

Do not use general replacement parts. The switching transistors and diodes used in SMPS must operate effectively at high currents and high frequencies. The most reliable results are achieved using the original manufacturer's parts.

Digital television dead set flowchart

The troubleshooting flowchart shown in Fig. 4-19 covers several Zenith digital television chasses. Several minor revisions and the inclusion of later digital receivers are the

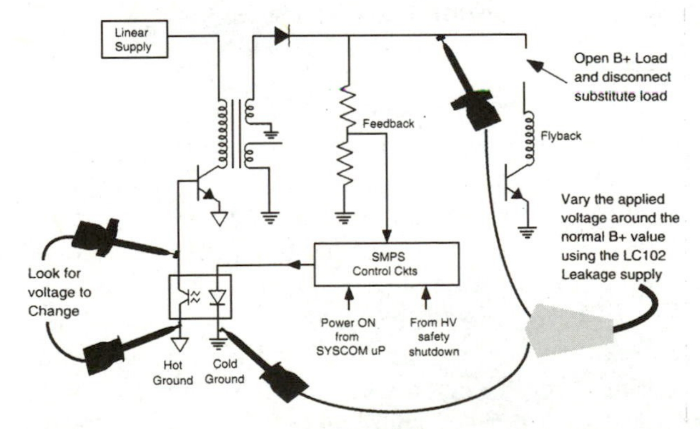

4-18 Applying an external dc voltage to the feedback circuit to confirm the proper operation of circuit 4. (*Courtesy of Sencore, Inc.*)

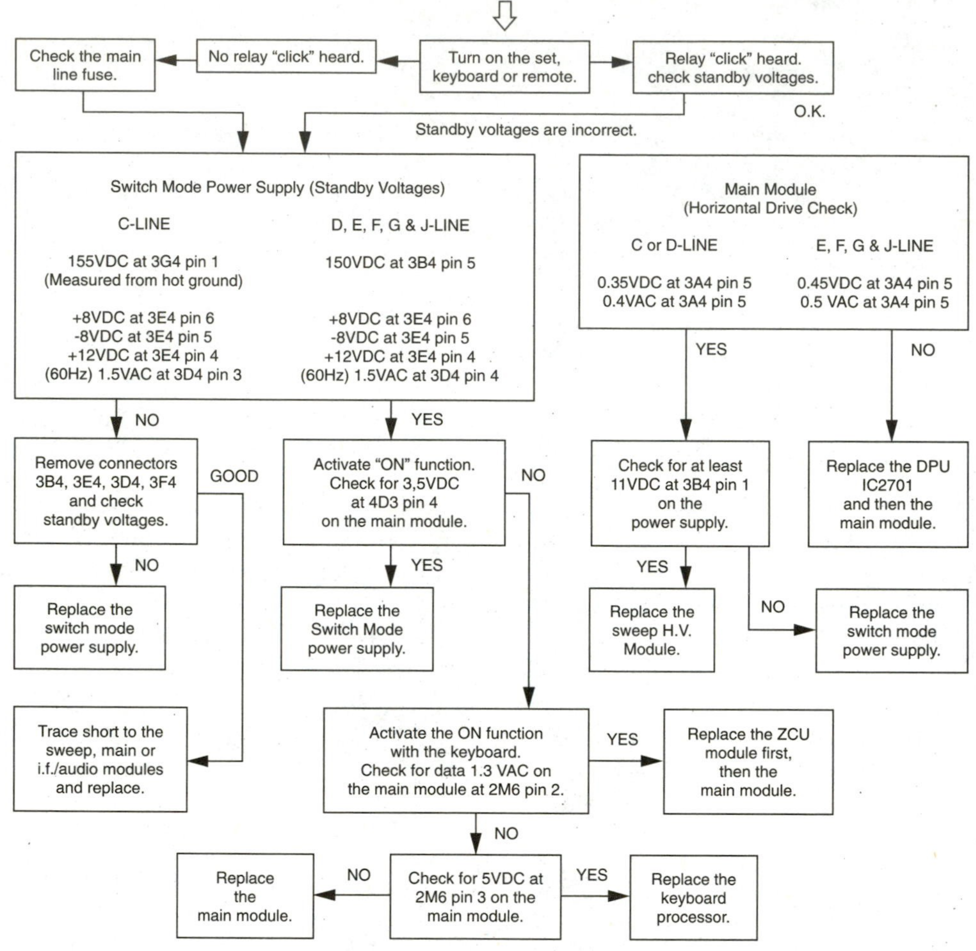

4-19 The ac rms and dc values will vary in these troubleshooting blocks depending on the failure and the type of DMM being used. Remember, you are checking for the presence of a signal and voltage, not just amplitude. (*Courtesy of Sencore, Inc.*)

only differences. This chart is designed to aid the technician in quickly locating a failure (other than video, audio, or sweep-related) in a module through the use of only a DMM.

Begin by attempting to turn the receiver on with the remote hand control and by the keyboard on the set. If the "click" of the degaussing relay is heard, the dc switch operating voltages are present. No click would mean that the main fuses should be checked out, followed by all standby voltages. Any time a receiver is plugged into an ac outlet, standby voltages appear that prepare the receiver for "turn on." When measuring these voltages, keep in mind that on a dead receiver they will be higher because of an abnormal load condition. Continuing on with the chart means nothing more than checking the "on-signal horizontal drive," "horizontal drive B+," and "keyboard serial data pulse."

5 CHAPTER

Color television and PC monitor sweep systems

This chapter begins with an explanation of television horizontal sweep and horizontal output stage circuits and their key components. It continues with a discussion of how the flyback transformer and timing capacitor operate. Various mysteries of these stages are revealed, and an explanation of horizontal and vertical yoke deflection is offered. You will learn how to conduct dynamic tests of sweep transformers and yokes, how to perform load tests, and what problems are experienced in high-voltage troubleshooting and breakdown. Next, there is a detailed discussion of sweep and circuit ringing troubleshooting using the Sencore VA62A Video Analyzer.

The chapter concludes with an in-depth explanation of the circuit operation of color televisions and personal computer (PC) monitors: how they work and how to fix them when they are not working. Included are some actual symptoms and case histories of problems that have occurred and their solutions. Finally, some horizontal sync problems, as well as problems with switch-mode power supply (SMPS) circuits, will be reviewed along with tips for troubleshooting them.

Understanding the horizontal output stages

This section will explain the mysteries of the horizontal sweep stages found in modern television sets and PC monitors. Please keep in mind that the basic concept of how these stages operate has not changed for many years. However, the use of scan-derived supplies and startup and shutdown circuits has made fixing horizontal output stages quite troublesome for many technicians. There is little practical information available on this subject, and that which is available can be very confusing to the reader. Hopefully, this chapter will fill the void.

I know that many service technicians have difficulty relating problematic symptoms of horizontal output stages to their possible causes and are often misled when interpreting circuit voltages and other troubleshooting clues. Let us begin by examining the operational theory of circuitry used in the horizontal output and flyback transformer and by looking at the currents and voltages in the horizontal output stage.

Key horizontal output stage components

The horizontal output stages operate virtually the same regardless of the make or model of the television set. All television receivers and video monitors apply a sawtooth current into the primary winding of the flyback transformer. The output stage receives power from the main B+ supply, typically about 130 V dc. The B+ supply can deliver peak currents of several amperes while maintaining a regulated voltage. The peak-to-peak current varies depending on the size of the cathode-ray tube (CRT), the number of scan-derived supplies, and whether or not the chassis is color or black and white.

Figure 5-1 shows a simplified horizontal output stage. It consists of six key components:

1. Horizontal output transistor
2. Flyback or sweep transformer
3. Retrace timing capacitor or "safety capacitor"
4. Damper diode
5. Horizontal deflection yoke
6. Yoke series capacitor

Let us now take a closer look at the role that each of these components plays.

Output transistor operation.

As you refer to the circuit drawing in Fig. 5-2, think of the horizontal output transistor (HOT) as a switch. It provides a path for current to flow through the pri-

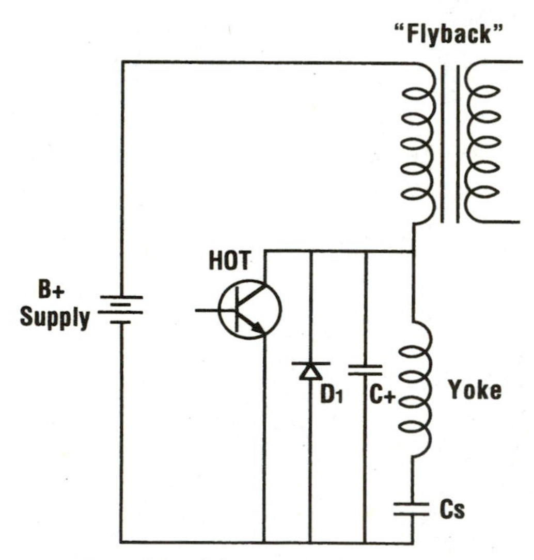

5-1 Simplified horizontal output stage. *(Courtesy of Sencore, Inc.)*

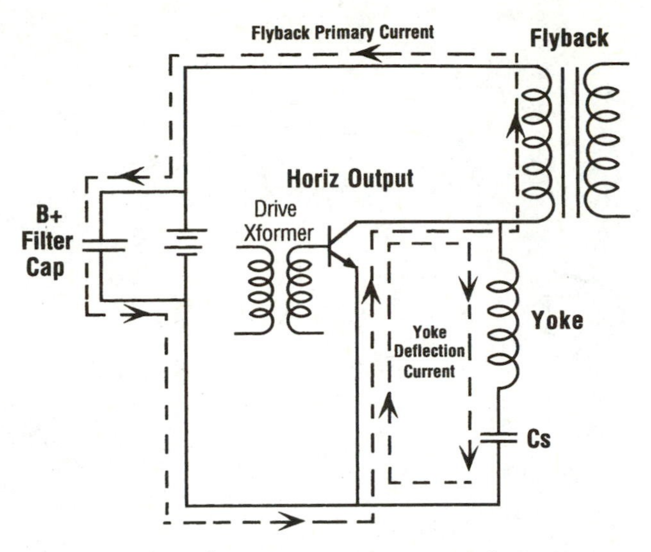

5-2 Horizontal output transistor conduction current paths. (*Courtesy of Sencore, Inc.*)

mary winding and horizontal yoke of the flyback. The transistor is switched on and off by a signal applied to its base. Because it is a power transistor, a hefty drive current is required. This drive current is supplied by the horizontal driver transistor via the driver transformer. The driver transformer steps up the current while providing impedance matching and isolation.

The horizontal output transistor passes a collector current at levels that range from 200 mA in a black-and-white chassis to 1.5 A in a large-screen color television chassis that has multiple scan-derived power supplies. These are the average current values that a dc current meter would indicate. Since the flyback and yoke are inductive, the HOT collector current has a sawtooth rise that reaches peaks of several amperes.

Transistor theory tells us that the collector current equals the base current multiplied by the current gain (beta) of the transistor. The base drive current must be sufficient to enable the transistor to pass the required collector currents. Therefore, base drive currents that range from 100 to 300 mA are needed. Low transistor gain (beta) or reduced base current drive will prevent the transistor from passing the required flyback primary current.

The horizontal output transistor is switched on and off at the horizontal frequency of 15,734 Hz. The horizontal oscillator, which controls the driver stage, is synchronized to turn on the horizontal output transistor approximately 30 to 35 μs before horizontal sync appears (see Fig. 5-3). The HOT conducts current until the start of the horizontal sync and then is turned off abruptly. This switching action must correspond with the resonant action of the flyback and yoke circuits, which will be explained in more detail later.

The time it takes to switch the horizontal output transistor between the off and on conditions is important. The drive current, applied to the base of the horizontal

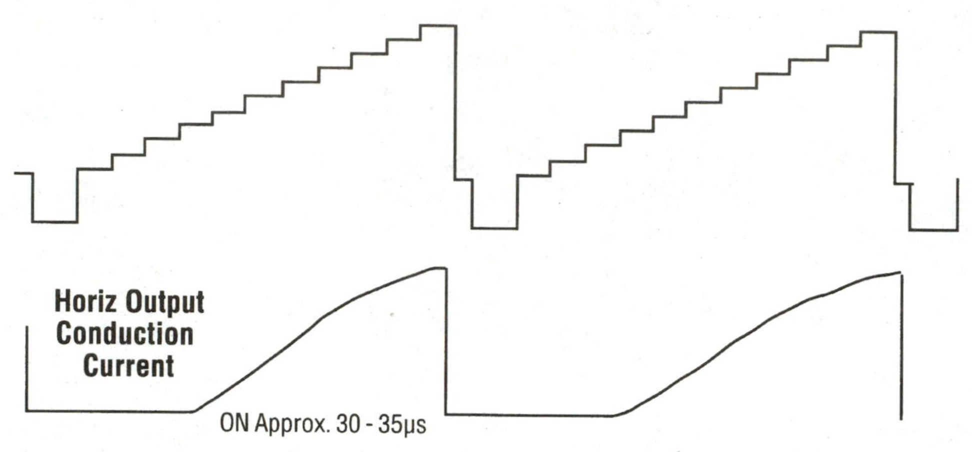

5-3 Horizontal output transistor. (*Courtesy of Sencore, Inc.*)

output transistor and to components in the base circuit, enables fast transistor switching. As the transistor is switched, the emitter-to-collector resistance changes from approximately 5 Ω (on) to about 10 MΩ (off). As the current flows through this changing resistance, it produces heat. The longer the transition occurs, the greater is the chance for thermal failure of the transistor from the heat.

The drive produces a waveform at the base of the horizontal output transistor similar to a square. The waveform has spikes that increase the peak-to-peak values indicated on schematics from 5 to 30 V. The waveform only confirms the presence of drive to the HOT; it cannot confirm whether or not the base drive current and switching transistors are adequate for normal operation of the horizontal output circuit. Improper drive can result in reduced deflection (width), image overlap, excessive transistor heating, or shortened HOT life.

Flyback transformer operation.

The horizontal output transformer is often called the *flyback or integrated high-voltage transformer* (IHVT). (An IHVT is a flyback that includes a high-voltage multiplier.) The flyback is primarily responsible for developing high voltage for the CRT. It is constructed with a powdered iron or ceramic core that allows it to work efficiently at high frequencies.

The flyback includes one primary winding and many secondary windings. The main secondary winding supplies voltage pulses to the voltage multiplier. Other secondary windings supply filament power, keying pulses, and scan-derived power supplies to the CRT.

The primary winding of the flyback is connected in a series with the HOT and B+ power supply (see Fig. 5-4). The switching action of the horizontal output transistor energizes the flyback to produce inductive voltage pulses. Even though the B+ power supply is only about 130 V dc, the flyback action of voltage pulses it produces is much larger—typically 700 to 1100 V peak to peak. To understand how this occurs, let us revisit some basic inductor theory.

When the HOT is conducting, the current in the primary winding of the flyback raises at a linear rate. This produces a constant amount of induced voltage in the flyback windings. However, when the HOT is turned off abruptly, the magnetic field in the flyback core rapidly collapses and induces a high voltage into both the primary and secondary windings.

The rate at which the magnetic field collapses in the output stage is controlled by timing components. If the rate were not controlled, it would induce voltage spikes of several thousand volts across the flyback primary. These spikes would exceed the breakdown ratings of both the horizontal output transistor and the flyback and produce excessively high voltage to all flyback windings.

Retrace timing capacitor.

The retrace timing capacitor plays a very important role in the timing of the horizontal output stage. Its main purpose is to slow the rate of the collapsing magnetic field in the flyback.

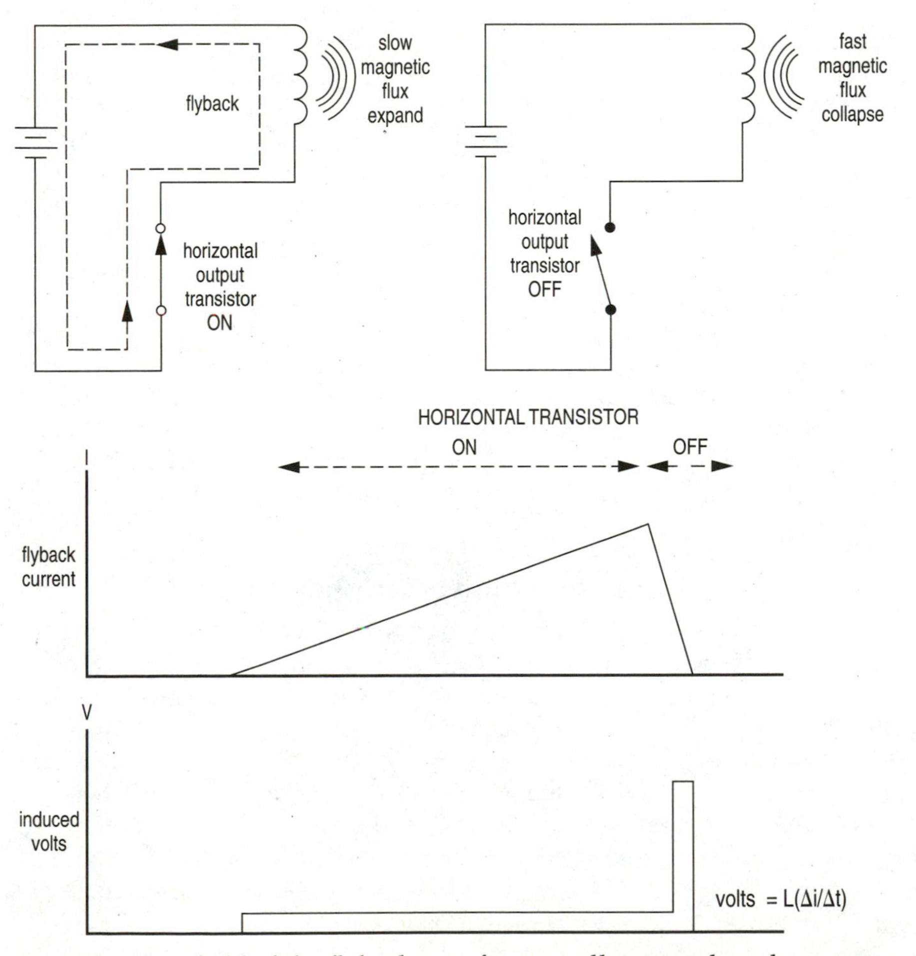

5-4 Magnetic field of the flyback transformer collapsing when the output transistor is switched off, developing an HV pulse. (*Courtesy of Sencore, Inc.*)

If the retrace capacitor value decreases, or if it opens, the pulse amplitude of the flyback will increase several thousand volts. To minimize this danger, several capacitors of smaller value usually are connected in parallel. Safety (shutdown) circuits also are added to disable the horizontal output stage if the high voltage should increase to an unsafe level. Because of the key role that this capacitor plays in controlling induced voltages (and the CRT high voltage), it is often called the *safety capacitor.*

Damper diode.
The damper diode serves to complete the path for resonant current for both the flyback primary and horizontal deflection yokes. The damper diode is a fast-switching diode capable of withstanding the addition of currents from the flyback and yoke. If the damper diode should open, it would force the horizontal output transistor to operate in reverse breakdown. This would add considerable heat to the transistor and likely lead to its failure.

Horizontal yoke.
The horizontal yoke causes the electron beam of the CRT to scan from left to right across the face of the CRT. Horizontal deflection is produced by a sawtooth rising and falling of current to the horizontal yoke windings. The horizontal yoke is driven by the current in the horizontal output stage. Since this component is part of the stage, it influences the retrace timing of the circuit. The yoke, or any one of its series of components, can have problems that will alter the operation of the horizontal output circuit and not be suspected as the cause of the operational breakdown.

Yoke series capacitor.
Capacitor Cs, which is in series with the yoke, has four functions. First, it is primarily responsible for matching the resonant timing of the yoke deflection current. Second, as stated earlier, it affects the retrace time. Third, it prevents a fixed dc bias from developing on the yoke and causing improper picture centering. Finally, it shapes the sawtooth rise in deflection current to match the slight curvature of the CRT screen.

Unraveling the mysteries of output stage operation

Now that we have looked at the key components, individually, we will put them together and see how the whole circuit operates. Then we will analyze the output stage in two parts, according to the major functions that the horizontal output stage performs:

1. Flyback primary current and retrace time
2. Horizontal deflection

Although these two functions interact, discussing them separately will help you better understand the operation of the horizontal output section. The first function, flyback primary current and retrace time, is responsible for producing the high voltage, focus voltage, and scan-derived supplies to the CRT. The second function, as its name implies, deals with deflection and the electron beam.

Flyback primary current and retrace time.

Let us begin by looking at the paths of current in the primary winding of the flyback transformer. Figure 5-5 shows the flyback action and current paths at four times during one output cycle, beginning with the turning on of the horizontal output transistor.

When the horizontal output transistor is turned on, current flows through the flyback primary from the B+ power supply. All the power required by the output stage and by the secondary windings of the flyback is delivered to the circuit from the B+ supply during this time. The current and the magnetic field in the flyback core continue to build until the transistor is turned off.

The magnetic field that was stored in the flyback core begins to collapse immediately after the horizontal output transistor is turned off. This is the beginning of retrace time and corresponds to the start of horizontal sync. When the horizontal output transistor is switched off, the retrace timing capacitor is effectively placed in parallel with the flyback primary. Thus a resonant circuit is formed, as shown in Fig. 5-4. The time constant of the resonant circuit is determined mainly by the value of the retrace capacitor and the inductance of the primary winding of the flyback.

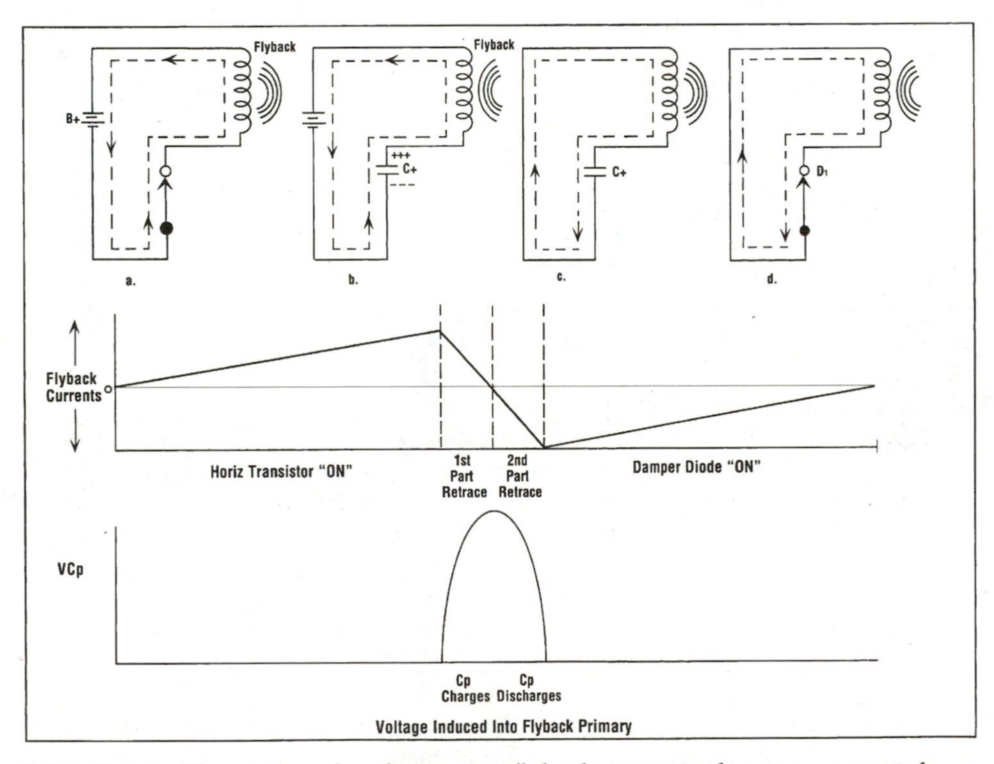

5-5 Circuit illustrating the alternating flyback current, the current waveform, and the flyback voltage pulse for one complete horizontal cycle. (*Courtesy of Sencore, Inc.*)

Note that the yoke components in parallel with capacitor Ct (the yoke and Cs) also affect retrace timing.

The collapsing magnetic field causes current to flow through the low-impedance filter capacitors of the B+ supply and into Ct. This current charges Ct. The rise in voltage across Ct is the flyback pulse formed at the collector of the horizontal output transistor. Properly operating horizontal output stages have retrace periods (flyback pulse duration) of 11.3 to 15.9 μs.

When Ct has discharged completely, the magnetic field again begins to collapse. The collapsing field induces a voltage with a polarity that forward biases the damper diode (DI). The damper diode serves as a switch and allows the magnetic current in the flyback and yoke to decay at a controlled rate. When the damper diode turns on, the circuit becomes highly inductive and once again produces a slowly increasing current in the primary winding of the flyback. Approximately 18 μs later, the horizontal output transistor is once again turned on, and the cycle repeats. Note that if the horizontal output transistor could conduct current of either polarity, the damper diode would not be needed.

Flyback power transfer.

The flyback transformer essentially works like any other transformer, in that the ac in the primary winding induces power (voltage and current) into the secondary windings (secondaries). If all the loads in the secondaries were open, most of the power stored in the magnetic field of the transformer would return to the primary circuit. However, the circuits in the secondaries draw power from the primary circuit. Thus, as the load on secondary windings increases, more current flows in the primary winding and more current is drawn from the B+ supply.

Some problems, such as a short in a secondary load circuit or in a flyback winding, produce such a strong current that the circuit cannot accommodate the power demand. These situations may cause the horizontal output transistor to overheat and short, the primary winding to open, or the B+ supply to fail.

Horizontal yoke deflection.

The second major function of the horizontal output stage is to provide the current needed for the deflection yoke to move the electron beam in the CRT from left to right across the screen. The collector current of the output transistor is split between the flyback and the horizontal yoke. Both current paths share the damper diode and retrace timing capacitor.

Figure 5-6 shows the yoke deflection current at four different times during one horizontal cycle. These four times are the same as those shown in Fig. 5-5.

When the horizontal output transistor is turned on, the bottom side of the yoke series capacitor (Cs) is connected to the top of the yoke. Because Cs is fully charged through the horizontal output transistor, the resulting current produces an expanding magnetic field in the yoke that moves the electron beam from the center of the screen toward the right side.

When the horizontal output transistor opens, the retrace timing capacitor is added to the circuit (Fig. 5-6*b*). This increases the resonant frequency and causes the magnetic field of the yoke to collapse rapidly. This is the beginning of retrace

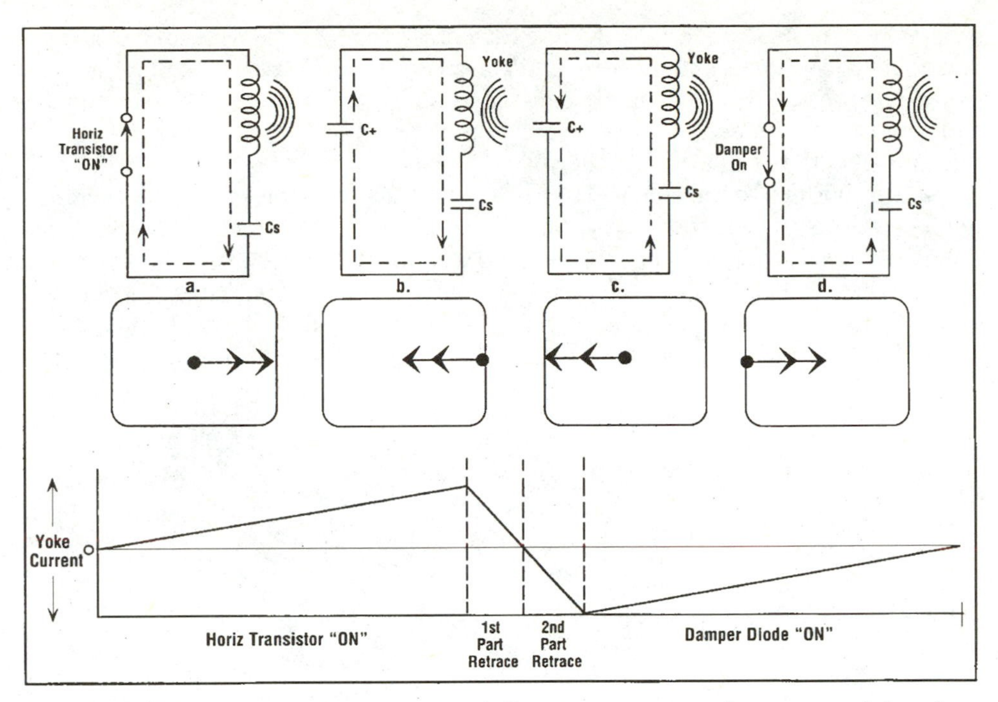

5-6 Simplified circuits depicting the deflection current and position of the electron beam for one complete horizontal cycle. (*Courtesy of Sencore, Inc.*)

time, during which the CRT beam is snapped from the right side of the screen back to the center. The induced voltage causes current to flow and returns the energy that was stored in the magnetic field of the yoke to capacitors Ct and Cs. The retrace timing capacitor is replenished with charging current from the flyback transformer and becomes the source of current for the yoke.

During the second part of retrace, Ct and Cs discharge and force the current to flow in the opposite direction (see Fig. 5-6*c*). The timing is identical to the first part of the retrace, and the CRT beam is moved quickly from the center of the screen to the left side.

When capacitors Ct and Cs are fully discharged, the magnetic field of the yoke begins to collapse (Fig. 5-6*b*). The circuit timing now is determined by the yoke and capacitor Cs and agrees with the timing during the right trace time. The collapsing magnetic field of the yoke produces current through the damper diode, which returns energy to the circuit and charges capacitor Cs.

When the magnetic field of the yoke collapses, the damper diode stops conducting. This must coincide with the turning on of the horizontal output transistor; otherwise, there will be nonlinear horizontal lines in the center of the raster.

Summary

To simplify the explanation of the horizontal output stage, we will analyze the flyback and yoke functions later in this chapter. Note that these circuits are not independent of each other. The flyback current is transferred to the yoke by the retrace timing capacitor, and the yoke and flyback currents share the conduction time of the horizontal output transistor, damper diode, and retrace timing capacitor. Because of this interaction, most problems in the horizontal output circuits affect the currents in both the flyback and the yoke.

It is important to remember that capacitor Ct effectively is charged by simultaneous currents of both the flyback and yoke during retrace time. This rising and falling retrace voltage pulse can be analyzed to gain information about the operation of the horizontal output circuit. The voltage pulse may be measured at the collector of the horizontal output transistor with respect to emitter ground; the B+ power supply voltage also may be measured at this point to confirm proper supply operation.

> **Caution:** Before making measurements in the horizontal output stage, be sure the test instrument is designed to withstand peak-to-peak voltage pulses of at least 1000 V.

Dynamic testing of sweep transformers and yokes

Too often valuable troubleshooting time is wasted in checking everything else in the circuit before suspecting that the flyback transformer or yoke may be defective. In this section we will discuss the use of the Sencore TVA92A Video Analyzer, which provides special features for dynamic testing of the flyback/IHVT, yoke, and other expensive television components.

The horizontal output stage interacts closely with the B+ power supply, horizontal oscillator/driver, and startup and shutdown circuits. Many television problem symptoms mislead service technicians to suspect the flyback. Consequently, many service technicians order replacement flybacks too quickly or spend time testing for defects when the horizontal output stage and flyback/IHVT are working fine.

Horizontal output load test

The Sencore TVA92A horizontal output load test provides a fast and easy method to determine if a problem exists in the horizontal output stage—without even turning on the receiver. Since the flyback and horizontal yoke are crucial components of this circuit, many defects in these components become evident during the test.

This horizontal output load test is performed with the television turned off. It supplies 15 V that substitutes for the television's power supply, as well as a transistor switched at the proper rate and time to simulate the action of the horizontal output stage. During the load test, alternating currents that closely mirror the operation of the horizontal output stage are produced in the flyback and yoke but at approximately $\frac{1}{10}$ the current and voltage amplitudes (Fig. 5-7).

The horizontal output load test requires three simple connections to the television chassis under test. The B+ test lead is connected to the B+ test point, either at the

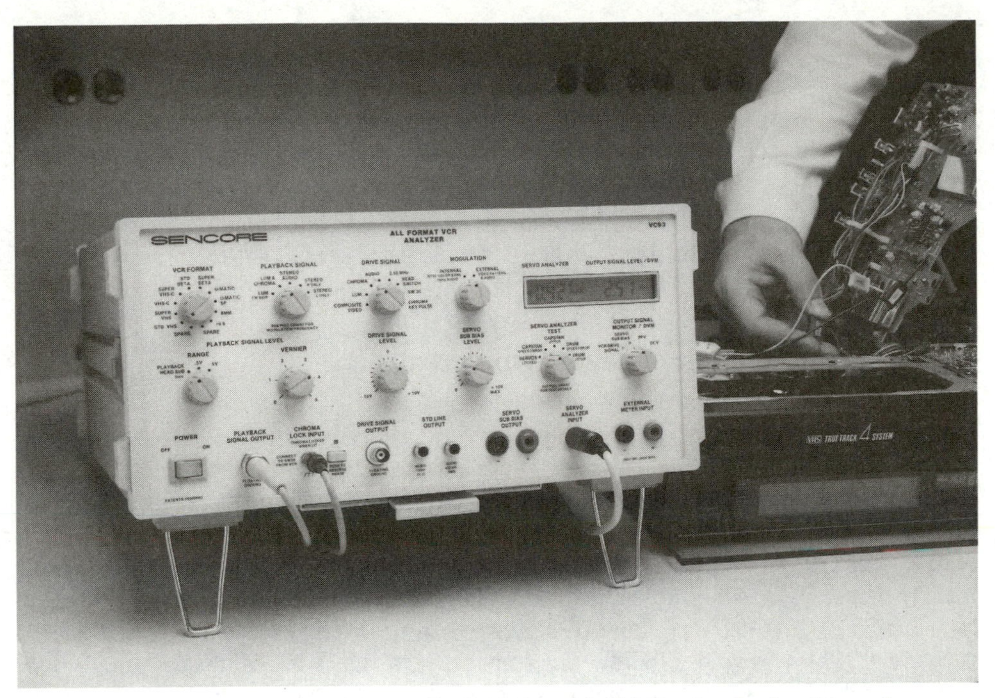

5-7 Horizontal output load test, which detects problems in the horizontal output stage and indicates when to test the flyback and yoke for defects. (*Courtesy of Sencore, Inc.*)

power supply or at the B+ terminal of the flyback transformer primary. The other connections are made to the emitter and the collector of the horizontal output transistor.

The horizontal output load test checks the operation of the horizontal output stage and reflects the load of current to the television's B+ supply. Two separate read-outs (milliamps and microseconds) are provided to detect abnormal conditions. Each read-out is accompanied by a good/bad indicator.

The meters on the TVA92A display the 15-V current supplied by the horizontal output load test. Read-outs between 5 and 80 mA represent a normal range of current for a wide variety of horizontal output stages and are considered "good" by the TVA92. Current levels less than 5 mA indicate improper test lead connections or an opening in the horizontal output circuits. Readings greater than 80 mA indicate heavy current load demand from the horizontal output, flyback, or other B+ circuits.

During the horizontal output load test, a flyback pulse is produced at the collector of the horizontal output transistor. The TVA92A measures and displays, in microseconds, the duration of the flyback pulse produced by the television's horizontal output stage. The duration, or time of the flyback pulse, depends on the operation of the horizontal output circuits of the television, primarily the flyback, retrace timing capacitors, yoke, and yoke capacitors.

The pulse time read-out indicates if proper timing and resonant action are occurring in the flyback and yoke circuits. Readings between 11.3 and 15.9 μs represent a normal range for horizontal output stages and are considered "good." If the microsecond read-out indicates "good," you can be assured that the horizontal output stage is developing normal flyback pulses.

If the horizontal output load test read-outs indicate an abnormal circuit condition, you will want to test the components in the horizontal output stage, including the flyback and horizontal yoke. If the read-outs show no serious loading or timing defects, the flyback/IHVT and yoke are probably fine. In this event, you do not want to waste time testing the flyback or yoke; rather, proceed with making "television on" measurements to pinpoint the problem. This will let you isolate the defect in the least amount of time.

Flyback transformer and yoke ringer test

A "bad" reading from the TVA92 horizontal output load test indicates a possible flyback or horizontal yoke defect. The most likely cause is a short between a winding or turn or between multiple turns. Even one shorted turn within the flyback or yoke greatly reduces the efficiency of the coil, causing additional power demand. This results in various chassis symptoms such as blown HOTs, overcurrent shutdown, or B+ supply loading. More often than not, the shorted turn is buried inside the transformer or yoke with no visible clue to the defect.

The TVA92 ringer test shows if a shorted turn or turns exist between adjacent turns of a flyback or yoke winding. The ringer test works by including the transformer or yoke coil into a tuned inductance and capacitance (LC) circuit and then applying energy pulses. The LC circuit oscillates, creating a number of waveform cycles before dampening down. How fast the cycles dampen depends on the quality, or Q, of the coil. An efficient coil or transformer will oscillate or ring many cycles before dampening.

The number of cycles or rings before dampening to a 25 percent level is counted and displayed on a digital meter by the TVA92. A good coil (high Q) will ring 10 or more rings, while one with a shorted turn will ring less than 10 times.

A shorted turn will cause all the other windings that share the common core of the transformer to also ring bad. Therefore, just ring the flyback primary winding. The flyback primary can be identified by locating the winding that connects to the collector of the horizontal output transistor and to the B+ power supply.

To "ring" a coil with the TVA92, first remove ac power to the television chassis, and then connect the ringer/load test leads to the flyback primary or the horizontal yoke winding. Note the drawing of these connections in (Fig. 5-8). The test results are calculated and displayed automatically on the TVA92 digital readout. A reading of 10 or more is "good" and indicates that the coil does not have shorted turns.

To save time, you can ring a flyback while it is connected in the circuit. However, other components could reduce the rings to a "bad" level. Just isolate the flyback primary winding from the circuit and open the flyback secondaries until the display reads "good." If the flyback still rings "bad" after all the circuit paths are open, it more than likely has a shorted turn. Remove the flyback from the chassis, and confirm it again with the ringer test.

The ringer test also can be used to detect shorted turns in the horizontal yoke or power supply switching transformer. Use the "Switch X former" position of the TVA92 ringer test when testing switch-mode power supply transformers.

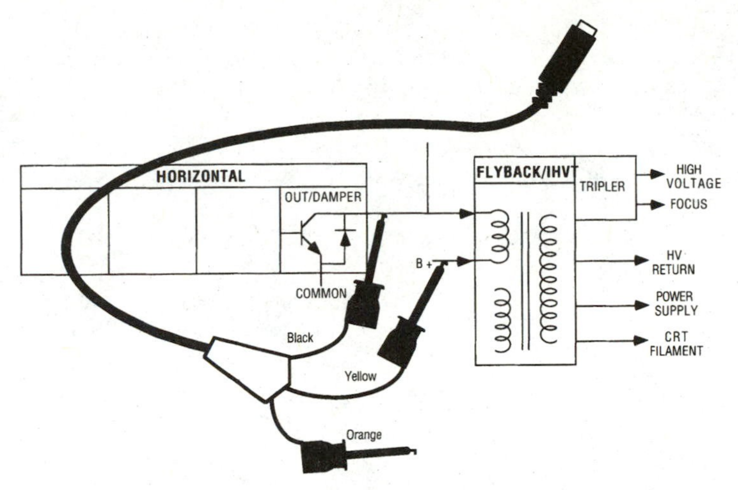

5-8 Ringer test pinpointing shorted turns in yokes and flyback transformers down to a single shorted turn on the coil. (*Courtesy of Sencore, Inc.*)

Checking the IHVT multiplier stage

An IHVT is similar to older flyback transformers with an important difference. The IHVT includes the high-voltage multiplier circuits with the flyback transformer in the same package. An IHVT is easily recognized by the high-voltage lead coming from it and going directly to the CRT.

The high-voltage diodes that are responsible for creating the high voltage and focus voltages are included in the IHVT. The diodes may short, open, or develop leakage and cause missing or low CRT anode and/or focus voltages. Shorted or open secondary windings within the multiplier section will cause similar symptoms.

You only need to test the multiplier section when the horizontal output stage is operating normally but the CRT anode and focus voltages are low or missing.

By driving the primary winding with a pulse similar to the horizontal output pulse, you can operate the IHVT dynamically to prove that the drive pulses are rectified and multiplied. A bad diode, winding, or flyback core will reduce the output voltage of the IHVT. A common 25-V peak-to-peak drive level from the TVA92 enables a comparison to the typical IHVT outputs that is based on flyback primary input pulses and expected high-voltage output.

To dynamically test the HV multiplier section of the IHVT with the TVA92, remove the IHVT from the circuit and apply a 25-V peak-to-peak horizontal key pulse drive signal to the flyback primary winding. Measure the voltage output from the anode lead of the CRT against the HV ground return or the aquadag connection to the flyback.

Note: Use the Sencore TP212X10 multiplier probe to prevent loading. To determine if the HV multiplier output of the IHVT is working properly, multiply the TVA92 reading by 10 to compensate for the ×10 probe, and compare the reading with the expected result (Table 5-1). Check

Table 5-1. Determining the minimum dc voltage for a good integrated flyback by using the ratio between the normal HOT collector peak-to-peak voltage and the normal high voltage

Normal HOT collector peak-to-peak voltage*	Normal chassis high voltage					
	10,000	15,000	20,000	25,000	30,000	35,000
100	2500	3750	5000	6250	7500	8750
200	1250	1875	2500	3125	3750	4375
300	833	1250	1667	2083	2500	2917
400	625	938	1250	1563	1875	2188
500	500	750	1000	1250	1500	1750
600	417	625	833	1042	1250	1458
700	357	536	714	893	1071	1250
800	313	469	625	781	938	1094
900	278	417	556	694	833	972
1000	250	375	500	625	750	875
1100	227	341	455	568	682	795

*Minimum dc voltage output.

the schematic and match the peak-to-peak voltage shown with the normal horizontal output flyback pulse of the chassis and with the listed CRT high voltage for the television chassis high voltage. The TVA92 reading should be equal to or greater than the value listed. If the value is lower, change the drive signal polarity of the TVA92 and repeat the measurement. Use the highest of the two output voltage readings for your comparisons.

Powering the IHVT to find breakdown and corona problems

Some high-voltage (HV) breakdown problems with flybacks/IHVTs or separate HV multipliers are evident only when HV potentials are present. The TVA92 horizontal output device substitute and drive enables you to power and test the horizontal output HV components at full operating voltages. It provides a substitute horizontal output transistor with a known good horizontal base drive signal. This lets you power and test the horizontal output, flyback, HV, and flyback scan–derived power circuits to full voltage potentials.

The TVA92 substitute HOT works as a replacement for the HOT of the television chassis. It switches on and off just like the HOT of the chassis, completing a current path for the flyback and yoke currents. Switching comes from a horizontal drive signal generated in the TVA92. The drive signal is synchronized to a video signal supplied to the television from the Sencore VG91 Universal Video Generator, which is a companion to the TVA92. When using the horizontal output device substitution and drive, a working chassis produces near-normal deflection, high voltage, and a video-pattern display on the CRT. Note the test setup in Fig. 5-9.

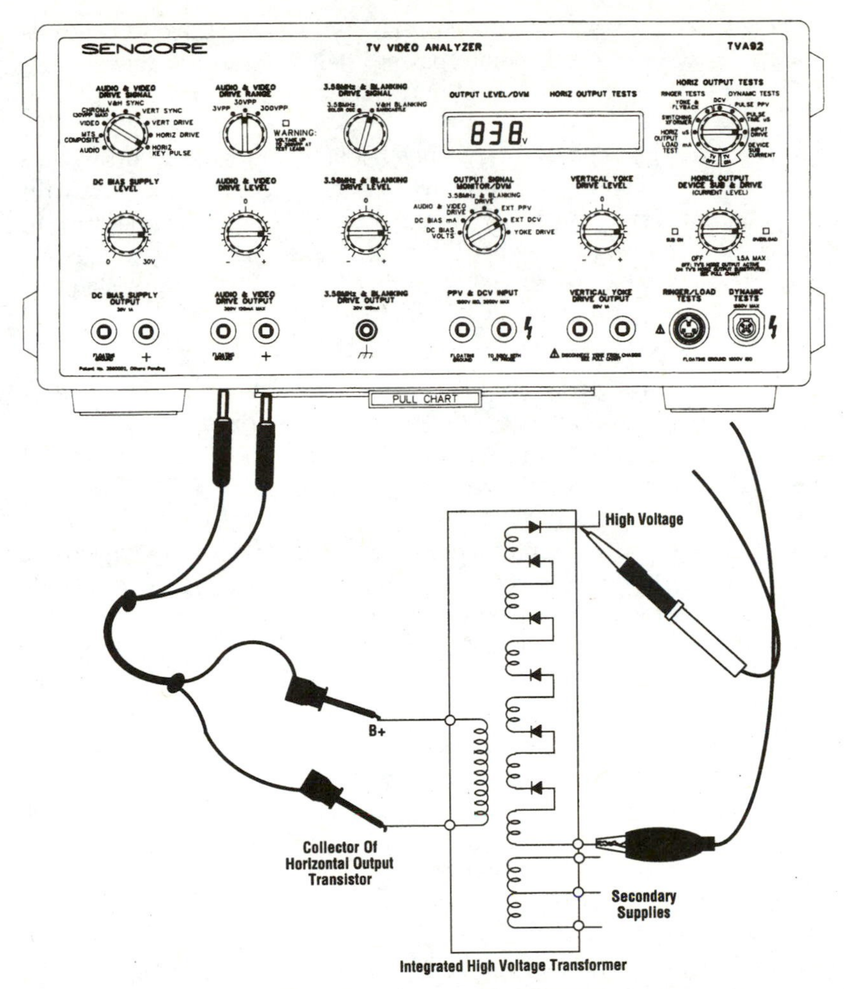

5-9 Setup to drive the HV section of an IHVT with the Sencore TVA92 Video Analyzer. (*Courtesy of Sencore, Inc.*)

When turned on, the horizontal output device substitution and drive control adjusts the duty cycle of the horizontal drive signal. This varies the conduction time of the TVA92 substitute HOT from approximately 5 μs (minimum) to 35 μs (maximum). By changing the conduction time of the substituting transistor, the current flow (power) to the horizontal output stage can be limited and increased slowly. This lets you slowly increase the pulse amplitude of the flyback and resulting HV to detect component or HV breakdown.

To assist in identifying breakdown or high-current conditions when using the horizontal output device substitution and drive, the TVA92 provides an indication of the current flowing through its substituting transistor. The device substitution current read-out of the TVA92 indicates the current through the collector of the TVA92 substituting transistor. This read-out reflects the average flyback primary plus yoke currents in the horizontal output stage of the chassis. The collector current provides the most accurate indication of the current (power) delivered to the flyback and yoke and, therefore, the best means to detect normal or abnormal conditions when substituting.

The device substitution current read-out, when using the horizontal output substitution and drive to test the television at full high voltage and deflection closely reflects how the conduction current of the horizontal output transistor is operating in the television chassis. The current through the substituting transistor is provided by the B+ supply, horizontal output, and flyback secondary circuits of the television.

Dynamic vertical yoke testing

The changing current through the windings of the deflection yoke produces a magnetic field that moves the electron beam vertically and horizontally across the CRT. Deflection yokes often develop shorted or open windings. An open or shorted winding can cause a complete loss of deflection, reduced deflection, image overlaps, or nonlinearities.

Vertical circuits are difficult to troubleshoot for several reasons: (1) stages are directly coupled; (2) a broad frequency response is required; and (3) a critical feedback loop delicately shapes the drive current of the yoke.

These problems, which cause only minor degradation, are especially difficult to see on an oscilloscope waveform. Also, when they are visible, all stages look bad because of the signal feedback. These difficulties lead technicians to a procedure of shotgunning parts until the only component left is the yoke. Too often it is assumed that the yoke must be bad without knowing for sure.

The TVA92 vertical yoke drive signal is designed to take the uncertainty out of vertical yoke testing. It is used as a substitute for the signal that normally feeds the vertical deflection yoke and provides a linear current ramp that tests the ability of the yoke to produce a full linear deflection at full operating current.

The vertical yoke drive signal is synchronized with the VG91 and the other TVA92 drive signals. Therefore, when using the TVA92 vertical yoke drive, the CRT displays a locked video pattern. You can use the CRT to determine whether or not the yoke produces full linear deflection.

To test the vertical yoke with the TVA92, simply disconnect the vertical yoke plug from the chassis and connect the TVA92 test leads. Turn the television on, and increase the vertical yoke drive level control to produce a pattern on the CRT. Use the crosshatch video pattern to judge deflection and linearity.

If the yoke is good, you will see a full and linear crosshatch pattern on the CRT. If the yoke is defective, the symptoms observed with the TVA92 vertical yoke drive signal will closely match the symptoms of the television chassis, indicating that the yoke is the defective component.

Horizontal output load test

As we have discovered previously in this chapter, many of the problems involving the horizontal output stage are difficult to troubleshoot. Problems in the horizontal output stage may cause startup or shutdown symptoms or may instantly destroy replacement output transistors or power supply components. Since these conditions occur only momentarily, normal troubleshooting measurements and procedures with the chassis operating are impossible.

Using the horizontal output load test

A severe load or timing problem in the horizontal output stage places immediate high-current stresses on the B+ power supply and output components when the chassis is turned on. This often results in immediate damage to the replacement horizontal output transistor and/or B+ power supply components.

The horizontal output stage and B+ power supply closely depend on each other for proper operation. The output stage needs a well-filtered and well-regulated voltage, but the B+ power supply only can provide regulated and filtered voltage when the current demand from the output stage is within a normal range. This makes it difficult, for example, to determine if a symptom of low B+ voltage is caused by a defect in the power supply or by the horizontal output stage demanding higher than normal current.

The startup and shutdown loops tie closely to the horizontal output stage. A problem in the output stage may prevent the chassis from starting, even when the startup circuits are good. Abnormal conditions in the output stage or B+ regulator may cause immediate shutdown. However, since these conditions occur momentarily, conventional voltage measurements cannot be performed.

Many chassis use a switch-mode power supply (SMPS) as the B+ supply. A popular shutdown method in these chassis is to defeat the SMPS when excessively high voltage is detected. This, of course, removes the B+ voltage from the output stage, making it difficult to determine whether the problem is in the B+ supply, the horizontal output stage, or the safety shutdown circuits. A further complication is that most SMPSs do not allow you to reduce the B+ voltage by lowering the ac line voltage. This prevents you from analyzing the horizontal output stage at a reduced voltage.

The horizontal output load test allows you to detect severe loading and timing problems in the horizontal output stage before you apply ac power to the chassis. This enables you to analyze the operation of the output stage, no matter what the symptoms, and quickly determine if it is causing B+ supply loading (low B+), improper startup, safety shutdown, or some other problem.

For any horizontal output circuit to operate, three things are required: (1) B+ voltage to the primary of the flyback transformer, (2) a switch (transistor) that completes the current path from the flyback primary to ground, and (3) a drive signal to turn the switching transistor on and off at a 15.734-kHz rate with an on time of approximately 30 μs.

The TVA92A horizontal output load test satisfies these basic requirements. It supplies a low-level B+ voltage to the output stage and simulates the switching action of the horizontal output transistor in the chassis. The circuit blocks and key components of the TVA92A internal horizontal output load test are shown in Fig. 5-10.

The horizontal output load test energizes the horizontal output stage of the chassis. If the horizontal output stage is good, the load test accurately simulates the

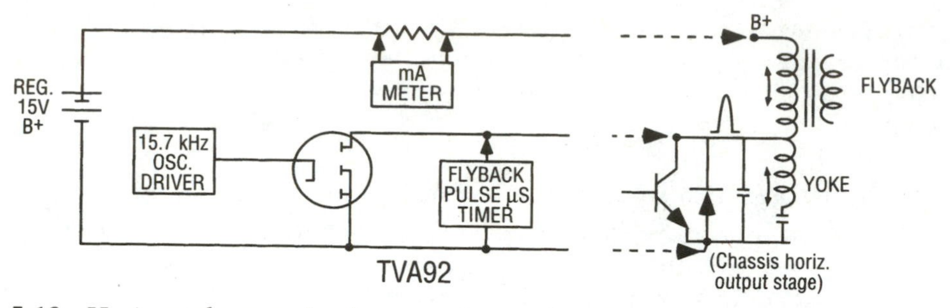

5-10 Horizontal output load test, applying a low B+ voltage and switching action to simulate operating conditions of the horizontal output. (*Courtesy of Sencore, Inc.*)

operating conditions of the horizontal output stage. Lead currents from the flyback, yoke, and secondaries are produced, just as they would be if the chassis were operating at its full potential. Flyback pulses also are produced, reflecting the critical timing of the resonant circuits.

The TVA92A monitors the horizontal output stage during the horizontal output load test. The microsecond load test measures the duration of the flyback pulse, while the milliamp test measures the current supplied by the 15-V B+ supply to the output stage. In a normal operating output stage, pulse time and a normal range of current of the flyback will be indicated "good" by the load test. However, if the output stage has severe problems, abnormal pulse time or current will be measured.

Horizontal output load test procedure

Three connections to the chassis are necessary to test the horizontal output stage with the horizontal output load test (see Fig. 5-11): (1) B+ side of the flyback primary, (2) collector of the horizontal output transistor, and (3) emitter of the horizontal output transistor. The clips on the ringer/load test lead are labeled and color coded for easy identification: orange/B+=B+ side of flyback; yellow/C=collector of HOT; and black/E=emitter of HOT.

If the read-out of a horizontal output load test shows dashed lines, the TVA92A is not receiving flyback pulses or delivering current to the output stage. Check to make sure you are connected to the proper test points.

Tests using the horizontal output load test are performed with ac power to the chassis removed. The TVA92A contains internal protection circuitry; however, to remove any chance of damaging either the analyzer or the chassis, always unplug the ac line cord to the television chassis before connecting the test leads.

Caution: Only perform the horizontal output load test with the ac power to the chassis removed. Remove the ac power by unplugging the ac line cord.

Note: The horizontal output load test may be performed with the horizontal output transistor of the chassis in-circuit, unless the transistor is shorted. If the horizontal output transistor in the television chassis is shorted, it must be removed.

Horizontal output load test sequence

1. Unplug the ac cord to the television chassis.
2. Connect the horizontal test leads to the proper circuit points.
3. Set the horizontal output test switch to the horizontal output load test "milliamps."
4. Read the results on the horizontal output test display.
5. Set the horizontal output test switch to the horizontal output load test "microseconds."
6. Read the results on the horizontal output test display.

> **Caution:** The horizontal output load test produces voltage pulses from the flyback at the collector of the horizontal output transistor and the flyback transformer secondaries in the television chassis. Do not come in contact with the energized circuit points during the load test.

Interpreting the test readings

The read-out of the horizontal output load test displays the two parameters that most accurately reflect the operation of the horizontal output stage under test. These parameters indicate if the horizontal output stage has a severe problem that will cause loading, startup, shutdown, or other problems with the B+ supply. Each test includes a numerical reading, as well as a "good/bad" indication. Table 5-2 summarizes the "good/bad" ranges for each test: (1) The milliamp reading is the amount of B+ current drawn from the TVA92A 15-V power supply. Readings between 5 and 80 mA represent a normal range of current and are considered "good." Current levels less than 5 mA usually indicate an invalid test condition, such as improper test lead connections or an open in the circuit of the horizontal output stage. Current readings greater than 80 mA are "bad" and indicate a heavy load in the circuitry of the horizontal output or flyback that will likely overload the power supply. (2) The microsecond read-out indicates the time duration (also called *pulse width*

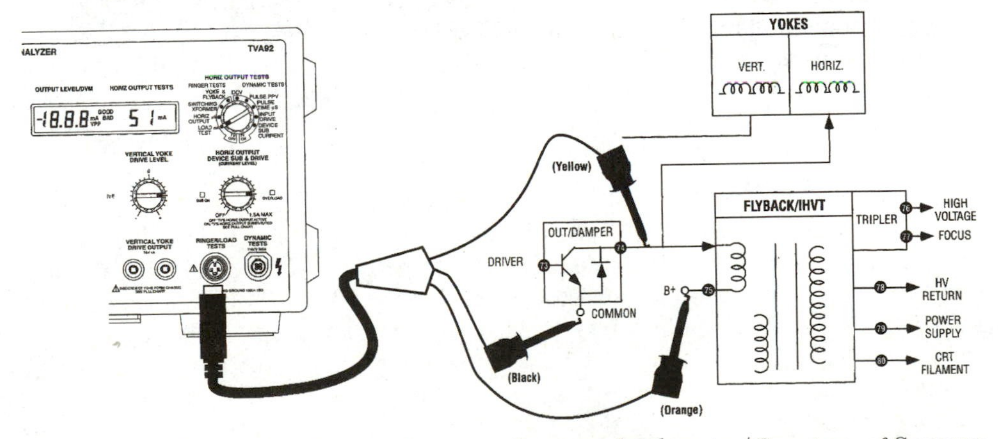

5-11 Setup for performing the horizontal output load tests. (*Courtesy of Sencore, Inc.*)

Table 5-2. Horizontal output load test good/bad ranges

Test	Normal range	Bad range
mA	5–80 mA	<5 or >80 mA
μs	11.3–15.9 μs	<11.3 or >16.9 μs

or *retrace time*) of the flyback pulse generated during the load test. This pulse time is set by the horizontal output circuits in the chassis, primarily the flyback, retrace timing capacitors, yoke, and yoke capacitor. Therefore, it is nearly identical to what the time would be if the chassis were powered to its full ac line voltage.

The pulse time indicates if proper timing and resonant action are occurring in the flyback and yoke circuits. Readings between 11.3 and 15.9 μs represent a normal range of pulse widths and are considered "good." Readings outside this range are "bad" and indicate improper timing, flyback defects, or severe loading problems.

The normal "good/bad" ranges shown in Table 5-2 take into account a wide variety of horizontal output stages. If a particular horizontal output stage reads "good" in both load tests, you may be certain that the horizontal output stage and flyback secondaries are not an immediate threat to an output transistor replacement or to the B+ power supply of the chassis when applying full ac power. You also know that the output stage is not the cause of a startup or shutdown condition.

Usually a "good" reading for both load tests indicates that the horizontal output circuit is 100 percent problem-free. However, some minor leakage paths in the output stage or a secondary load may not drastically change circuit parameters. In these rare cases, the horizontal output load test may not be altered significantly and will remain in the "good" range. When these problems occur, the output stage may be thoroughly analyzed, with ac power applied to the chassis, using the TVA92A dynamic test.

A "bad" reading in one or both of the horizontal output load tests indicates a problem in the horizontal output circuit, flyback, or flyback secondary. Use the "bad" read-out to help determine what to look for when isolating the problem. Some of the likely causes for different combinations of load test results are outlined in Table 5-3.

Some horizontal output circuit problems will cause the microsecond readings to fluctuate between "good" and "bad" values during the test. Fluctuating pulse time readings indicate multiple pulses or severe pulse ringing in the flyback. In either case, a problem of abnormal loading or timing that needs to be corrected is indicated.

Loading and leakage tests

The most likely cause of a dc short on the B+ supply is a shorted horizontal output transistor. Disconnect the B+ (orange) ringer/load test lead from the chassis and remove the horizontal output transistor. Then reconnect the B+ ringer/load test lead and note the milliamp reading. If the milliamp read-out is 10 mA or less, you have confirmed that the horizontal output transistor is shorted. If opening the horizontal output transistor does not remove the short, continue opening the possible dc short-circuit paths shown in Fig. 5-12 until you have isolated the defective components.

Table 5-3. Possible horizontal output load test readings and likely causes

Test readings mA	μs	Most likely causes
—	—	Improper connections Open flyback Open output stage circuit paths
Bad	—	Severe B+ supply short or leakage path <5 mA=open flyback or circuit path
Good	—	Open flyback Improper "collector" connection Open ringer/load fuse
Good	Good	No severe loading or timing defects
Bad	Good	Severe B+ leakage and/or flyback Secondary short or leakage path Flyback transformer
Good	Bad	Defective output timing components Flyback transformer Severe flyback secondary short or leakage path
Bad	Bad	Severe B+ leakage Flyback secondary short or leakage path Flyback transformer Defective output timing components

Note: Fluctuating microsecond read-out values indicate abnormal flyback pulse ringing or timing.

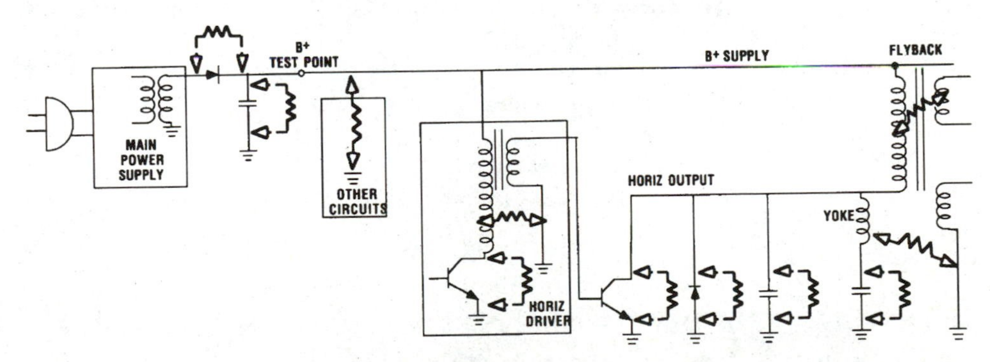

5-12 Some possible dc short or leakage paths that can drain the B+ power supply. (*Courtesy of Sencore, Inc.*)

Note: The horizontal output load test can be performed with the horizontal output transistor and damper in or out of the circuit. These components, if good, will not change the milliamp or microsecond readings.

Loading problems other than a low-resistance short produce current readings ranging from 80 to 200 mA. The first step in isolating these loading problems is to determine if the added current is caused by a dc load or by an ac load. Do so by disconnecting the collector (yellow) ringer/load test lead from the horizontal output stage, as shown in Fig. 5-13. Removing the collector lead stops the switching action of the output stage and removes all the ac currents in the flyback and yoke. All power transfer to the circuits on the flyback secondaries also stops.

Check the milliamp read-out with the collector lead disconnected. The current now is only the dc current to circuits that are powered by the B+ supply. This includes the horizontal output stage and perhaps the horizontal driver stage and oscillator. These stages typically draw less than 10 mA during the horizontal output load test. If the current is much higher than this, suspect a dc short or leakage path on the B+ power supply. If the current drops to normal, the excessive load is caused by ac leakage.

There are many possible dc leakage paths, as shown in Fig. 5-12. A dc short or leakage path can be caused by a leaky filter or diode in the B+ supply or by shorts or leakage paths in the horizontal output stage or other stages connected to the B+ power supply. To isolate a dc short or leakage path, open the circuit path of the B+ supply and hook the B+ (orange) test lead directly to the B+ side of the flyback transformer. Leave the collector (yellow) lead open. Now, through a process of elimination, disconnect each possible leakage path while you monitor the current. Begin by removing the horizontal output transistor and damper diode. Repeat the horizontal output load test and compare the milliamp read-out with the previous reading. If the milliamp reading decreases, the transistor or damper diode is leaky.

To isolate shorts on the flyback secondaries using the TVA92A, use the external peak-to-peak voltage and dc voltage input to measure the dc voltages and peak-to-peak pulse amplitudes to the secondary windings or the flyback. Remember that the load test simulates the normal horizontal output circuit operation of the chassis at approximately $\frac{1}{10}$ normal. Therefore, the voltages will be approximately $\frac{1}{10}$ the normal values that are shown on the schematic.

Ac or dc voltages that are considerably lower than $\frac{1}{10}$ normal or that are completely missing indicate a shorted component or scan-derived circuit associated with that flyback winding. If you suspect a short circuit, open the current path. Then repeat the load test and compare the milliamp read-outs. A current reading on the high side indicates that a component is breaking down. Use the ringer test to check the flyback and yoke for a shorted turn, and use the Sencore Z-Meter to check capacitors and other components for breakdown at their full-rated voltage.

Understanding the microsecond read-out

The horizontal output load test accurately simulates the normal switching action of the horizontal output stage. During the test, a pulse is produced at the collector of

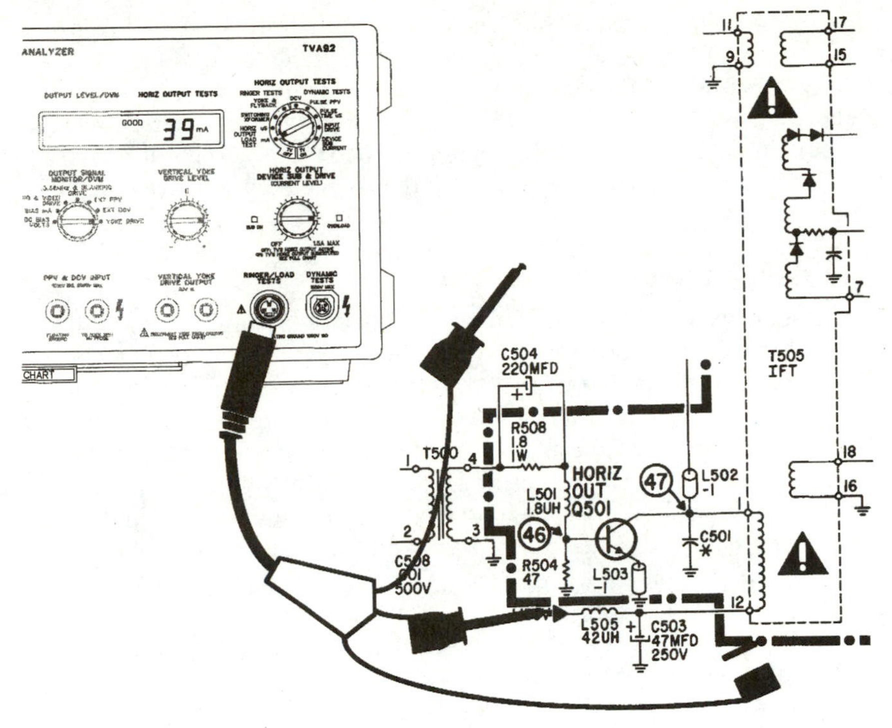

5-13 Determine if the loading is caused by dc or ac leakage. (*Courtesy of Sencore, Inc.*)

the horizontal output transistor. The shape and width of this flyback pulse reflect the normal resonant timing in the chassis. Normal duration (time in microseconds) of the flyback pulse varies from about 11.3 μs to as long as 16 μs. Any stable microsecond reading in this range indicates normal output stage timing and is accompanied by a "good" indication. If the read-out indicates "good," you know that the horizontal output stage is developing normal flyback pulses of the proper duration and is not the cause of a dead or shutdown symptom.

However, a "bad" read-out indicates that there is a timing or minor loading problem associated with the circuit of the horizontal output stage.

Ringer tests/checks with Sencore VA62A Analyzer

Various ringing tests and checks can be performed using the Sencore VA62A Deluxe Analyzer. These tests include in-circuit tests of the horizontal sweep transformer, checks of horizontal and vertical deflection yoke windings, and ringing checks of other transformer windings and coils. Aiding in these tests and checks is a built-in digital voltmeter and current meter, also discussed in this section.

The ringer test, built into the VA62A, is designed only for checks of yokes and horizontal sweep transformers. The digital read-out is updated continuously (unlike the LC53, which updates only once each time the ringer test button is pressed or the impedance match switch is rotated). This continuous updating allows the external shorted-turn test to be used for testing secondary windings of a flyback transformer. The same 10-count "good/bad" reference is used, as on previous Sencore ringers.

Horizontal sweep circuit checks

If the drive signal is correct or you have driven the sweep output stage with the VA62A and still have horizontal sweep/HV problems, more circuit checks are required. The next step would be to test the horizontal amplifier components.

Sencore engineers gave much thought to the possibility of providing a universal amplifier to drive the horizontal output load directly. They decided not to include a horizontal output amplifier for several reasons.

1. Impedance matching became a real problem with the many yokes and sweep/HV transformers in use today, to say nothing of the many tubes, transistors, and silicon-controlled rectifiers. Every attempt became misleading and was left with constant interpretations of the results.
2. A test lead of 5 kV would have been exposed at the front panel of the VA62A, endangering shop personnel.
3. All three types of amplifiers would have to be built-in, including tube, transistor, and SCR. Mistaken use of the wrong amplifier would result in damage to components, such as expensive flyback transformers.

It was decided that the very best way to check these three types of horizontal amplifiers was by using a tube tester (or substitute) in a known good tube, a transistor tester for solid-state sweep systems, and a substitute for the SCR.

A much easier and universal way of troubleshooting the horizontal sweep output circuit is to eliminate the loading. The most vital components we want to eliminate are the sweep transformer and the deflection yoke. Moreover, shorted capacitors should not be overlooked. The coil checks can be performed with the VA62A by doing the ringer test to see if the Q values of the transformer and yoke windings are good. This critical but easy-to-do test has been used with the Sencore VA48 Analyzer and the YF33 Ringer Yoke/Flyback Checker and has proven to be correct 100 times out of 100 with the thousands of these checkers in use today.

This same test is incorporated in the VA62A when the digital meter is switched over to serve as the ringer test indicator. All one has to do is connect the leads of the ringer test to the sweep transformer, switch over to the ringer test, and rotate the impedance-matching switch through its six positions. Figure 5-14 shows the analyzer ringer test setup. Like all impedance-matching devices, the output (thus the meter reading) will increase when you have matched the load correctly. The digital read-out shows the number of rings before decay to the 25 percent level, the same as if you were viewing the waveform on a scope (Fig. 5-15).

The beauty of the ringer test is that it will detect even one shorted turn, while opens are indicated by no read-out at all. You can check small sections of the sweep transformer too, giving you a positive check on the numerous small takeoff coil

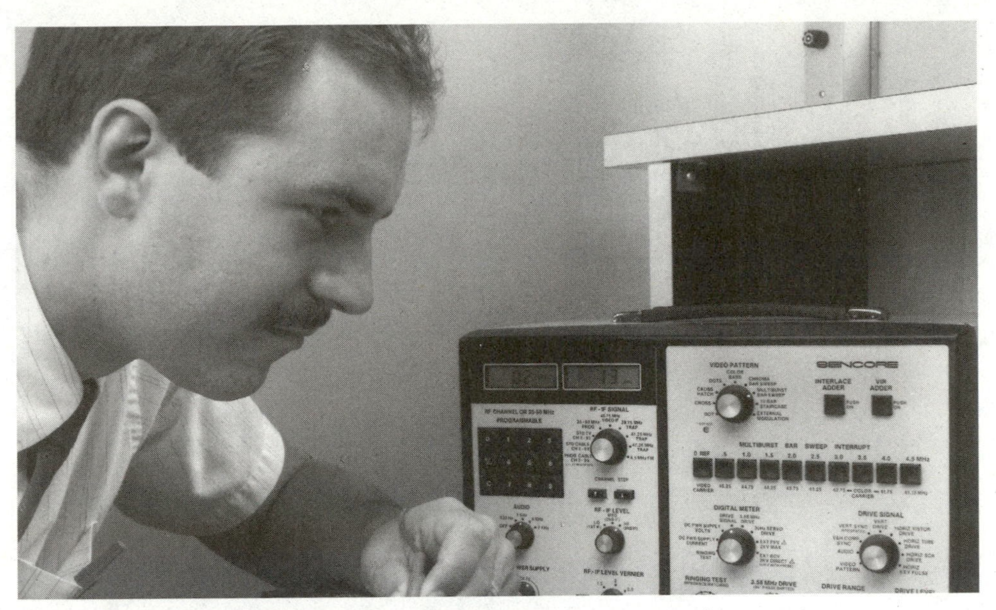

5-14 Analyzer ringer test setup. (*Courtesy of Sencore, Inc.*)

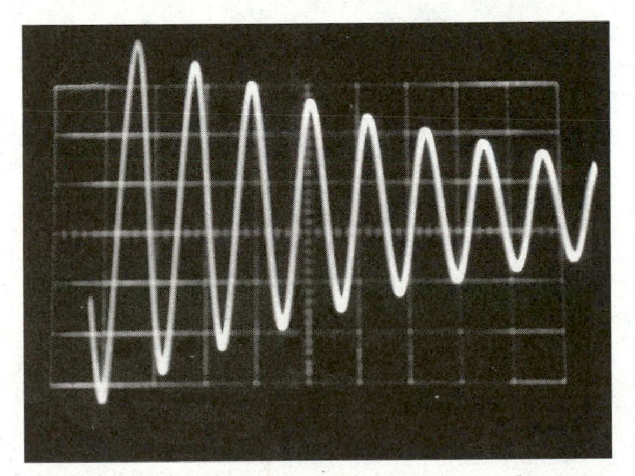

5-15 Damped scope waveform seen during a ringing check of a good coil. (*Courtesy of Sencore, Inc.*)

windings—coils that you probably have had no way to check before. Note the small windings being checked in Fig. 5-16.

If the horizontal sweep transformer and takeoff winding taps check okay by the ringer test, do the same test on the deflection windings of the yoke. The procedure is exactly the same, and the results can be just as conclusive. The ringer test also can be used to check pincushion and horizontal efficiency coils.

If you have eliminated all components with these easily performed checks and still have trouble, you are down to straightforward dc paths, and it is time to use the

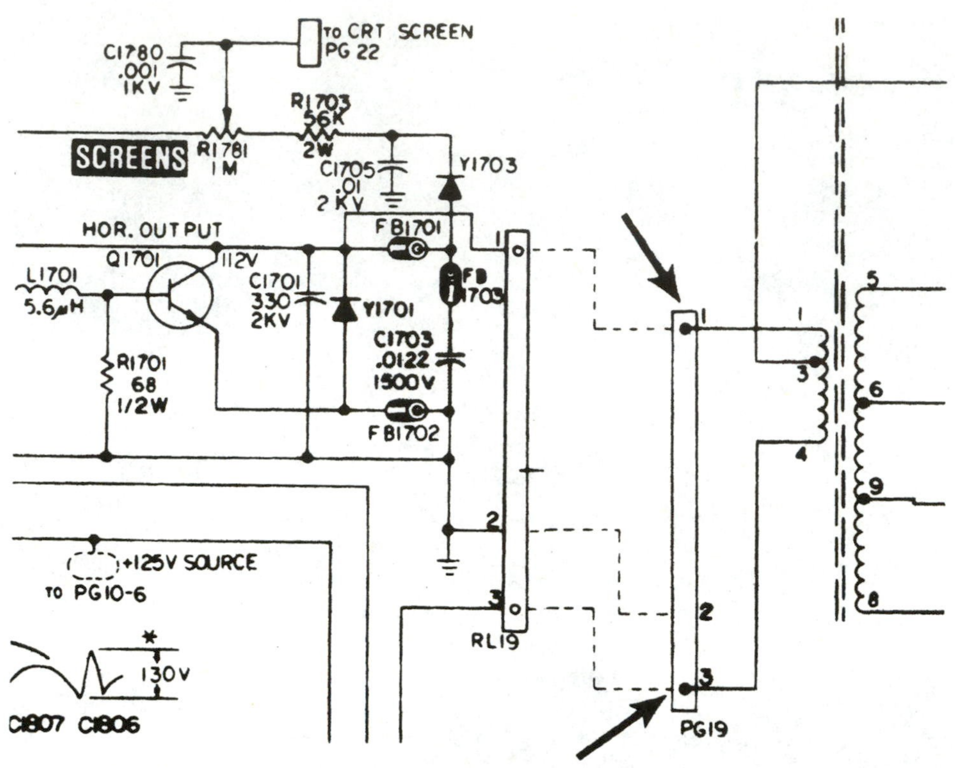

5-16 Takeoff windings of a sweep transformer being checked with the ringer. (*Courtesy of Sencore, Inc.*)

built-in digital voltmeter (and the 50-kV high-voltage probe) to check some of the dc voltages in the sweep circuit.

The boost (Fig. 5-17) is an important check at this point, because proper drive, a good output amplifier stage, and a good yoke and sweep transformer should produce the boost voltage. If it does not, either the damper is defective or the boost voltage circuit is loaded down by a shorted component.

If the boost voltage is okay, you should now know that the problem is in the HV section of the sweep system. If the set has an HV tripler, this can be disconnected to see if this unloads the sweep transformer. Use the HV probe to check the dc developed from the sweep transformer pulses; this should give you some clues about the condition of the horizontal sweep stages.

You may have noted that all these tests are performed without interpretation or reference to a single chart and that all television receivers are made to look alike for faster, easier troubleshooting of horizontal sweep circuits.

The ringer test

The VA62A ringer test allows you to test any deflection yoke or sweep transformer in or out of circuit. This test involves connecting two leads to the coil under test and rotating the ringer test impedance-matching switch through its six positions. If a "good" read-out is obtained on the digital meter for one or more of the switch positions, the coil under test is good. If none of the positions provides a "good" test, the ringer test will locate the cause of the "bad" test results.

Before a test of a yoke or flyback transformer is performed in a tube-type or hybrid set, the HV rectifier should be removed from its socket in the HV compartment. If this tube remains in circuit, its filaments may act like a short across the sweep transformer, causing it to test as defective.

Normally, the in-circuit ringer test is performed first. If this test results in a "good" digital meter reading, the sweep transformer and horizontal yoke windings both have been tested and are operating properly.

A backup out-of-circuit test is performed if (1) the in-circuit test resulted in a "bad" meter reading or (2) an open is suspected in one of the takeoff coils (other than the HV winding) of the sweep transformer.

Tube-type (sweep transformer) in-circuit test

In-circuit ringer tests on a tube-operated set can be performed with the test leads connected to the horizontal output tube-plate lead and chassis.

> **Caution:** Do not make any ringer connections until the power has been removed from the set.

Ringer test procedure

1 Connect the common test lead from the VA62A analyzer to the chassis of the set.
2. Connect a lead from the ringer test jack to the cap connections of the HC rectifier or plate cap of the horizontal output tube.
3. Switch the selector switch of the digital meter to the ringer test position.
4. Switch the ringer test impedance-matching switch through all six of its positions while observing the digital meter read-out.

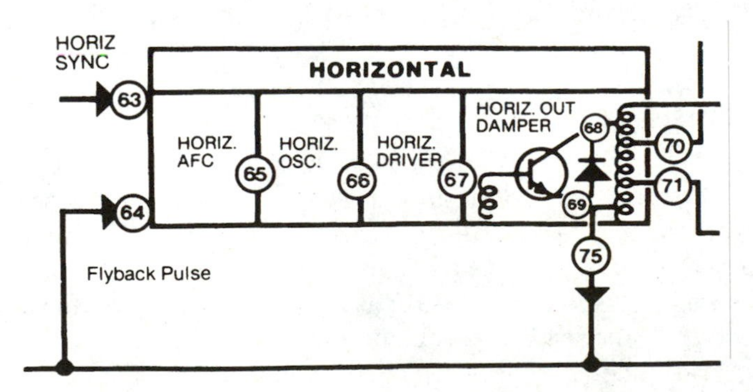

5-17 Simplified diagram of a typical horizontal sweep output system. (*Courtesy of Sencore, Inc.*)

5. If one or more of the positions results in a "good" reading on the meter, the HV windings of the sweep transformer are good, indicating that none of the coil turns is shorted.
6. If the test reads "bad" in all six positions, disconnect the yoke connections or plug and repeat step 4.
7. If the ringer test still shows "bad," see "Extended In-Circuit Transformer Tests."

Solid-state in-circuit test

An in-circuit ringer test on a transistor of the horizontal output system can be performed with the test lead connected to the collector of the output transistor and the common lead connected to the chassis.

Caution: Do not make any ringer connections until power has been removed from the set.

Ringer test procedure

1. Connect the common test lead from the analyzer to the chassis of the television set.
2. Connect a test lead from the ringer test jack to the lead running from the sweep transformer to the HV tripler or to the lead coming from the collector of the horizontal output transistor to the sweep transformer.
3. Switch the selector switch of the digital meter to the ringer test position.
4. Switch the ringer test impedance-matching switch through all six of its positions while observing the digital meter read-out.
5. If one or more of the positions results in a "good" reading on the meter, the sweep transformer is good with no shorted coils.
6. If the test reads "bad," disconnect the yoke plug and the convergence unit plug and repeat steps 4 and 5.
7. If the test still reads "bad," disconnect one end of the damper diode and repeat steps 4 and 5.
8. If the test still reads "bad," see "Extended In-Circuit Transformer Test."

An out-of-circuit sweep transformer is being checked with the ringer test in Fig. 5-18.

Extended in-circuit transformer test

At the conclusion of the in-circuit test listed above, a "bad" test almost always indicates a defective sweep transformer. To be absolutely sure the sweep transformer is defective, the ringer test may be repeated as each of the sweep transformer loads is disconnected. If a flyback that produced a "bad" indication begins to read "good" after a load has been disconnected, that load path should be checked for a possible shorted component. If all the loads are now disconnected and the flyback transformer is still reading "bad," one-half of the job of replacing the transformer is done, since all the load connections already have been unsoldered.

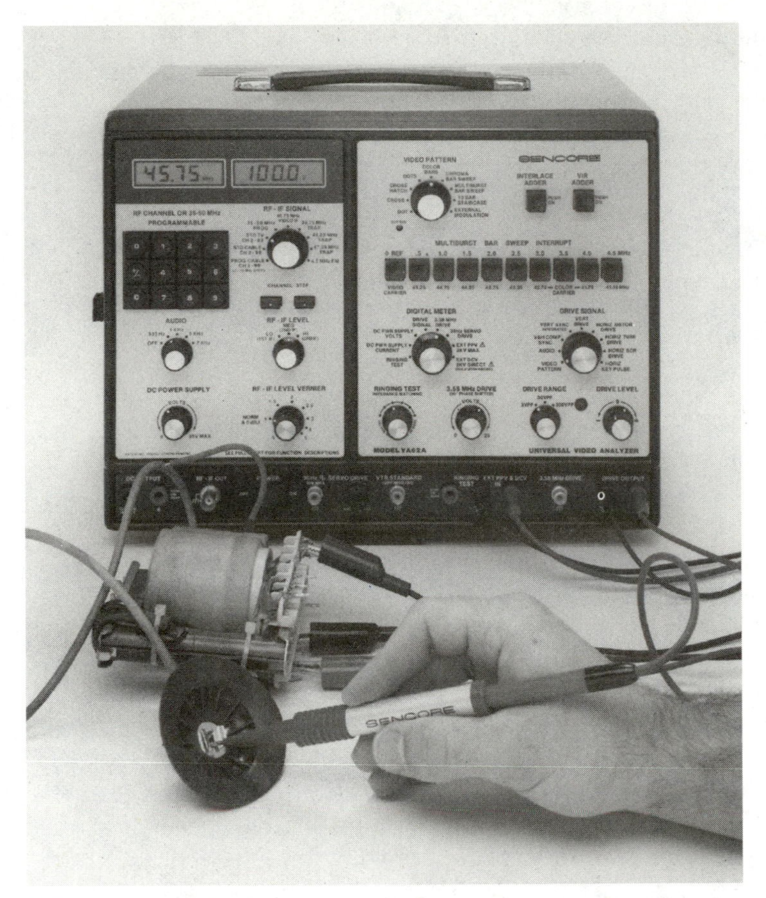

5-18 Ringer check on an out-of-circuit sweep transformer. (*Courtesy of Sencore, Inc.*)

Sweep transformer takeoff winding test

The previously listed in-circuit checks of the sweep transformer will find most defective sweep transformers. The test will show if a coil is open or if any of the coils in the transformer are shorted (since a shorted coil will lower the Q for all of the windings in the transformer).

In a few rare cases, however, one of the other windings in the transformer (such as the automatic gain control keying-pulse coil) may be open. If this condition is indicated (by operation of the set), each coil may be ring tested individually for a complete sweep transformer test.

Caution: Do not make any ringer connections until the power has been removed from the chassis under test.

Procedure for additional sweep transformer tests

1. Determine the connections of the coil to be tested by checking the service data for the transformer on the television chassis schematic.
2. Connect the ringer test lead to the coil under test and the common lead to the other end of the coil.

> **Note:** A coil with several taps may be checked with a single test if the coil is ring tested from end to end. All series-connected sections of the coil will be tested, and an open in any section will produce a "bad" test.

3. Switch the digital meter control to the ringer test position.
4. Switch the ringer test impedance-matching switch through all six of its positions while observing the meter.
5. If the first coil section reads "good," move the test lead to the next suspected "bad" coil and repeat steps 3 and 4.
6. If all suspected coils test "good," the sweep transformer is not the cause of the horizontal sweep problem. The HV secondary winding of a sweep transformer is being checked in Fig. 5-19.

Deflection yoke testing

The impedance-matching range of the ringer test allows for virtually all yokes and sweep transformers to be tested with the same basic check. The coil to be tested is connected between the common and the ringer test jacks. The impedance-matching switch is rotated through all its positions, and the quality of the coils is indicated by the counts on the digital read-out meter.

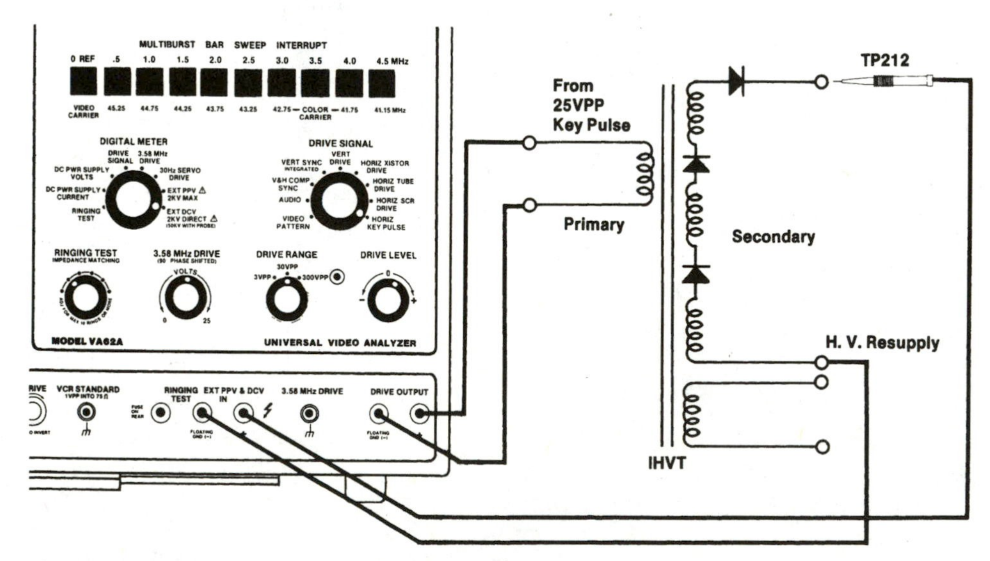

5-19 Ringer check of the HV coil. (*Courtesy of Sencore, Inc.*)

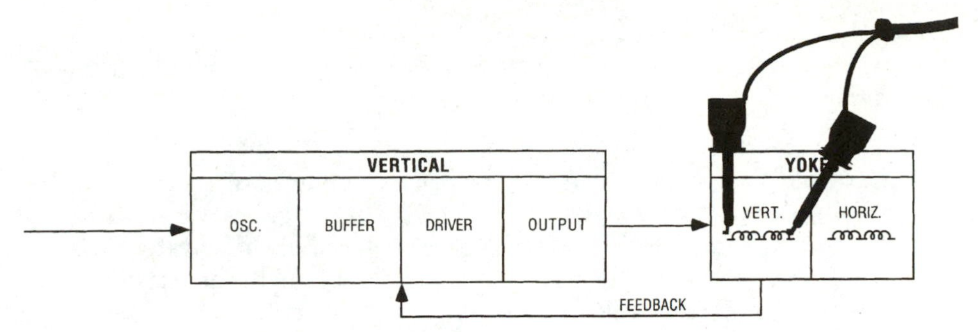

5-20 Ringer check for yoke deflection. (*Courtesy of Sencore, Inc.*)

Horizontal yoke testing procedure

1. Disconnect the horizontal yoke from the television circuit. Notice the test leads connected to the flyback yoke being tested in Fig. 5-20.
2. Connect the coil to be tested between the common and the ringer test jacks on the VA62A analyzer.
3. Select the ringer test position of the digital meter selector.
4. Rotate the ringer test impedance-matching switch through all its positions while observing the read-out on the digital meter.
5. If one or more of the positions produces the proper count on the digital meter, the coil is good.
6. If two coils are series connected, make sure both coils are tested. If not, reconnect the ringer test leads across the second coil and repeat the test.

Vertical yoke testing procedure

1. Disconnect the vertical yoke from the television circuit. Unsolder the leads or, if they are plug-in type, remove the yoke plug.
2. Determine if the vertical damping resistors for the yoke are located in the yoke housing or in the television chassis. If they are in the chassis, they were disconnected when the yoke was unplugged. If they are mounted on the coil housing, one end of each should be disconnected temporarily from the coil.
3. Connect the ringer test leads across one of the two series-connected coils and test it by rotating the ringer test impedance-matching switch through all its positions. Notice which position gives the highest reading.
4. Move the test leads to the other half of the series-connected coils. Repeat the test, and again, notice which position gives the highest reading and the number of ringing counts.
5. If none of the switch positions produces a "good" count on the digital read-out meter or one side of the yoke gives a different number of ringing counts than the other, the deflection yoke should be considered defective.

If a schematic is not available for the yoke coil to be tested, remember that the vertical yoke coils are located at the sides of the picture tube and the horizontal coils are located on the top and bottom of the yoke assembly.

Deflection yokes should be tested, whenever possible, while still mounted on the CRT. Some yokes may have intermittent shorts. Removing the clamp on the yoke mounting hardware will release the pressure, holding two shorted turns or coil wires together. The result is a deflection yoke that tests "bad" when mounted but "good" when it is removed from the neck of the picture tube.

Caution: Do not test a yoke or horizontal sweep transformer on a metal surface, such as a metallic-topped workbench. The metal surface will act like a shorted coil and cause the coil to test as defective. If coils must be tested near a metal surface, be sure to separate the coil under test from the metal surface with at least 1 in of nonmetallic material, such as wood, plastic, or cardboard.

Digital meter

The digital read-out meter (Fig. 5-21) is used for measuring voltage (peak-to-peak and dc) and dc current and for counting the number of rings during the ringer test. The meter can be switched to monitor all drive signals and built-in dc power supply voltage or current or can be used as an external voltmeter for circuit troubleshooting. The digital meter read-out is fully autoranged.

Dc power supply monitor

The digital meter measures either the dc voltage or the dc current obtained from the adjustable built-in power supply. This feature allows you to set the dc voltage

5-21 Digital meter section of the VA62 analyzer. (*Courtesy of Sencore, Inc.*)

output much more accurately than you can with control-panel markings. Also, the digital measurements of current provide so much troubleshooting information that using another meter is not necessary. (The fuse for the power supply output, by the way, is located on the back panel of the VA62A and not in the test lead. This ensures that the fuse will always be used if different test leads are used.)

Drive monitors

Three separate drive monitor switch positions allow the drive signal functions, the servo drive signal, or the 3.58-MHz drive signal to be monitored by the digital meter, which is autoranged to provide direct readings.

External meter modes

The external modes function on the digital meter allows you to read peak-to-peak and dc voltages directly, just as you would with a separate meter. You also will find that the frequency response and accuracy of the peak-to-peak function are more than adequate for any troubleshooting procedure.

The external dc function is important for troubleshooting or testing triplers and integrated sweep transformers. The input impedance of the dc function is 15 MΩ so that the TP212 10-kV probe or the HP200 50-kV probe can be used for HV tests. Both peak-to-peak and dc volts are autoranged.

The common test lead connection for the meter is floating and is not common to the drive output or the dc power supply. This allows the meter to be used to measure voltages across components, such as resistors, or to make measurements in non-grounded circuits.

Horizontal and vertical sweep stages

There are a number of procedures available when troubleshooting vertical and horizontal sweep circuit operations of both color televisions and PC monitors.

Vertical sweep troubleshooting

For vertical sweep troubleshooting of color televisions and PC monitors, the following information should be noted when isolating problems with the vertical oscillator, driver, and sweep output. The vertical driver and output stages amplify the oscillator signal and provide the current drive needed for the vertical deflection yoke.

A defective driver, output, or yoke can cause loss of deflection, reduced height, or loss of vertical linearity. Before using signal injection to troubleshoot a vertical problem, use the DVM or DMM of the analyzer to confirm the proper bias on the output components. The vertical stages are usually dc coupled to get good linearity. A wrong dc voltage affects all the components in the oscillator, driver, and output stages. Therefore, a dc bias problem must be repaired before signal injection can be used effectively in the vertical stages. Use an analyzer, such as the Sencore VA62A or TVA92A, to inject vertical and horizontal sweep signals into the circuit (see Fig. 5-22).

Collapsed vertical raster (a thin horizontal line across the screen)

Inject the vertical drive signal of the analyzer into the output of the vertical driver circuit (Fig. 5-23).

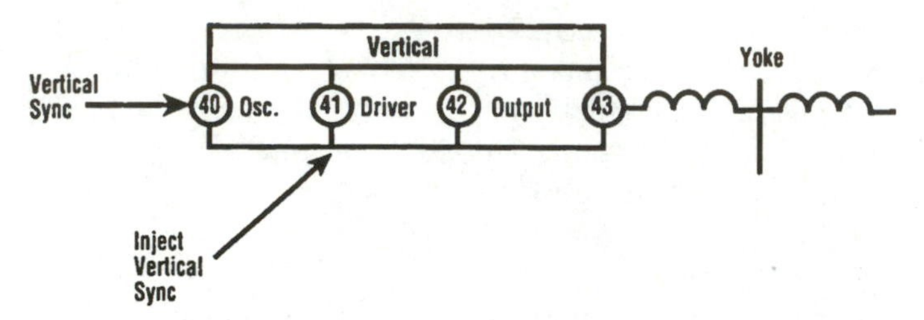

5-22 Using an analyzer to inject vertical sync signals. If the picture returns with signal injection into any of these stages, troubleshoot the vertical oscillator, driver, or sweep output stage. (*Courtesy of Sencore, Inc.*)

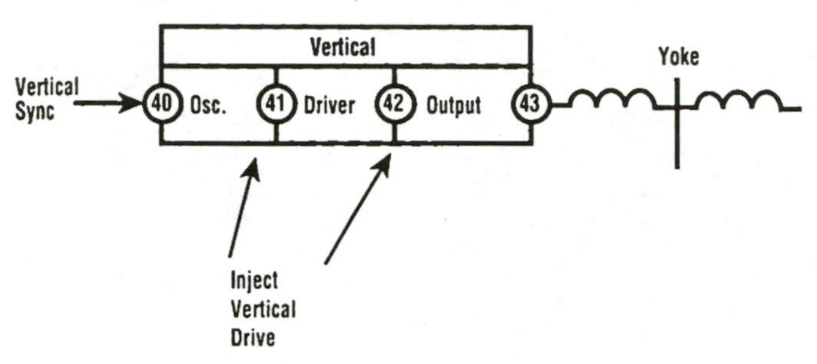

5-23 Injecting the vertical drive signal into the vertical driver and output stages. (*Courtesy of Sencore, Inc.*)

Note: Injecting into the vertical stages will not always produce full vertical deflection because most signals are uniquely shaped by feedback loops and wave-shaping circuits.

Look for the sweep to expand, but remember, it may not be a full raster. If the sweep expands, either partially or fully, the circuits from the injection point to the output are good. If the sweep does not expand, check the output components or ring the deflection yoke.

Note: The vertical driver signal will not drive the vertical yoke directly.

Isolating horizontal sync problems

The horizontal sync pulses control the timing of the horizontal oscillator. Many monitors receive horizontal sync directly. Television sets and some PC monitors have a composite sync ("sync on video") input and require the use of sync separators. Sync pulses that are low in amplitude, the wrong frequency, or missing will cause the monitor to lose horizontal hold.

Loss of horizontal sync symptom

Inject the horizontal sync drive signal of the video analyzer into the input of the horizontal oscillator, as shown in Fig. 5-24. If the television or PC monitor regains horizontal hold control and gives full horizontal deflection, the driver and output stage are working properly. Troubleshoot the horizontal sync part. If the same symptoms are displayed with the drive signal applied, troubleshoot the horizontal oscillator circuit.

Checking yokes for shorts or opens

The changing current through the windings of the deflection yoke produces a magnetic field that scans the electron beam across the face plate of the CRT. Yokes often develop shorted or open windings. An open or shorted winding may cause reduced vertical or horizontal raster size or a complete loss of deflection.

The ringer test will find a defective yoke, even if it has a single shorted turn. Readings of 10 rings or more are accompanied by a "good" display and show that the winding does not have a shorted turn. "Bad" readings, less than 10 rings, indicate a shorted turn.

Collapsed raster symptom

For this symptom, ring the horizontal and vertical yoke windings. Always disconnect the yoke from the circuit and unsolder any damping resistors (leave the yoke mounted on the picture tube).

If the horizontal and vertical yoke windings ring above 10 rings, the yoke is good. If any of the windings ring below 10, the yoke is defective and needs to be replaced.

Making meaningful peak HV and B+ voltage readings

A wealth of troubleshooting information can be gained about the operation of televisions by measuring the dc and peak-to-peak voltage at the collector of the horizontal output transistor. The video analyzer has a dc and peak-to-peak voltmeter with the input protection required for measuring signals at this test point. The dc

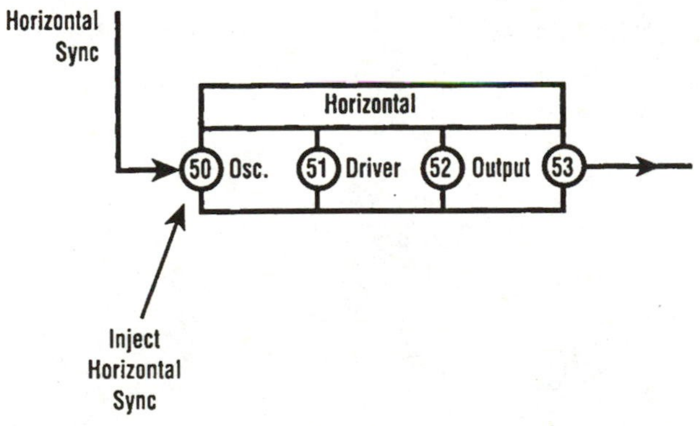

5-24 Injecting the horizontal sync signal into the horizontal sync path. (*Courtesy of Sencore, Inc.*)

reading tells you if the B+ supply is working correctly, while the peak-to-peak reading tells you if the output circuits are creating the needed high voltage.

Dead monitor case-history problem

With the video analyzer, measure the dc voltage at the collector of the horizontal output transistor. If the B+ voltage is low or missing, unload the power supply by disconnecting the collector of the horizontal output transistor from the circuit. Measure the voltage at the output of the power supply regulator again. If the voltage is still low or missing, troubleshoot the power supply. If the voltage to that noted on the schematic also is low, then something is loading down the supply. In this case, troubleshoot the output transistor, flyback transformer, or yoke (see Fig. 5-25).

Ring the flyback for shorts or opens

The flyback transformer in a PC monitor is responsible for creating the focus, high voltage, and other scan-derived power supply voltages. The flyback is a high-

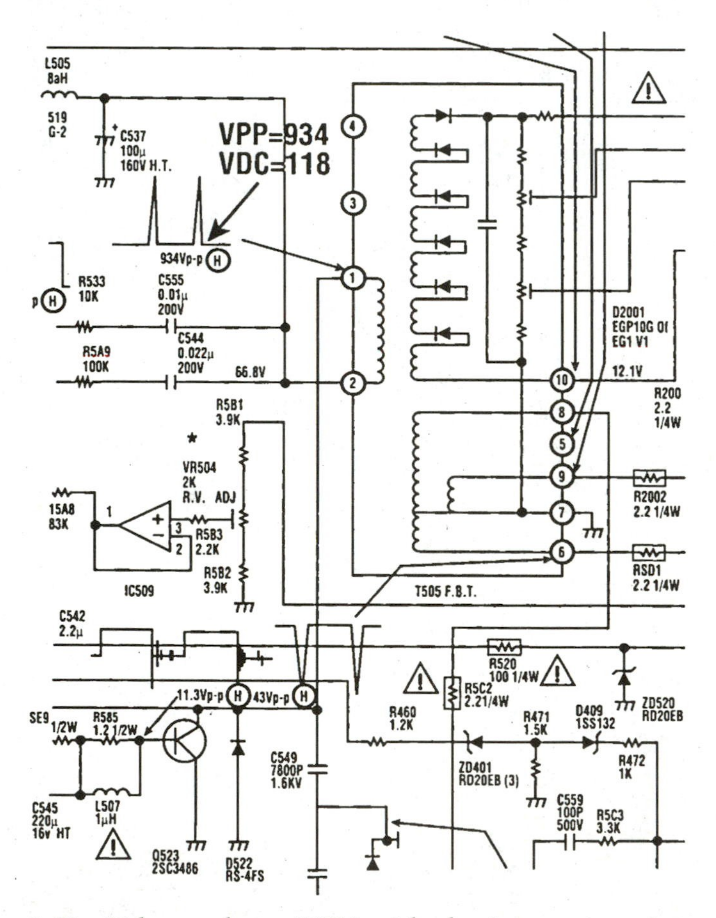

5-25 Video analyzer DVM with the input protection required to measure the pulse at the horizontal output collector. (*Courtesy of Sencore, Inc.*)

failure component; it also is one of the most expensive parts in the monitor, except for the CRT.

While a transformer winding that is open is easy to identify using an ohm meter, the more common problem of a transformer winding that is shorted is nearly impossible to find using conventional testing methods. The video analyzer has a patented ringer test that gives you an easy-to-use fail-safe method of finding opens and shorts in flyback transformers.

Another DOA monitor problem

Connect the ringer across the winding of the flyback primary, and ring the transformer. A "good" reading of 10 rings or more indicates that none of the windings in the flyback has a short or open. You do not need to ring any other winding. A shorted turn in any other winding will cause the primary to ring bad (Fig. 5-26).

A "bad" reading, fewer than 10 rings, may be caused by a circuit connected to the flyback that is loading the ringer test. Sequentially disconnect and retest the flyback of the most likely circuits in the following order: (1) deflection yoke, (2) CRT filament (unplug the CRT socket), (3) HOT collector, and (4) scan-derived supplies.

If, after disconnecting and retesting each circuit, the flyback rings "good," it does not have a shorted winding. If the flyback still checks "bad," after you have disconnected each circuit, unsolder it and completely remove it from the circuit. If the flyback primary still rings fewer than 10, the flyback is bad and must be replaced.

Testing the HV diode multipliers and IHVTs

During normal television and PC monitor operation, a large pulse appears at the collector of the horizontal output transistor. The output connects to the primary of the flyback transformer, and the pulses are induced into the flyback secondaries. The pulses are stepped up and rectified to produce the focus and high voltages. These voltage pulses are rectified by HV diodes contained in the flyback package or in a stand-alone multiplier package.

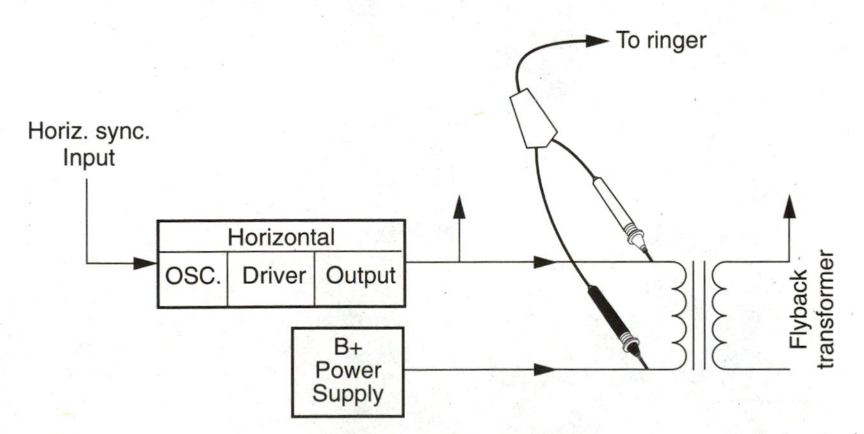

5-26 Ringer test hookup for finding shorts or opens in a flyback transformer. (*Courtesy of Sencore, Inc.*)

Because these are HV components, it is often difficult to determine dynamically if the diodes will break down under HV conditions. The video analyzer has a special test for determining if these diodes are good or bad.

Low or no high-voltage symptom (IHVT)

> **Note:** It is only necessary to do this test if all the following conditions occur:

1. Voltage is low or high voltage or focus voltage is missing.
2. The B+ and peak-to-peak voltages at the horizontal output transistor are normal.
3. The flyback transformer passes the ringer test.

With the analyzer, feed a 25-V peak-to-peak horizontal sync drive signal into the primary winding of the flyback transformer. The step-up action of the transformer and the HV diodes should create a dc voltage between the second anode and HV re-supply pin on the flyback. Measure this voltage with a dc voltmeter. Look up this voltage on a reference chart to decide if the HV diodes are good or bad.

Locating problems in the horizontal oscillator, driver, and output

If the horizontal yoke, flyback, multiplier, output transistor, and B+ supply have tested "good" but the television still lacks deflection or high voltage, the horizontal driver circuit may be defective. A missing or reduced amplitude horizontal drive signal could prevent the receiver from starting and operating properly. Use the horizontal sync drive signal of the Sencore video analyzer to isolate problems in the horizontal drive circuit. Refer to signal injection points in Fig. 5-27.

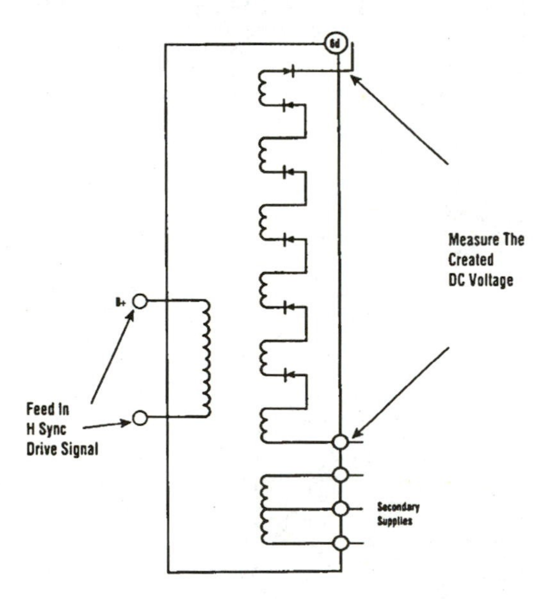

5-27 Testing the multiplier diodes by feeding the video analyzer drive signal into the primary and monitoring the dc voltage across the secondary. (*Courtesy of Sencore, Inc.*)

Television receiver will not start

> **Note 1:** Before injecting into the horizontal drive circuit, test the flyback, the yoke, the HV multiplier, the horizontal output transistor, and the B+ supply.
>
> **Note 2:** When injecting at the output transistor, disconnect the secondary winding of the driver transformer from the base.

Inject the horizontal drive signal into the driver circuit. Watch for horizontal deflection on the CRT. If it returns, you are injecting after the defective stage. If nothing happens, inject the drive signal at the base of the horizontal output transistor. Figure 5-28 shows these injection points.

Measuring high voltage

The CRT requires a very high dc voltage to accelerate the electrons toward the screen. This voltage develops in the secondary winding of the flyback transformer and is amplified and rectified by the integrated diodes in the flyback or by a separate multiplier circuit.

Measuring the high voltage at the second anode of the CRT lets you know if the output circuit, flyback, HV multiplier, and power supply regulation circuits are working correctly. Additionally, some televisions and PC monitors have adjustments to set the high voltage and focus voltage.

Dim, bloomed-picture symptom

Start troubleshooting by measuring the high voltage with the Sencore analyzer dc voltmeter and the HP200 high-voltage probe. Compare the voltage reading with that shown on the schematic diagram.

Testing switching transformers in switch-mode power supplies

Switching transformers are used in power supply circuits to step voltages up or down. They are one of the most common failure components in switch-mode power supplies. Open windings are easy to find with an ohm meter, but shorted turns are nearly impossible to find using conventional test methods. The ringer test of the video analyzer easily finds both open and shorted windings in switching transformers.

Dead power supply symptom

> **Note:** For this test, the switching transformer must be removed from the circuit.

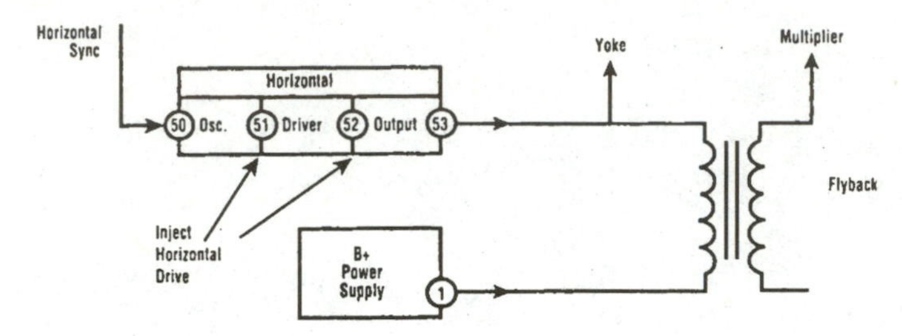

5-28 Injecting the horizontal drive signal into the horizontal driver and output stages. (*Courtesy of Sencore, Inc.*)

Connect the analyzer ringer test leads across a winding on the switching transformer. A reading of 10 rings or more will show that the winding does not have a shorted turn. Perform this same test on all windings of the switching transformer.

Television vertical sweep systems

In my conversations with technicians, I have found that many believe that vertical sweep systems are the most difficult circuits to troubleshoot in a PC monitor or television receiver. Even a minute change in component parameters can cause reduced deflection, nonlinear deflection, or image overlaps. These symptoms can be due to either a small circuit component or an expensive vertical yoke. Thus you must carefully plan a strategy to take the guesswork out of isolating vertical sweep problems.

How the vertical yoke deflects the CRT beam

To know how the vertical sweep stages operate requires an understanding of CRT beam deflection. These beam electrons travel to the face of the CRT, striking the phosphor surface to produce light on the face of the picture tube.

If the stream of electrons travels to the face plate of the CRT without control from a magnetic or electrostatic field, the electrons will strike the center of the CRT screen, producing a white dot. To move the dot across the face plate of the CRT screen requires that the electrons be influenced by the electrostatic or magnetic field.

In video display CRTs, a magnetic field is produced by the vertical coils of a yoke mounted around the neck of the CRT. The yoke is constructed with coils wound around a magnetic core material.

When current flows through the vertical yoke coils, a magnetic field is produced. The yoke core concentrates the magnetic field inward through the neck of the CRT. As the electrons pass through the magnetic field on the way to the CRT face plate, they are deflected (pulled upward or downward) by the yoke. This causes the electrons to strike the CRT face plate at points above or below the screen center.

To understand how electrons are deflected requires a review of the interaction of magnetic fields. As you refer to Fig. 5-29, you may recall that an individual electron in motion is surrounded by a magnetic field. The magnetic field has a circular motion surrounding the electron. As electrons travel through the magnetic field of the yoke, the magnetic fields interact. Magnetic lines of force in the same direction create a stronger field, while magnetic lines in opposite directions produce a weaker field. The electron is pulled toward the weaker field.

The direction of the current in the yoke coil determines the polarity of the magnetic field of the yoke, that is, whether the electron beam is deflected upward or downward. Current flow in the direction shown in Fig. 5-29*a* causes the electron beam to deflect upward. Current in the opposite direction through the yoke coils (Fig. 5-29*b*) reverses the magnetic field, causing the beam to deflect downward.

How far the electrons are repelled when passing through the magnetic field of the yoke is determined by the design of the yoke and the level of current flowing in the vertical coils. The higher the current, the stronger is the magnetic field and resulting electron deflection.

A requirement of vertical deflection in a television set or PC monitor is that the current in the coils of the vertical yoke increases an equal amount for specific time intervals. This linear current change causes the deflection of the electron beam to make a uniform, or smooth, movement from the top to the center and from the center to the bottom of the CRT face plate.

The waveforms shown in Fig. 5-29 represent a current increasing and decreasing in level with respect to time. In Fig. 5-29*a*, the waveform shows the current increasing quickly and then decreasing slowly to zero. This would cause the electron beam to quickly jump to the top of the CRT and then slowly drop to the center.

Looking at the waveform in Fig. 5-29*b*, the current is increasing slowly in the opposite direction and then decreasing quickly to zero. This would cause the electron beam to move slowly from the center to the bottom of the CRT face plate and then return quickly to the center.

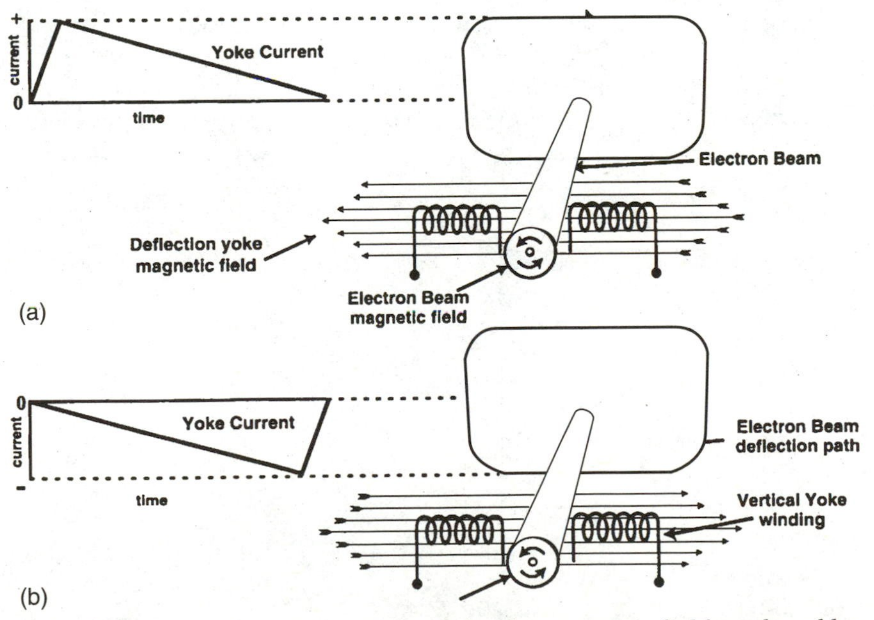

5-29 Deflection of the electron beam by the magnetic field produced by the yoke mounted on the CRT neck. (*Courtesy of Sencore, Inc.*)

During normal television or PC monitor operation, the yoke current increases and decreases as shown in the waveforms in Fig. 5-29. The current changes directions, alternating as shown in Fig. 5-29*a* and *b*, at approximately 60 times a second. The alternating current moves the electron beam from the top of the CRT face plate to the bottom and quickly back to the top of the screen.

Producing the vertical drive signal

The vertical stages of the television receiver are responsible for developing the vertical drive signal. This signal is fed to an output amplifier that produces alternating current in the vertical yoke.

The vertical section consists of four basic circuits, or blocks (Fig. 5-30). These include the following sections: (1) oscillator or digital divider, (2) buffer/predriver amplifier, (3) driver amplifier, and (4) output amplifier. The circuitry for these stages may be discrete components on the circuit board or may be included as part of one or more integrated circuits.

The vertical oscillator generates the vertical sweep signal. This signal is then fed to the amplifiers and drives the yoke to produce deflection. Vertical oscillators may be free-running circuit oscillators or the more modern digital divider generators.

Free-running oscillators use an amplifier with regenerative feedback to self-generate a signal. More common types are resistor-capacitor (RC) oscillators associated with ICs or with discrete multivibrator or blocking oscillator circuits.

A digital divider generator uses a crystal oscillator. The crystal produces a stable frequency at a multiple of the vertical frequency. Digital divider stages divide the signal down to the vertical frequency. Usually you will find most of the digital divider oscillator circuitry located inside an IC.

The output of a vertical oscillator must be a sawtooth-shaped waveform. A ramp generator is often used to shape the output waveform of a free-running oscillator or digital divider. A ramp generator switches a transistor off and on, alternately charging and discharging a capacitor. When the transistor is off, the capacitor charges to the supply voltage through a resistor. When the transistor is switched on, the capacitor is discharged.

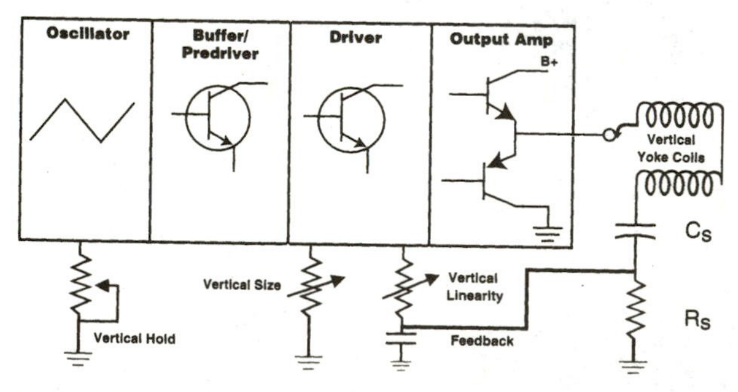

5-30 Vertical section of a television, consisting of an oscillator, buffer, output amplifier, and deflection yoke. (*Courtesy of Sencore, Inc.*)

The vertical oscillator must then be synchronized with the video signal so that a locked-in picture can be viewed on the CRT display. The oscillator frequency is controlled in two ways: (1) a vertical hold control may be used to set the free-running oscillator close to the vertical frequency, or (2) vertical sync pulses, removed from the video signal, may be applied to the vertical oscillator to lock it into the proper frequency and phase. If the oscillator is not synchronized, the CRT picture will roll vertically. The picture rolls upward when the oscillator frequency is too low and downward when the frequency is too high.

There are several intermediate amplifier stages between the output of the vertical oscillator and output amplifier stage. Some common stages are the buffer and the predriver and/or driver. The purpose of the buffer amplifier stage is to prevent loading of the oscillator, which may cause frequency instability or wave-shape changes. The predriver and/or driver stages shape and amplify the signal to provide sufficient base drive current to the output amplifier stage. Feedback maintains the proper dc bias and wave shape to ensure that the current drive to the yoke remains constant as components, temperatures, and power supply voltages drift. These stages are dc coupled and use ac and dc feedback similar to audio amplifier stages.

Ac feedback in most vertical circuits is obtained by a voltage waveform derived from a resistor placed in series with the yoke. The small resistor is typically placed from one side of the yoke to ground. A sawtooth waveform is developed across the resistor as yoke current alternates through it. This resistor provides feedback to widen the frequency response, reduce distortion, and stabilize the output current drive to the yoke. The feedback is often adjusted with gain or shaping controls referred to as the vertical height or the size and vertical linearity controls.

Dc feedback is used to stabilize the dc voltages in the vertical output amplifiers. Dc voltage from the output amplifier stage is used as feedback to an earlier amplifier stage. Any slight increase or decrease in the balance of the output amplifiers is offset by slightly changing the bias. Since the amplifiers are directed, the change in bias slightly shifts the bias on the output transistors to bring the stage back into balance.

Much of the difficulty in troubleshooting vertical stages is due to the feedback and dc coupling between stages. A problem in any amplifier stage, yoke, or yoke series component alters all the waveforms and/or dc voltages, making it difficult to trace the problem.

How the vertical output stage causes picture scanning

The vertical output stage produces yoke current that pulls the electron beam up and down the face plate of the CRT. The vertical yoke may require up to 500 mA of alternating current to produce full picture tube deflection. A power output stage is now required to produce this level of current.

A vertical output stage commonly consists of a complementary symmetry circuit with matching power transistors (see Fig. 5-31). The transistors conduct alternately in a push-pull arrangement. The top transistor conducts current in one direction to scan the top half of the picture. The bottom transistor conducts current in the opposite direction to scan the bottom half of the picture.

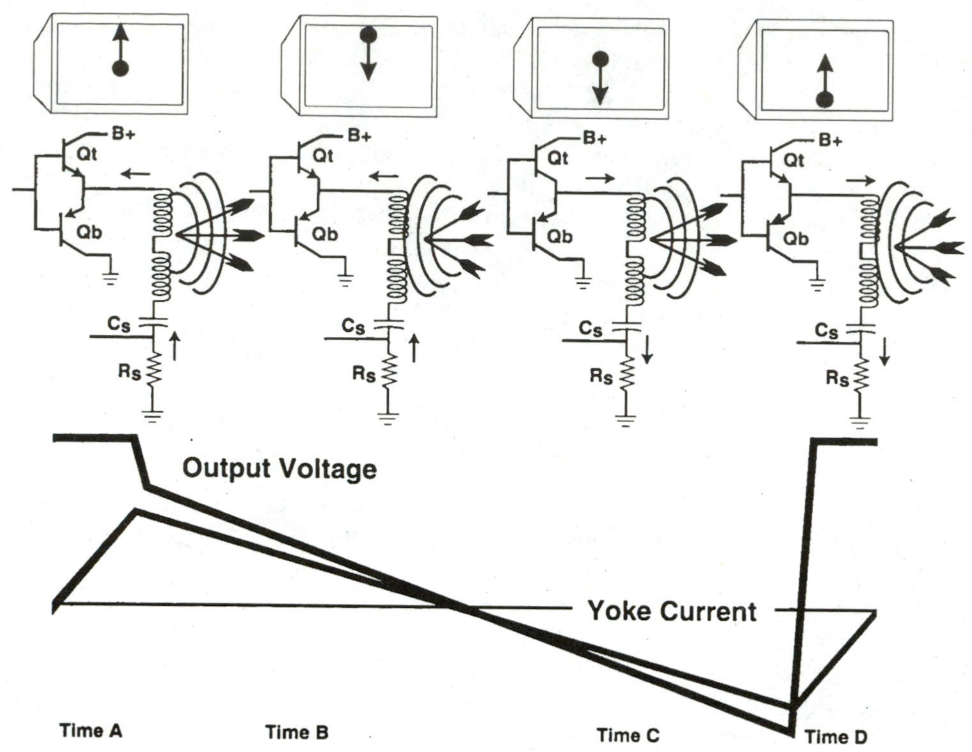

5-31 Deflection, currents, and waveforms during four periods of time in the vertical sweep cycle. (*Courtesy of Sencore, Inc.*)

Most vertical output stages are now part of an IC package and are powered with a single, positive power supply voltage. The voltage is applied to the collector on the top transistor. In this balanced arrangement, the emitter junction of the transistors should measure about one-half the supply voltage. In series with the vertical yoke coils is a large electrolytic capacitor. This capacitor passes the ac to the yoke but blocks dc to maintain a balanced dc bias on the output amplifier transistors.

To better understand the typical vertical output stage, let us now work through the current paths at four points in time during the vertical cycle, as illustrated in Fig. 5-31. Starting with time *A*, the top transistor Qt is turned "on" by the drive signal to its base. The transistor is biased "on," resulting in a low-conduction resistance from collector to emitter that provides a high level of collector current. This puts a high positive voltage potential at the top of the yoke, resulting in a fast-ringing current in the yoke.

During time *A*, capacitor Cs charges toward positive voltage and current flows through the yoke and the top transistor Qt. This pulls the electron beam of the CRT from the center of the CRT quickly to the top. During time *A*, an oscilloscope connected at the emitter junction displays a voltage peak, shown as the voltage output waveform in Fig. 5-31. The inductive voltage from the fast-changing current in the yoke, along with retrace speedup components, cause the voltage peak to be higher than the positive voltage supply.

The current flowing in the yoke during time *A* produces a waveform as viewed from the bottom of the yoke to ground. This is the voltage drop across Rs, which is a reflection of the current flowing through the yoke.

During time *B*, the drive signal to Qt slowly increases the transistor emitter-to-collector (E-C) resistance. Current in the yoke steadily decreases as the signal increases its E-C resistance and reduces its collector current. The voltage at the emitter junction falls during this time, and capacitor Cs discharges. A decreasing current through the yoke causes the electron beam of the CRT to move from the top to the center of the screen.

The creation of a linear fall in current through the yoke during time *B* demands a critically shaped drive waveform to the base of Qt to meet its linear transistor operating characteristics. The drive waveform must decrease the base current of the transistor at a constant rate. Thus the transistor must operate with base-to-collector linear current characteristics. These reductions in base current must result in proportional changes in collector current.

At the end of time *B*, transistor Qt E-C resistance is high, and the transistor is approaching the same E-C resistance as the bottom transistor Qb. Capacitor Cs has been slowing discharging to the falling voltage at the emitter junction of the output transistors. Just as the voltage at the emitter junction is near one-half the positive supply voltage, the bottom transistor begins to be biased "on" to begin time *C*. This transition requires that the conduction of Qt and Qb at this point be balanced to eliminate any distortion at center screen of the CRT.

During time *C*, the C-E resistance of transistor Qb slowly is decreased by the base drive signal and the increase in collector current. Capacitor Cs begins to discharge, producing current through the yoke and through Qb. As Qb resistance decreases and Qb collector current increases, the voltage at the emitter junction decreases. This can be seen on the voltage output waveform as it goes from one-half positive supply voltage toward ground in time *C*. The current increases at a linear rate through the yoke, as shown in the yoke current (voltage across Rs waveform) in Fig. 5-31.

The resistance decrease of Qb must mirror the opposite of transistor Qt during time *B*. If not, the yoke current would be different in amplitude and/or rate, causing a difference in CRT beam deflection between the top trace and bottom trace times. At the end of time *C*, the E-C resistance of Qb is low, and the current in the yoke approaches a maximum level.

At the start of time *D*, the E-C resistance of Qb is increased quickly, and collector current is decreased. This quickly slows the discharging current from capacitor Cs through the yoke and transistor. As the current is reduced, the trace is pulled quickly from the bottom to the center of the screen. Time A begins again, and the cycle is repeated again.

6
CHAPTER
Video and chroma systems

This chapter begins with information for troubleshooting color television 3.58-MHz oscillators, demodulators, bandpass amplifiers, and chroma integrated circuits (ICs). Then use of the video analyzer to quickly isolate problems in the video and chroma sections of a color television chassis is discussed. Next, a simplified block diagram of a color television chassis is used to illustrate the "divide and conquer" servicing technique for quick circuit analysis. Some actual color television circuits also will be explained.

The chapter continues with a review of how the color picture tube operates and how it can fail in various ways. This section discusses how to check out and determine if a picture tube is defective and then shows some ways to repair certain picture tube failures. The chapter concludes with some picture tube adjustment notes.

Troubleshooting overview

Many faults in the tuner, i.f. stages, video detector, and chroma takeoff coil and comb filter can cause symptoms nearly identical to problems found in the color circuits. You can lose time troubleshooting circuits, unsoldering ICs, and ordering parts that are not defective. After these parts are replaced, you may find that only a trap is not tuned properly or that a coil is bad in the chroma takeoff circuits.

The color bars pattern signal from a video analyzer can be used to inject at the chroma input (just after the 3.58-MHz bandpass filter) to check the operation of the entire chroma circuit. Use the video pattern position of the drive signal switch, while the video analyzer's radiofrequency (rf) signal (connected to the antenna input) holds the vertical, horizontal, and video circuits in sync. The analyzer's 10-bar gated rainbow color bars pattern resembles the pure chroma signal normally present after the chroma takeoff coil.

You do not need to disconnect any components when substituting signals with the video analyzer. The analyzer swamps out the weaker received signals and replaces them with its own good test signals. You need only adjust the analyzer's drive level control until the built-in digital peak-to-peak meter shows the same amplitude as the schematic's scope waveform at your test point.

If substituting at the chroma input restores color, do not troubleshoot the color circuits; you now know that they are working properly. The problem will be before the chroma input. Move your substitute signal back to the video detector and then back through the i.f. stages, if necessary, to find out where the color signal disappears.

If substituting a known good signal at the chroma input does not restore normal color, you know the problem is in the chroma circuits. Missing color is often caused by problems in the color oscillator, so check out the 3.58-MHz path next. Note the test points shown in Fig. 6-1.

Color oscillator check

Check out the 3.58-MHz oscillator path by substituting the 3.58-MHz drive signal at the output of the color oscillator (at the crystal input on an IC). If the substituted 3.58-MHz signal restores color, you know the problem is in the 3.58-MHz oscillator path. If substituting a known good signal at the output of the color oscillator does not restore color, you know the problem is in the bandpass amplifiers or demodulators.

Color demodulators

Check the color demodulators by substituting the color bars drive signal from the analyzer at the demodulator's chroma input while still substituting the 3.58-MHz drive at the output of the color oscillator. This supplies good signals to both the demodulators' inputs. If color is not restored, you know the problem is in the demodu-

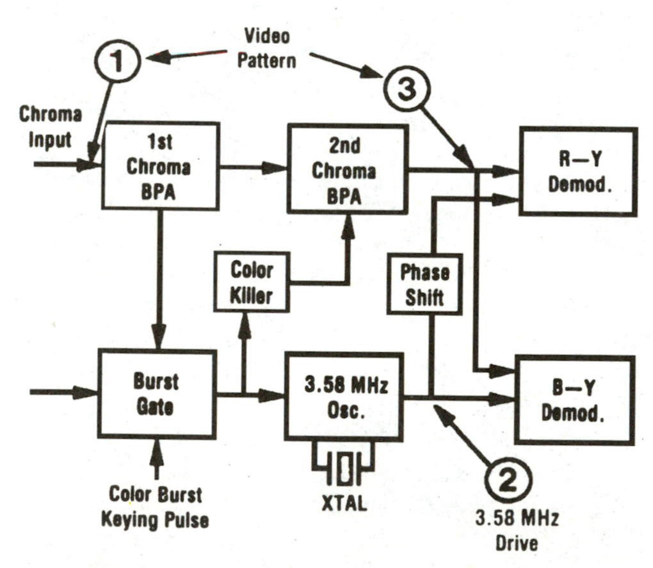

6-1 Signal injections into the color circuits to verify which stages are operating correctly. (*Courtesy of Sencore, Inc.*)

lators. If the substitute input signals restore color to the picture, you know the demodulators are working properly.

Color bandpass amplifiers

Check out the bandpass amplifiers by stepping back through the amplifier stages, substituting color bars drive signals to pinpoint the stage(s) where the color signal is not being properly processed. If the defective stage is controlled by the color killer, use a dc power supply to substitute the proper dc voltage onto the color killer input to the amplifier stage. If substituting the proper color killer voltage restores color, check for proper voltages and components in the color killer stage.

Chroma IC considerations

When you have color problems in a color chassis using a combined video/chroma IC, be sure to check the B+ supply to the IC. You may think that a B+ problem would kill the luminance signals if it affected the color. While this is very true of missing B+, it is not always the case when the supply voltage is low in value or may have a filtering problem that causes a less than pure dc voltage. Then the color is often affected before the luminance signal is.

The changing electronics service industry

You may not have noticed, but the "symptom cure" days have been numbered since the first IC was introduced in consumer products. If you have not noticed how the service industry has changed, your service procedures may not be as efficient as they should be. Today there are fewer and fewer easy-to-repair sets. Improved circuit designs, efficient ICs, and better manufacturing processes have drastically reduced the number of failures. Nowadays, easy repeat failures are becoming rare. Not many sets have the same circuits, even from one model to the next or from the same manufacturer, so a repeated failure only applies to a few sets that come across your bench. This makes "symptom cure" a secondary source of profit. Like most technicians, you find yourself making more and more in-circuit tests.

As you make these circuit tests, you probably find—unless you are already using a video analyzer—that there is not a "sure fire" way to test the special large-scale ICs used in televisions and VCRs. To make matters worse, service schematics are showing less and less troubleshooting information. Fewer test points show dc voltages. There are fewer scope waveforms, and the ones that are included are harder to interpret. In most service manuals, the circuit descriptions have either been cut back to a minimum, or there are none at all. And the parts problem with ICs is not good.

There are fewer "universal" parts that work with many different chassis, so you end up custom ordering the parts you need to repair a particular set. It is just not practical to have a big parts inventory. If you try to stock every part you may need, you will see more than 90 percent of them sitting on the parts room shelf for years. The overhead of stocking slow-moving parts can cancel all your service profits.

ICs can be checked out very easily with a video analyzer, as shown in Fig. 6-2. The analyzer tests are the same on this year's model, last year's model, or a set 15 or

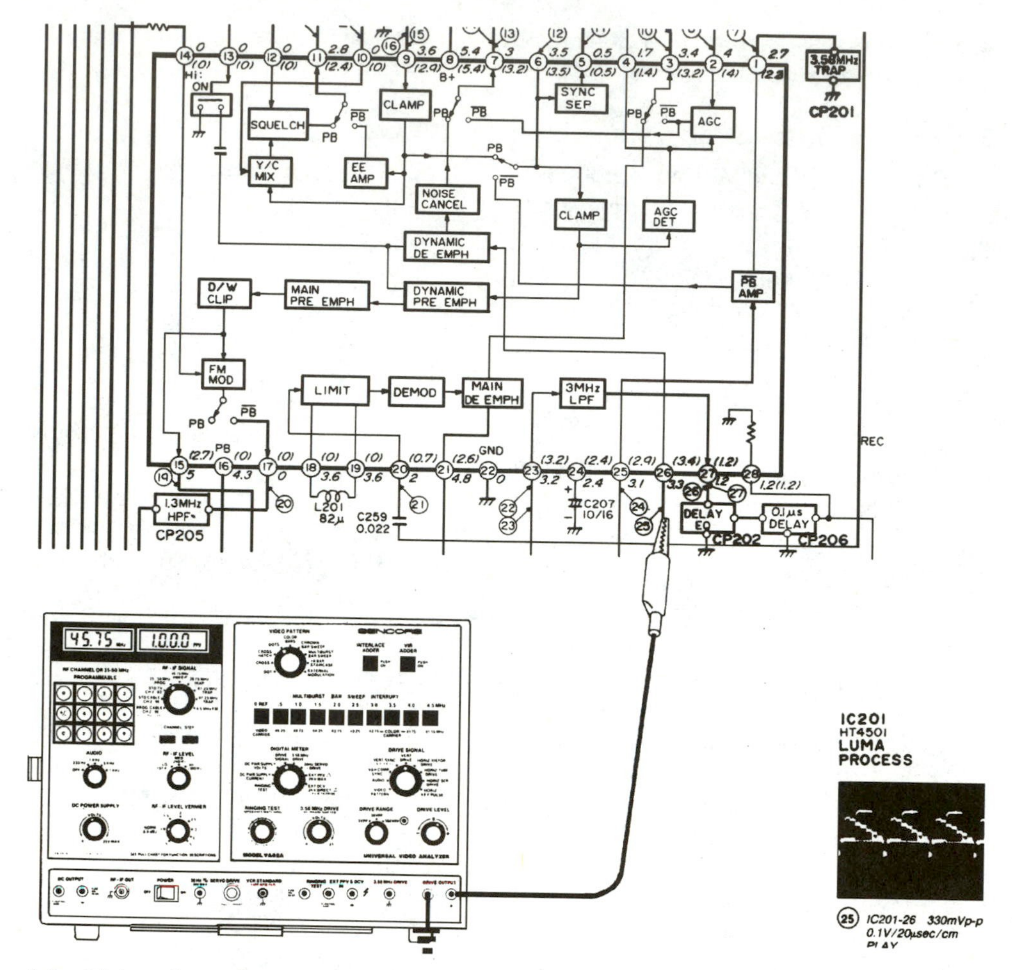

6-2 Using the video analyzer for dynamic in-circuit tests. (*Courtesy of Sencore, Inc.*)

more years old. Your schematic does not need nearly as much detail as you need when using a meter or a scope, since most of the tests involve substitution of "good" signals rather than measurements of "bad" ones.

Video analyzer troubleshooting steps

Before you start troubleshooting a television set, you want to identify as many symptoms as possible. For a certain set you could hear the crackle of high voltage on the CRT and a tone at the speaker output; thus you would reason that the power supply, high voltage, and sound circuits are operating normally. The major symptom at this point is simply "no picture."

Let us now look at a list of potentially faulty circuits. This "no picture" symptom could be caused by a break in the signal path anywhere after the 4.5-MHz sound takeoff. This would indicate that the list of suspects includes every circuit from the sound takeoff forward (except the power supply and high voltage). This is a pretty long list.

The major functional division point in the luminance circuits, referring to Fig. 6-3, is the video detector (right after the 4.5-MHz sound takeoff). This is where the signal changes from modulated to unmodulated, and this is where you need to begin using signal substitution with the analyzer's drive signals. This "divide and conquer" technique allows you to narrow down the circuit and center on the very component that is faulty, quickly and accurately.

> **Note:** In this particular chassis, the video detector is contained in the U300 IF/AGC processor ship, so you need to inject the signal at the nearest test point, the emitter of Q301, the video buffer transistor shown in Fig. 6-4.

Step 1

Rotate the analyzer drive signal knob to the video pattern position, and connect the drive output floating ground lead to ground and the positive lead to the main

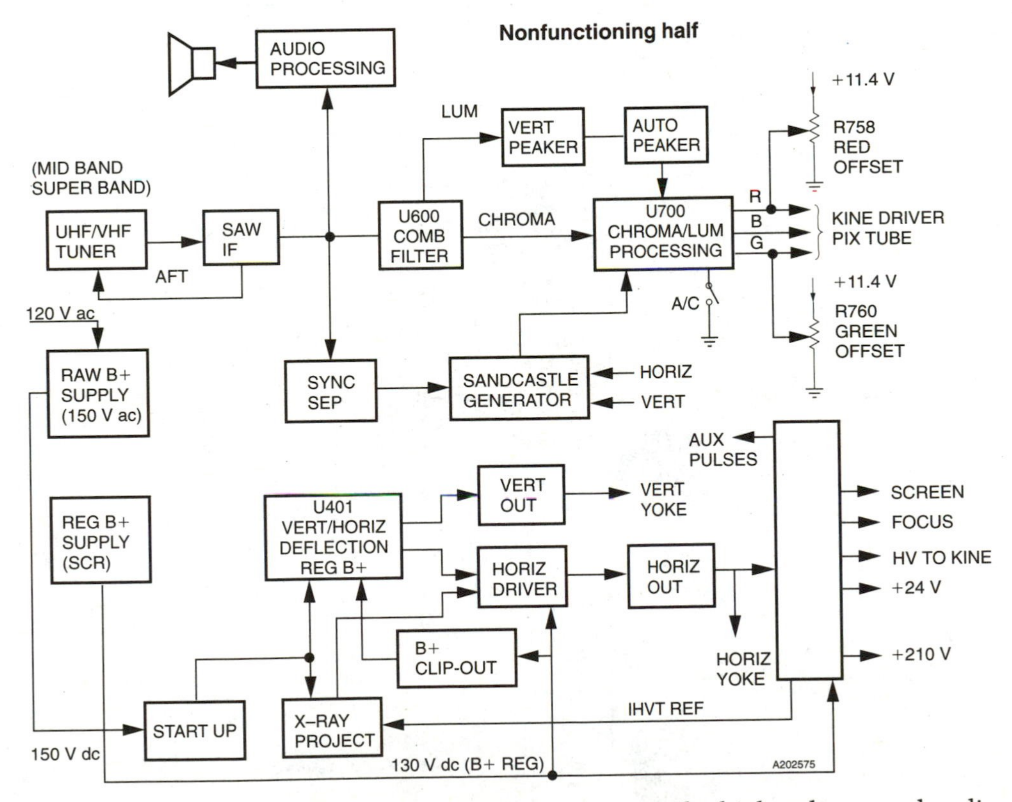

6-3 Block diagram used to determine that power supply, high-voltage, and audio circuits are operating properly. (*Courtesy of Sencore, Inc.*)

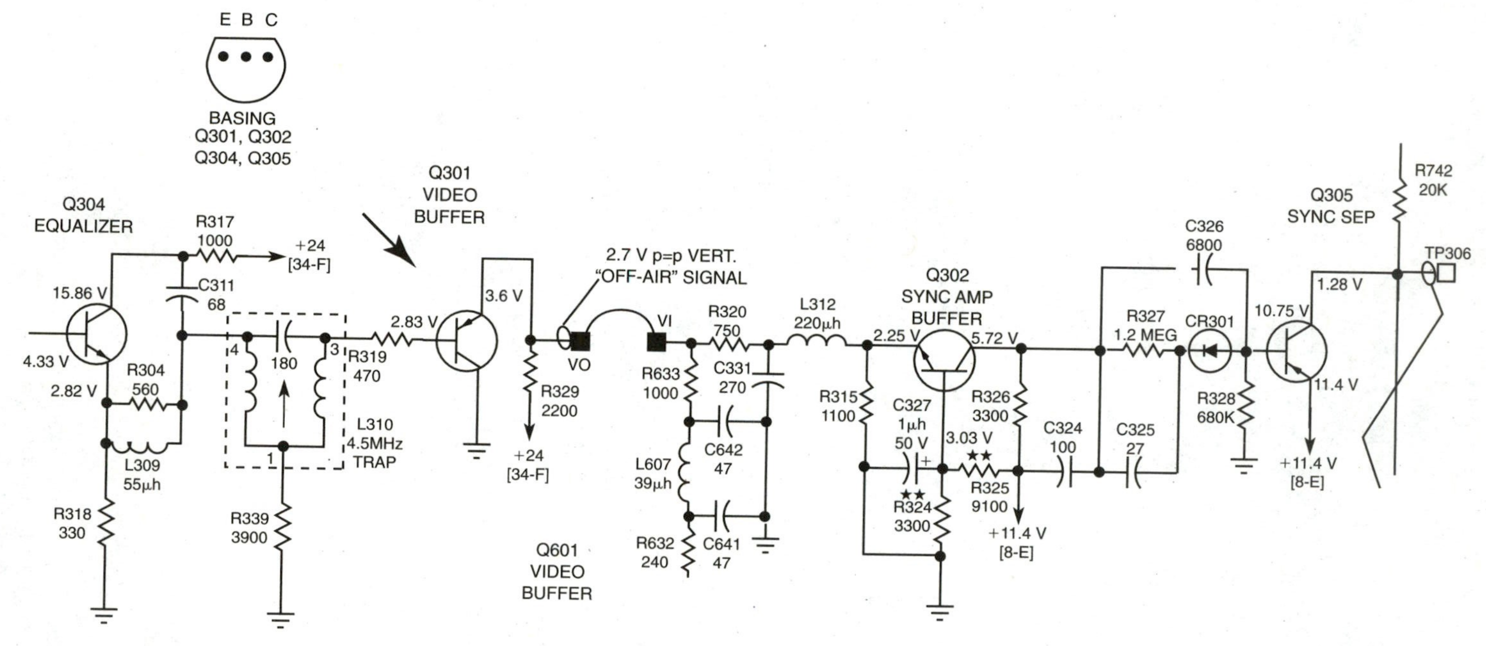

6-4 Video circuit used to confirm a defect between the point of injection (*arrow*) and the CRT. (*Courtesy of Sencore, Inc.*)

division point in the luminance circuits (video detector output). Now switch the analyzer drive range knob to the 3-V peak-to-peak position, and adjust the drive level knob until the same peak-to-peak amplitude shown on the schematic at the video detector (Q301) appears on the digital meter. If this does not improve the situation, then continue your analysis.

Step 2

Use the video analyzer to provide a known good signal to the television set's input. Use the video pattern drive signal, which will allow you to swamp out the existing video signal and use a known good composite video signal to drive the stages after the detector without disconnecting any components.

If the television picture symptoms do not improve, then it is because a known good signal is being blocked by a bad stage. After you confirm that the defect is between your point of injection and the CRT, you can divide the chassis into two parts—the functioning half and the nonfunctioning half.

Step 3

By now you have shortened the list of potentially faulty components by using the functional analyzing technique. The next logical place for signal injection would be at the Sandcastle, chroma, and luminance inputs at the lum/chroma IC chip. One of the main symptoms associated with a Sandcastle signal problem is a "no picture" symptom. Therefore, this appears to be a good place to start troubleshooting. At this point, try substituting a test signal for the Sandcastle signal with the video analyzer horizontal keying pulse, and then note any changes on the CRT. This test setup is illustrated in Fig. 6-5.

For this test, rotate the analyzer's drive signal knob to the horizontal key pulse position, connecting the drive output floating ground to ground and the positive test lead to the input of the Sandcastle circuit. Then switch the analyzer's drive range

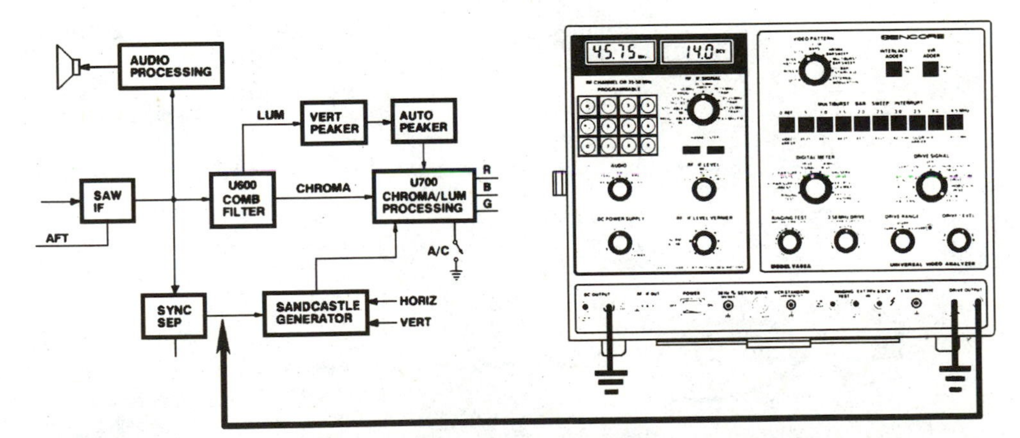

6-5 Injecting the horizontal keying pulse at the Sandcastle input to restore picture, proving that the video processing stages were okay all the way to the CRT. *(Courtesy of Sencore, Inc.)*

knob to the 30-V peak-to-peak position, and adjust the drive level knob until the same peak-to-peak amplitude indicated on the schematic appears on the digital meter. At this time you should see video information on the television screen. However, it may be out of sync and will not lock in. Let us now see why this might happen.

The analyzer's horizontal keying pulse has swamped out the existing Sandcastle signal and provided a known good signal to the Sandcastle decoder. When the symptom improved, it confirmed that the defect was between the first point of injection and the Sandcastle input stage. You may now safely assume that the circuits after the injection point are functioning normally. At this point you have eliminated the lum/chroma IC and all stages after the IC as potentially defective. The new problem with the out-of-sync video information could very well be caused by a fault in the sync stages. So now you should inject a test signal into the output of the sync separator (see Fig. 6-6). In Fig. 6-6, the symptom improved, ensuring that the stages past the point of injection were good.

Step 4

The defective stage can now be isolated to the video analyzer. Move the drive signal switch to the V&H comp sync position. Then connect the drive output floating ground lead to ground and the positive lead to the sync separator output to swamp out the existing sync signal. Rotate the drive range switch to the 30-V peak-to-peak position and the drive level until the same peak-to-peak amplitude shown on the schematic data (12 V peak to peak) appears on the digital meter. If the symptoms disappear and the picture returns completely in sync, then go on to the analysis information.

When you injected the video pattern drive signal at the emitter of Q301 in step 1, you effectively drove the sync separator input with a known good signal. The symptoms did not improve. However, when you drove the output of the sync separator in step 4, the symptoms disappeared. You have now isolated the defective stage—the sync separator circuits.

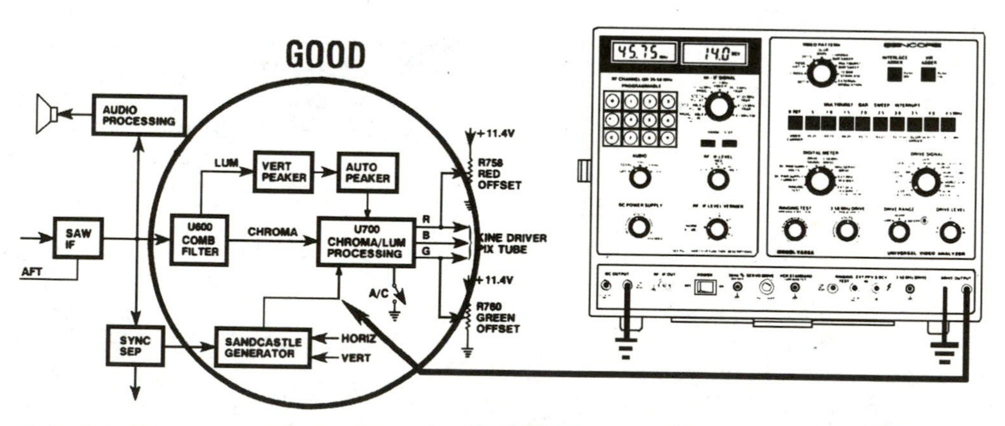

6-6 Injecting a test signal into the output of a sync separator. (*Courtesy of Sencore, Inc.*)

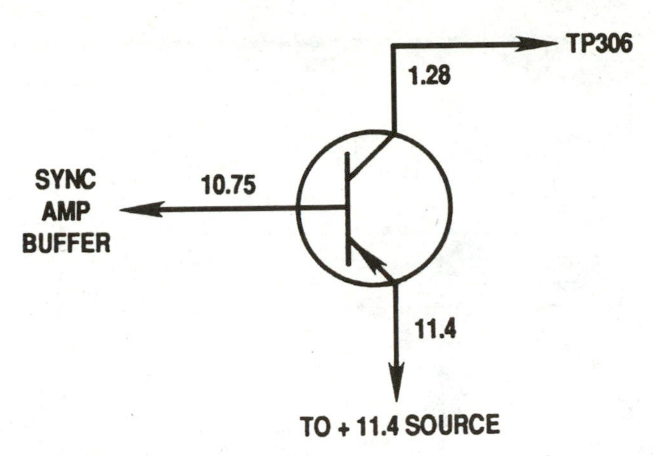

6-7 A shorted base-emitter junction at the sync separator transistor stage. (*Courtesy of Sencore, Inc.*)

Step 5

A quick check for proper signals at Q305 in the sync separator circuit showed no signal present at the collector. When the dc volt button on the analyzer was pressed, the digital read-out showed the dc bias voltage to be 0.00 V. The base and emitter voltages now read 11.88 V dc. The transistor was found to be shorted. There was no output and, of course, no sync pulse present. Refer to the sync circuit in Fig. 6-7.

The sync pulse is one of the three signals that make up the Sandcastle. Without the horizontal sync pulses from the sync separator, the luminance/chroma IC could not produce an output signal, and this is why the picture tube went blank.

Zenith TVR1920 main circuit board

This circuit board contains all the circuitry to drive the picture tube. Refer to circuit diagram in Fig. 6-8. IC1501 is a 52-pin DIP video processor. It handles all the video, sync separation, horizontal/vertical drive, and red, green, and blue (RGB) mixer and drive for all on-screen displays. The large-scale integrated (LSI) circuitry allows a great many functions to be combined into one chip. All customer controls are addressed through the microprocessor and are sent to the video processor as dc control signals.

The luminance signal comes in on connector P6T04 pin 7. It is then coupled to the base of transistor Q1202. The emitter output of Q1202 is coupled to the base of Q1501, Q1203, and Q1204. The chroma information taken off the emitter of Q1501 is coupled to pin 42 of IC1501. The luminance information taken off the emitter of Q1203 is coupled to pin 46 of IC1501.

The luminance and chroma signals are processed by the IC before being sent to the CRT module. The luminance signal exits the IC on pin 23. The R-Y, G-Y, and B-Y signals exit on pins 26, 25, and 24.

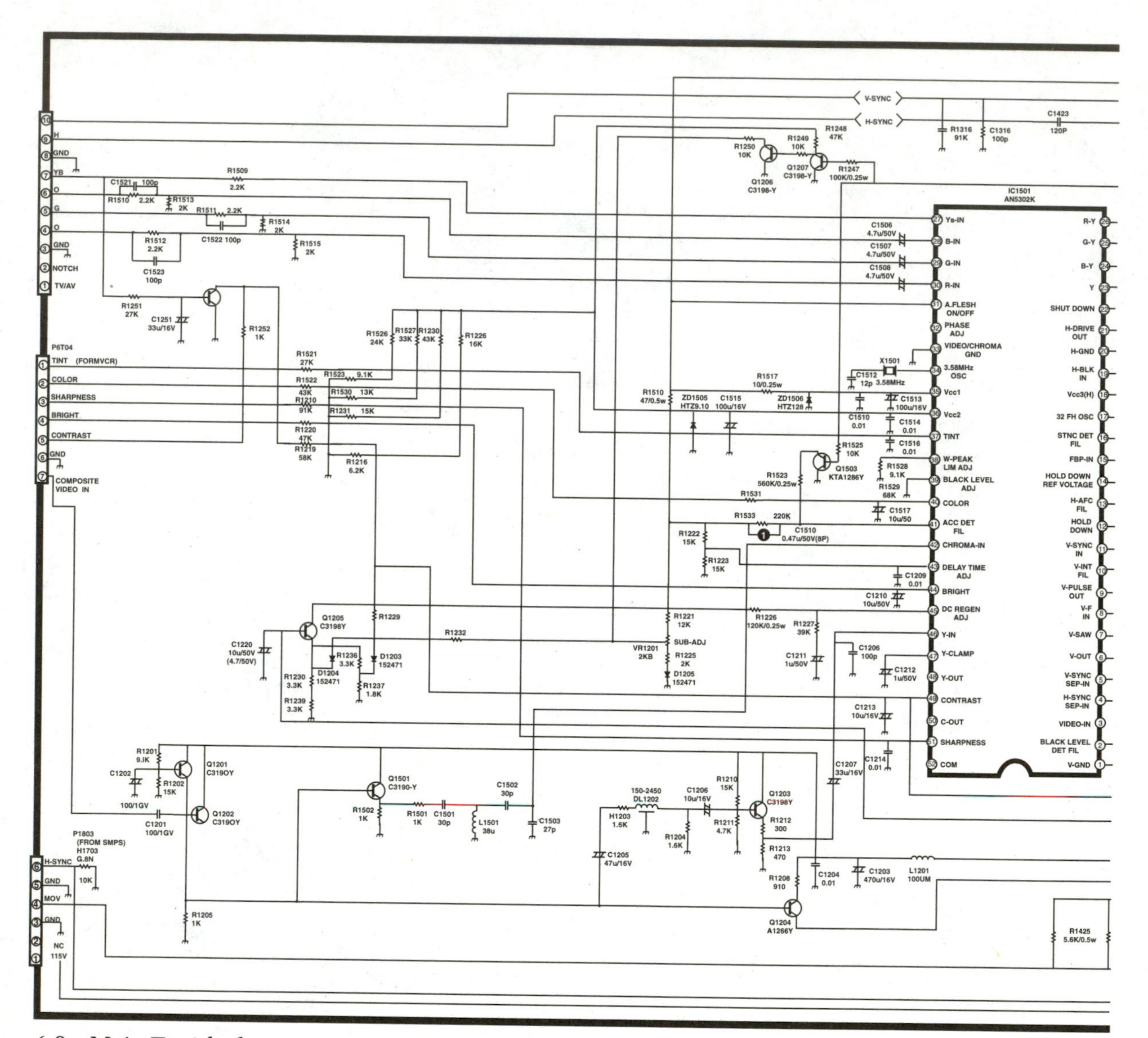

6-8 Main Zenith chroma circuits and microprocessor control system. (*Courtesy of Zenith Corporation.*)

This same video signal from Q1202 is coupled to the base of Q1204. The output from the collector is coupled to pins 4 and 5 of IC1501 as the horizontal and vertical sync input.

CRT output module (Zenith)

The CRT module, also shown in Fig. 6-8, mixes the luminance and chroma signals in output transistors Q1901, Q1902, and Q1903. The luminance signal enters on pin 8 of connector P1501 and is applied to the base of the three output transistors. R-Y chroma is applied to the base of Q1901. G-Y chroma is applied to the

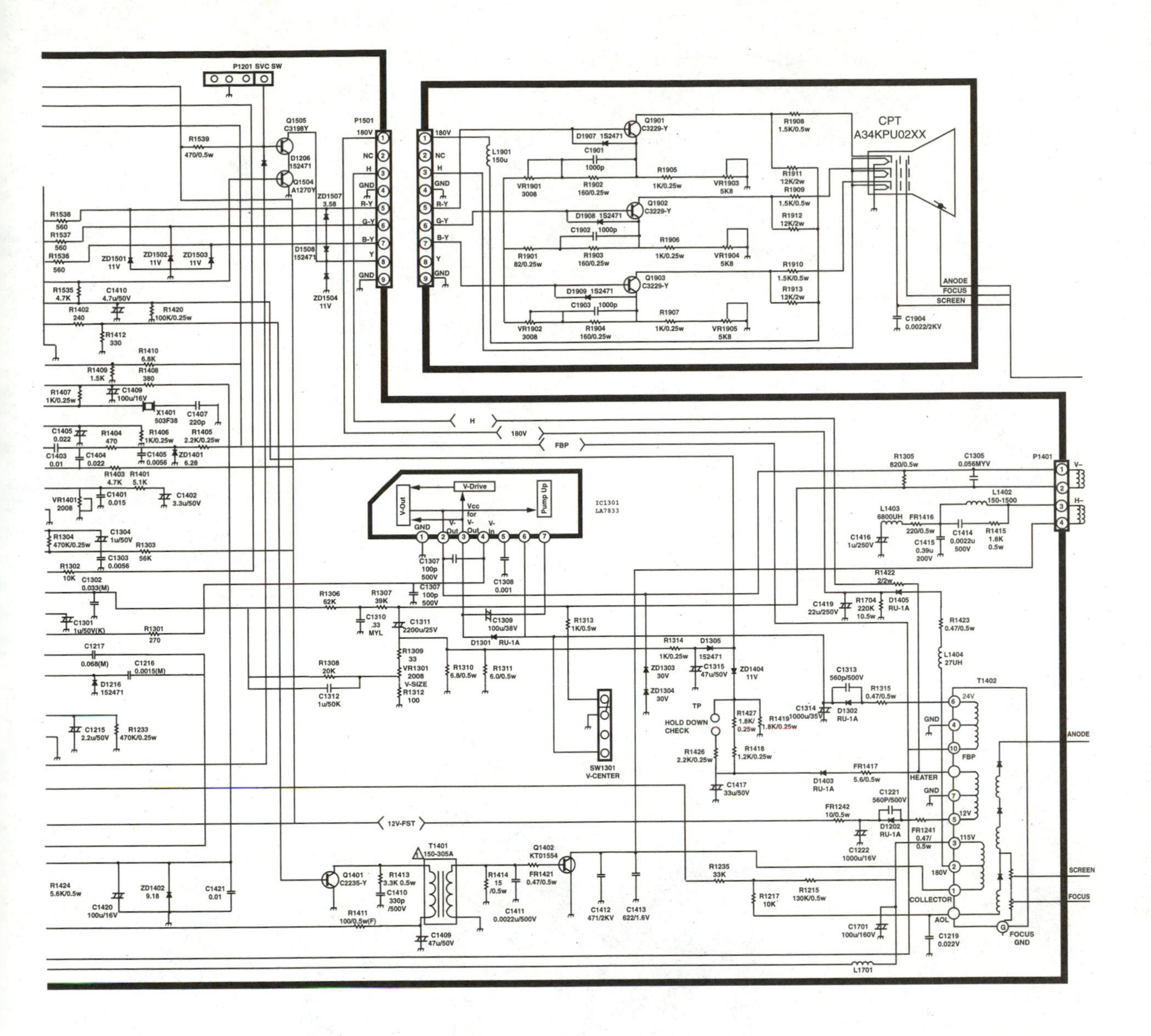

base of Q1903. Black-and-white tracking is accomplished using the red (VR901) and blue (VR902) drive controls together with the red (VR1903), green (VR1904), and blue (VR1905) bias controls. The B+ of +180 V is supplied to connector P1501 pin 1. Switch SW1201 is the black-and-white setup switch and is located on the main module.

CRT testing and restoration

The cathode-ray tube (CRT) is one of the oldest electronic technologies in everyday use. Its beginnings can be traced to 1879 when William Crookes deflected cathode rays inside a vacuum tube with a magnet. Modern CRTs have changed considerably from the first Crookes tube, but the basic operation remains unchanged: a hot cathode emits electrons that form a beam that strikes a phosphor screen to produce light. Because the ability of the cathode to emit electrons decreases with age, all CRTs eventually wear out (become weak) or other defects in the electron gun occur.

Every television technician who services video equipment today must be prepared to answer two very important questions concerning CRTs: "How do I know if this CRT is good or bad?" and "Is there a proven reliable alternative to replacing a worn-out CRT?" This part of the chapter will address the technical aspects of CRTs—how they work, how they fail, and how restoration may give some of them a new life.

Inner workings of a CRT

A CRT has three major sections:

1. Electron gun
2. Accelerator grids
3. Phosphor screen

The electron gun forms and controls an electron beam that is accelerated by grids. The amount of light produced by the screen depends on the intensity (current) of the electron beam striking it—more current produces more light. Deflection plates or deflection yokes move the beam to produce a scanned raster on the screen face plate. Refer to the CRT cut-away drawing in Fig. 6-9.

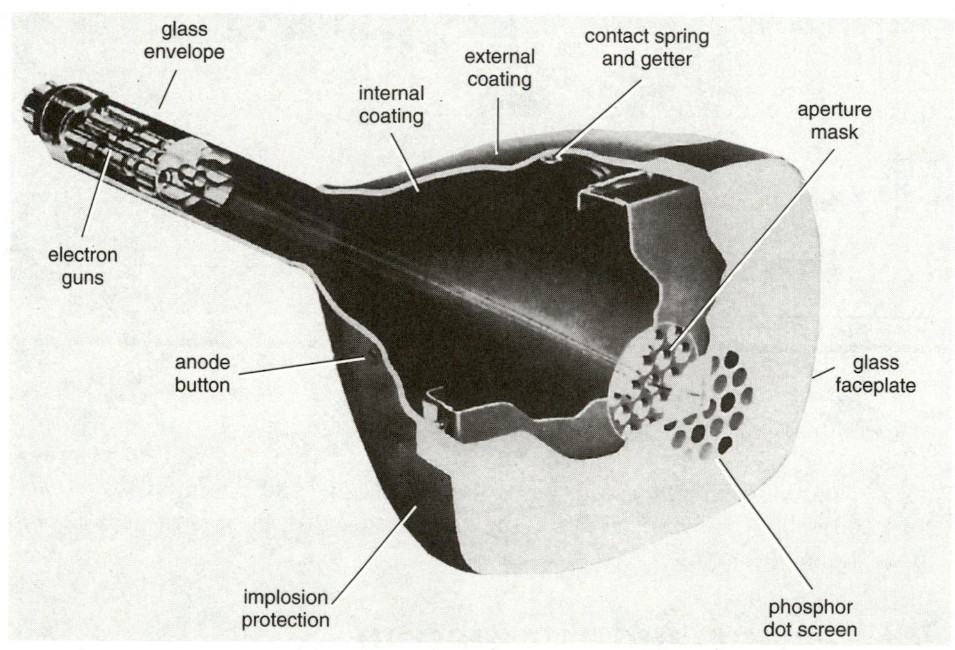

6-9 Cut-away view of color television CRT construction. (*Courtesy of Sencore, Inc.*)

A cathode in the electron gun emits electrons when it is heated. The electrons are formed into a beam by other grids in the electron gun. These grids, G1 nd G2, are cylinders with tiny holes. The G1 *control grid* determines the intensity of the electron beam by changing the amount of its negative voltage (bias) compared with the cathode—more negative bias equals less beam current. (Keep in mind that electrons are negatively charged.) Applying a video signal to either the cathode or the control grid changes the beam current in accordance with the picture information.

Positive voltage on the G2 *screen grid* pulls the electrons through G1 to form a thin, threadlike electron beam. The beam is accelerated by the high voltage on the accelerator grids and reaches a velocity great enough to cause light emission when it strikes the phosphor screen.

How CRTs fail

Most technicians think of CRT failures in terms of only two possibilities—shorts and low emission—because many CRT testers lump all tests into a "shorts" test and an "emission" test. However, each of these broad categories includes several distinct failure modes that need to be identified for reliable diagnosis and restoration.

Open filament.

An open filament cannot heat the cathode. The filaments, however, are pretty durable, so an open filament is not a very common problem. Open filaments cannot be repaired.

Heater-to-cathode short.

A heater-to-cathode (H-K) short occurs when the two elements physically touch or if a flake of conductive material from inside the tube shorts them. The symptoms of an H-K short depend on how the filaments are powered. Filaments powered directly from a 60-Hz power line cause "hum bars," poor contrast, and possible retrace lines if an H-K short exists. If the filament voltage is scan-derived, an H-K short will cause no visible problem if the flyback winding is floating. If the flyback winding is tied to ground, the CRT bias may be affected.

Grid 1 shorts.

Most G1 shorts occur when a conductive flake of material lodges between the cathode and the control grid, as illustrated in Fig. 6-10. Shorts between G1 and G2 are possible, but less common. A G1 short usually causes loss of beam control, allowing the beam to run wide open. The result is a bright white, red, green, or blue raster. The excessive beam current may even be enough to cause the chassis to shut down.

Poor gamma performance.

A picture tube with a gamma problem produces overdriven whites and deep blacks but few shades of gray in between. Although you may refer to this problem as a "gassy" CRT, the problem is caused by a defective cathode. Poor gamma occurs when the center of the cathode wears to the point it can no longer produce sufficient

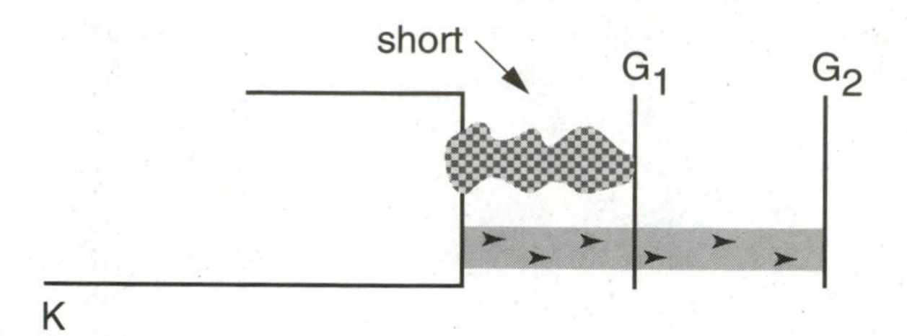

6-10 G1 short causing loss of beam control, which results in a bright white, red, green, or blue raster. (*Courtesy of Sencore, Inc.*)

beam current for the gray shades. The center of the cathode often wears down first because it is always contributing electrons to the beam, while the edges of the cathode only contribute electrons during white picture portions. Figure 6-11 shows how poor gamma results from a nonlinear change in beam current versus bias.

Weak emission.
A CRT with reduced brightness usually has a layer of contamination coating the cathode surface. The contamination, which is caused by minute amounts of air reacting with the hot cathode material, acts like a blanket to prevent the electrons from leaving the cathode's surface. If the contamination covers the entire cathode surface, the CRT will have reduced brightness over its entire range. Sometimes the contamination will develop only around the edges of the cathode because the center portion is always conducting heavily. This results in normal blacks and grays but reduced whites (the opposite of poor gamma), which causes the CRT to have poor contrast.

Stripped cathode.
A stripped cathode has lost most or all of its emitting material and produces little or no beam current. The cause of stripped cathodes is not normal wear. (A cathode will fail from contamination long before the emitting material wears out.) A stripped cathode is caused by excessive restoration, which removes good emitting material along with the contamination.

Temperature-sensitive cathode.
Some CRTs show good emission under normal operation but lose emission quickly with small reductions in filament voltage. All cathodes have lower emissions at reduced filament voltage, but a normal cathode produces more electrons than are needed for the electron beam. Thus a small drop in filament voltage produces no change in beam current as electrons are borrowed from the "reserve," as shown in Fig. 6-12. Less emitting material and a thin layer of contamination buildup cause the dropoff to be more severe than normal. Either condition reduces the amount of reserve electrons and will eventually hinder the electron beam at normal filament voltage. Therefore, a temperature-sensitive cathode is a sure sign that the cathode is failing.

Color tracking.

A tracking problem occurs when the three guns of a color CRT (or the separate CRTs in a projection system) do not balance with each other to produce white or pure shades of gray. Instead, black-and-white picture portions show a hint of color, and the color portions are the wrong tint and cannot be set correctly. A color tracking problem can occur even though each gun has "good" emission. CRT manufacturers specify that no gun in a color CRT or projection system should produce more or less than 55 percent of the current of any other gun. Any gun beyond this limit can fall outside the adjustment range of the screen and drive controls and cannot be balanced properly.

As you can now see, there are two types of shorts and several different types of emission problems. The Sencore CR70 beam builder can repair some of these CRT problems. However, before you restore a tube, you need to test it so that you can use restoration to its maximum effectiveness.

Testing for CRT defects

A CRT tester must provide reliable tests. Without a reliable test, you might restore a CRT that does not need restoration and end up with a damaged tube. The CR70 beam builder CRT analyzer and restorer is designed to show you exactly what is wrong with a CRT so that you know what restoration steps to take. Let us see what types of CRT tests the CR70 performs.

H-K shorts

The CR70 has two different shorts tests, H-K and G1, as shown in Fig. 6-13, so that you can determine which type of short the CRT has developed. Most CRT

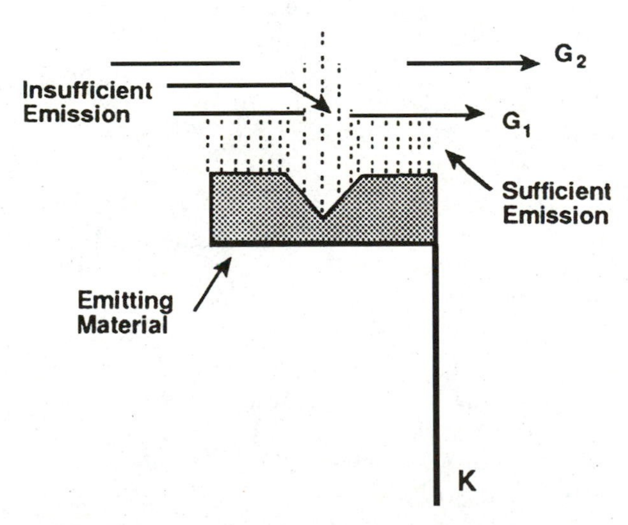

6-11 Poor gamma caused by a cathode that is severely worn in the center portion. The tube is able to produce blacks and whites but cannot produce gray shades. (*Courtesy of Sencore, Inc.*)

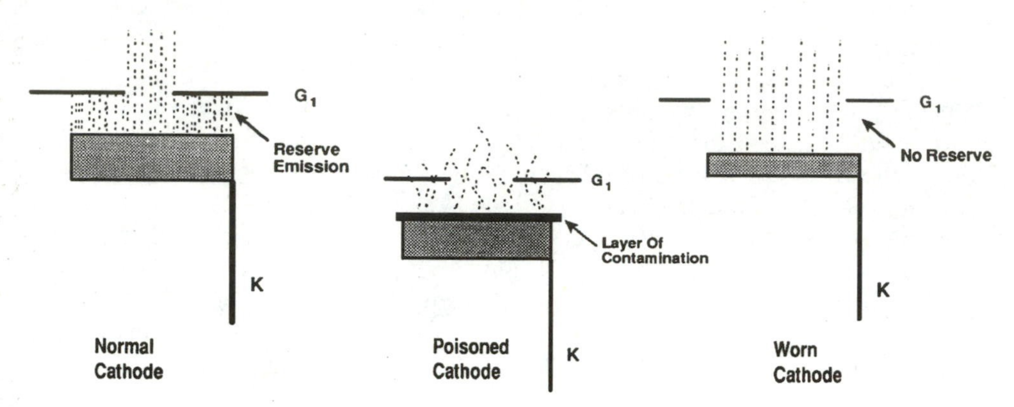

6-12 Reduced emitting material or a thin layer of contamination producing a temperature-sensitive cathode. (*Courtesy of Sencore, Inc.*)

checkers have only one indicator to show all types of shorts. It is best not to try to remove H-K shorts because you could easily damage or open up the filaments. (Most H-K shorts can be isolated by using a special filament transformer.) The H-K shorts test reads out on the same good/bad meter scale that is used for all the CRT checker's quality tests. A meter is used on the CR70 instead of indicator lights found on some other testers because the meter provides an accurate indication of the severity of the shorts.

G1 shorts

The G1 shorts test reads directly on the good/bad meter scale. In the G1 shorts test, the CR70 connects the cathode to G2 and looks for leakage between this connection and G1. Thus shorts or leakage between K and G1 and between G1 and G2 are detected. A meter reading far down in the "bad" area indicates a direct metal-to-metal short, which can be difficult to remove. Readings farther up in the "bad" area indicate resistive shorts, which are much easier to remove.

Dynamic emission test

The CR70's test of CRT emission is done in two parts. First, the cutoff is tested. During the cutoff test, the CRT is biased at the black (beam cutoff) point, which is the lower current output of the tube. This tests for the gamma problem explained earlier. If the CRT passes the cutoff test, you should now perform the emission test. The emission test checks the tube's current output at the white level. The current is measured at G2, rather than at G1, as other testers do. Measuring the true beam current at G2 ensures that the opening in G1 is not clogged by a flake of contamination. By checking both the black and white current levels, the CR70 CRT tester dynamically checks the tube over its entire operating range (Fig. 6-14).

Emission life test

Temperature-sensitive cathodes are identified with the checker's emission life test. During this test, the filament voltage is lowered by 25 percent. A normal cathode will not show a drop off in emission current, but a tube that has insufficient emitting material or one in which contamination is beginning to form will show a decrease in emission. Other testers remove the filament voltage completely. This leads to much interpretation because you must estimate how fast the dropoff occurs and take into account that smaller cathodes normally drop off faster than the cathodes in larger picture tubes.

Color tracking

Many CRT testers have three meters for comparing the emission of all three guns. The CR70 beam builder compares all guns automatically and gives you a simple "good/bad" determination. As you perform the emission test for each gun, the CR70 remembers the gun's emission level. Then, when you switch to the color tracking function, the emission level of each gun is compared with that of the other two guns as you switch to the gun select control. If the gun's emission current is less than 55 percent of either of the other two guns, the meter reads "bad."

CRT rejuvenation

The equipment for CRT rejuvenation and restoration has been around for many years. All restorers use the principle of increasing the cathode temperature in order to clean off the contamination. Increasing the cathode temperatures involves increasing both filament voltage and cathode current. Older types of rejuvenators do not work very reliably on today's CRTs because they often apply too much current and damage more tubes than they improve successfully.

The CR70 provides five levels of restoration and shorts removal. These five levels are referred to as *progressive restoration* because you start with the lowest (and safest) level and progress to higher levels only as required. The restoration

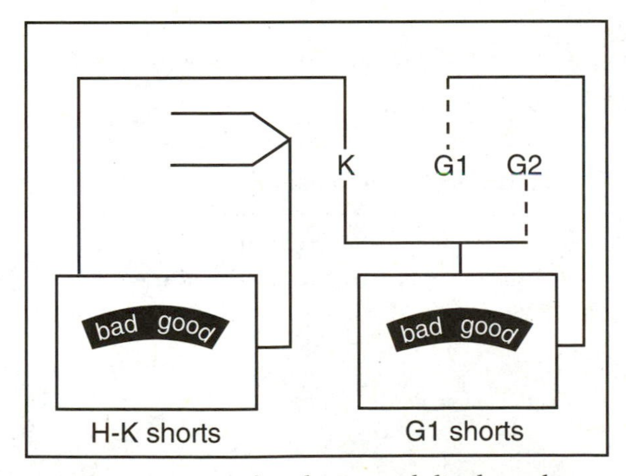

6-13 CR70 tests for shorts and displays the results on two separate meters. (*Courtesy of Sencore, Inc.*)

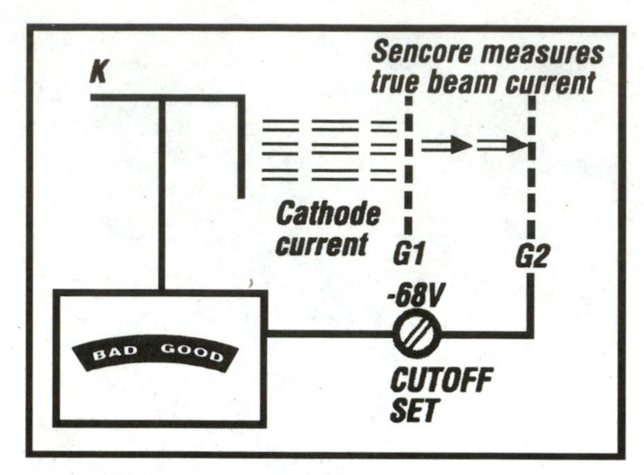

6-14 CR70 measuring the current at the black level and the "true beam current" at the white level. (*Courtesy of Sencore, Inc.*)

procedure you use for each CRT failure is given in Table 6-1. Let us now look at the CR70's method of shorts removal.

Shorts removal

A G1 short almost always can be vaporized by passing a high current through it. The CR70 quickly discharges a 450-V capacitor through the shorting material, as shown in Fig. 6-15. During the discharge, the filament voltage is removed to prevent possible damage to it or the cathode. Tougher shorts, which need more current to vaporize, draw more current from the capacitor than resistive shorts. When the short is gone, the current stops. Thus the CR70 is safe and effective.

Cathode restoration

Contamination can be removed from the cathode by heating it to a much higher than normal temperature. However, restoration is a subtractive process—it does not

Table 6-1. CR70 CRT restoration procedures

Test results				CR70
Cutoff	**Emission**	**Life**	**Tracking**	**restoration procedure**
Good	Bad			Auto cycle, then MAN1 if still weak
Bad	Good			Auto cycle once
Bad	Bad			Auto cycle. REJUV if less than 20 mA restore current
Good	Good	Bad		Auto cycle once
Good	Good	Good	Bad	Auto cycle lowest gun(s)

deposit new material back onto the cathode, it merely cleans off the contamination without removing any of the good emitting material. Let's now look at how the progressive restoration system works.

The tester provides three levels of restoration: auto restore, manual 1, and manual 2. Each is similar in operation and effect on the cathode. Restoration "boils off" the contamination and exposes fresh emitting material on the cathode's surface, as shown in Fig. 6-16. The levels differ in the intensity of the restoring current. Auto restoration is the least intense and always should be used first. The restore current is limited to 100 mA and is cycled on and off three times to prevent the cathode from overheating. Auto restore is sufficient to restore most cathode-related problems.

Manual 1 is used on tubes that are not adequately restored by auto restore. The restore current is again limited to 100 mA, but the current is allowed to flow for as long as you press the "restore" button.

Manual 2 is the highest level of cathode restoration. This is a "last resort" level and is used when repeated attempts at restoration at the lower levels prove unsuccessful. The current is limited to 150 mA and flows for as long as you press the "restore" button.

The rejuvenate function is used when the CRT cathode is so totally encrusted that no restore current can be drawn by the other restore functions. In rejuv, a capacitor is discharged between the cathode and control grid with the filament voltage applied. The sudden positive voltage from the capacitor discharge causes the electrons to break through the contamination. Once the layer of contamination has been cracked, auto restore usually brings the tube back to proper operation.

Color CRT purity

Demagnetize the CRT with an external demagnetizer coil before you perform a complete purity and convergence operation. Set the contrast and brightness controls to maximum. Attach a color bar–cross-hatch generator to the television receiver

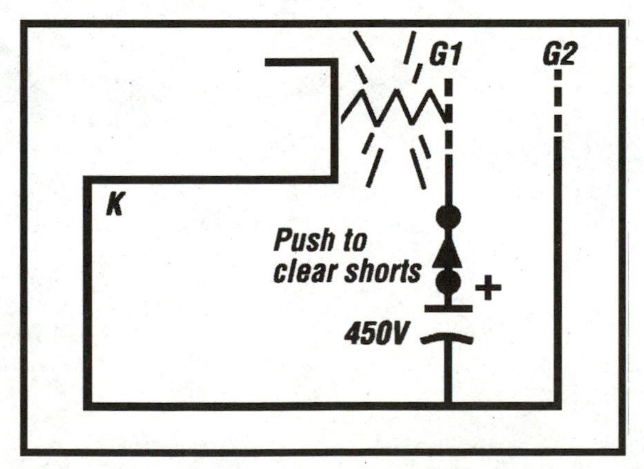

6-15 Remove G1 shorts function, in which a 450-V capacitor is discharged through the short. (*Courtesy of Sencore, Inc.*)

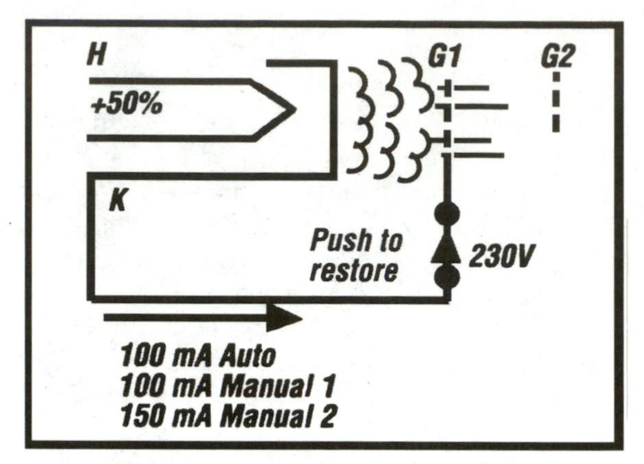

6-16 CR70 restore function. (*Courtesy of Sencore, Inc.*)

and set it for a green field. If a generator is not available, set the red and blue bias controls to minimum and the green to maximum. Then loosen the yoke clamp screw holding the yoke, and slide the yoke backward to provide a vertical green bar. Remove the rubber wedges. Rotate and spread the tabs of the purity magnet assembly (Fig. 6-17) until the green bar is in the center of the screen. Now move the yoke slowly forward until a uniform green screen is obtained. Tighten the clamp screw of the yoke temporarily. Check the purity of the red and blue fields. If every field is OK, tighten the yoke clamp, and proceed with the convergence adjustments.

> **Note:** Before attempting any convergence adjustments, the television receiver should be operated for at least 15 minutes.

Center convergence

Attach a color bar–cross-hatch generator to the television receiver, and adjust the brightness and contrast for a good picture (Fig. 6-18). Adjust the angle separating the two tabs of the four-pole magnets to superimpose the red and blue vertical lines in the center area of the picture. While keeping the angle separating the tabs constant, rotate them at the same time to superimpose the red and blue horizontal lines at the center of the screen. Adjust the two tabs of the six-pole magnets to superimpose the red/blue line on the green line. Adjusting the angle affects the vertical lines, and rotating both magnets affects the horizontal lines. Repeat these adjustments to obtain the best center convergence.

Edge convergence

> **Note:** This adjustment requires the use of rubber wedges.

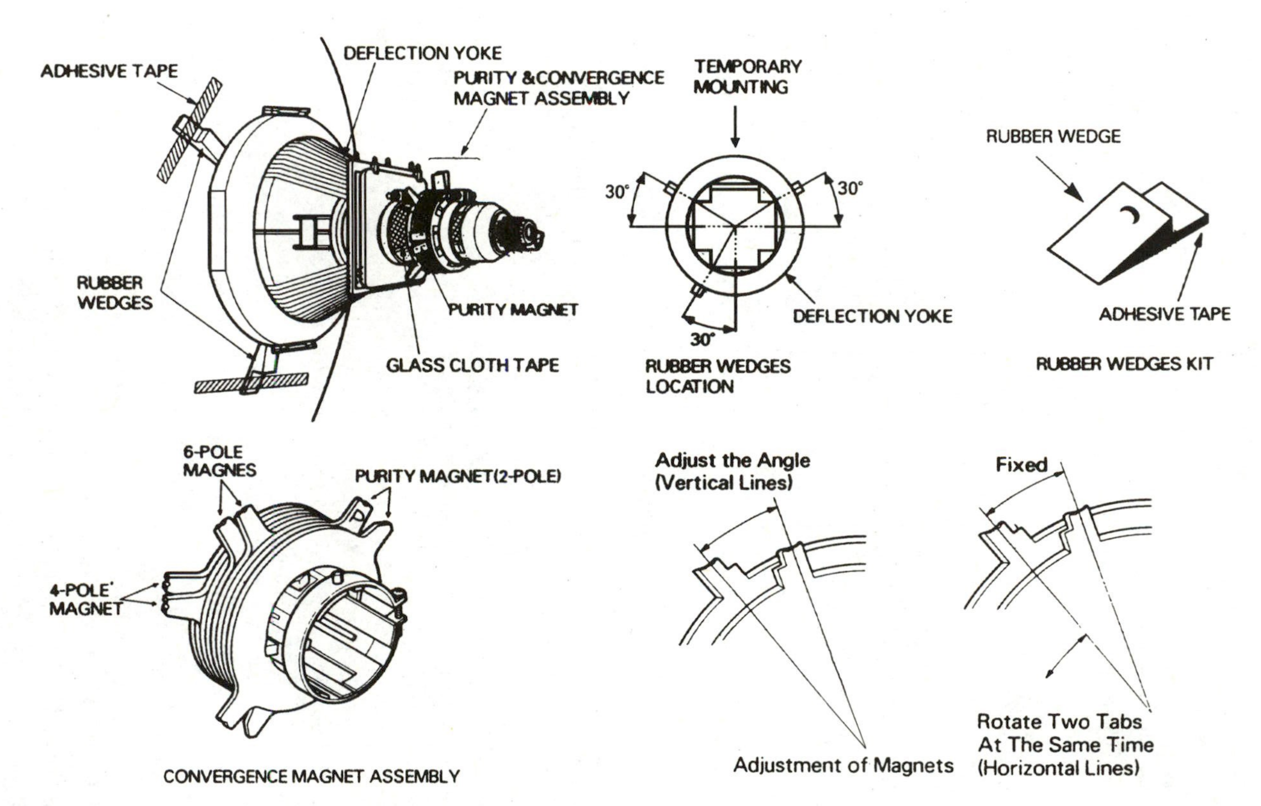

6-17 Picture tube adjustment components. (*Courtesy of Zenith Corporation.*)

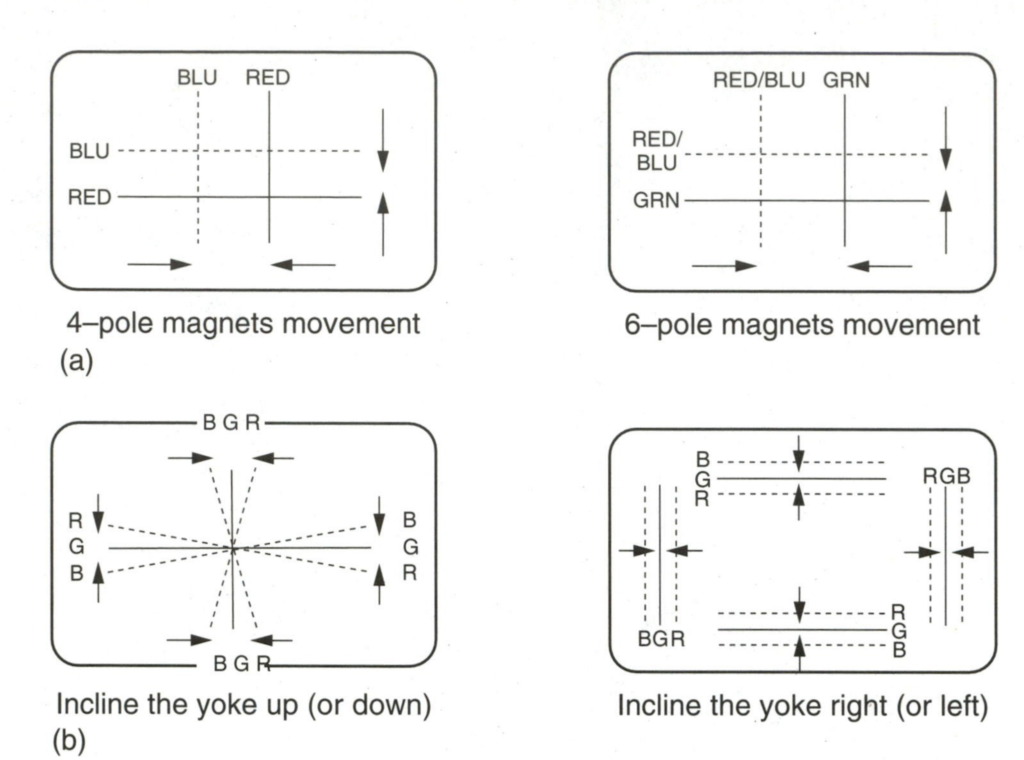

6-18 (*a*) Center and (*b*) convergence adjustments. (*Courtesy of Zenith Corporation.*)

Loosen the yoke clamping screw only enough to allow the deflection yoke to tilt. Tilt the front of the deflection yoke up or down to obtain the best edge convergence. Push the wedge into the space between the yoke and the picture tube at the top to find the position of the best edge convergence. Place the other wedge at the bottom, and do the same thing. Remove the cover paper of the bottom wedge, and stick it on the picture tube to hold the yoke in position. Tilt the yoke to the left and right to obtain the best convergence at the sides. Hold this position, and put another wedge in either upper space. Remove the cover paper, and stick the wedge on the picture tube to hold the yoke. Now remove the temporarily mounted wedge at the top and put it in the other upper space. Remove the cover paper, and stick the wedge on the picture tube to hold the yoke. After placing three wedges, recheck overall convergence. Tighten the yoke clamp tightly in place. Stick three adhesive tapes on the wedges, as shown in Fig. 6-19. The correct location of the pole magnet assembly on the neck of the picture tube is also shown in Fig. 6-19.

Horizontal tilt wedge adjustment

The vertical lines at 3 and 9 o'clock are converged by horizontally tilting the yoke and inserting a wedge, *first,* at 4 or 8 o'clock position, whichever has the larger space, until it is firmly seated between the CRT glass and yoke coils. Then insert the third wedge in the remaining horizontal tilt position until it is firmly seated between the CRT glass and yoke coils. Convergences at 3 and 9 o'clock should be maintained during operation.

When the three wedges are firmly installed for acceptable convergence, lock the wedges in place by applying a strip of tape across the tabs of each wedge, firmly against the CRT glass. Glass surfaces should be clean and free of dust and other foreign material.

Improving CRT corner purity

CRTs that display corner purity problems even after following servicing procedures can be modified with a correction kit. The purity can be improved by placing a picture correction magnet on the CRT funnel. Refer to the following modification steps and Fig. 6-20 to see how to place the magnet properly.

1. Place the magnet on the CRT funnel in the quadrant exhibiting impurity.
2. Rotate the magnet in place to the position shown for best purity.
3. Place a piece of ½-in Fiberglas tape over the magnet to hold it in place.
4. Degauss the CRT once the magnet is in place to ensure that the magnet is not over the internal magnetic shield.

> **Note:** If the magnet is placed over the internal magnet shield, any apparent purity correction will disappear after degaussing. Reposition the correction magnet off the internal shield and degauss again.

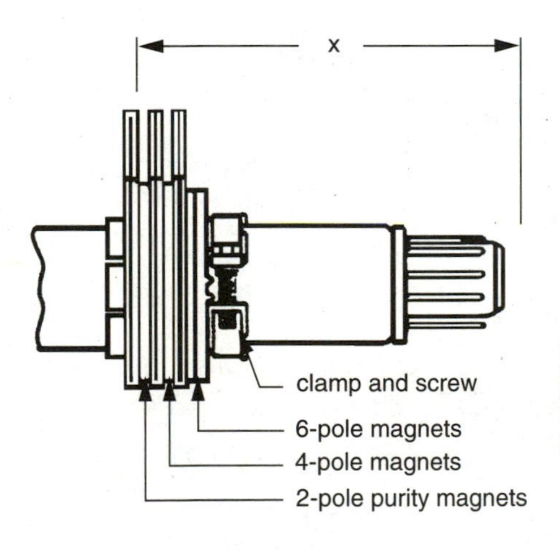

6-19
Location of magnet assembly on the CRT neck. (*Courtesy of Zenith Corporation.*)

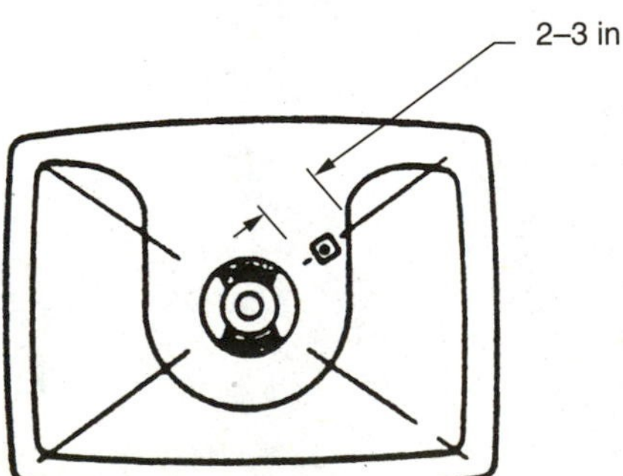

6-20
Improving CRT purity by using a corner purity magnet. (*Courtesy of Zenith Corporation.*)

7 CHAPTER

Electronic tuner and i.f. systems performance

This chapter shows you how to use test instruments to evaluate the performance of electronic tuners and i.f. systems. In addition, this chapter shows you how to correct problems when all tuner and i.f. systems are not "go." The Sencore VG91 universal video generator will be used to track down tuner and i.f. problems. This chapter also looks at some special checks required for cable-ready television tuners and some special problems that may develop. The chapter concludes with some troubleshooting tree checks and key test and checkpoint information.

Tuner and i.f. quick checks

The tuner of a television set is a section that can cause technicians to lose their minds very quickly. To replace a modern electronic tuner is very expensive and can approach the cost of buying a new television once you add labor and replacement parts into the repair costs. Once you have repaired any problems in the power supply, horizontal, or vertical stages, be sure to make a quick performance test of the tuner using the Sencore VG91 video analyzer.

With the number of channels that are now available from most cable television systems, you will need to test the set on the lower, midband, and hyperband channels. Testing the tuner section with the video analyzer also allows you to review the picture clarity, color, and hue from a known good signal before you make your estimates to any customer. This could save you from callbacks once the customer gets the television back home and also provides a valuable customer service.

The video analyzer provides accurate reference test signals and adjustable levels to fully test any NTSC video system. You can observe the operation of the video system by viewing the CRT or by using an external video monitor or waveform analyzer to monitor the video output.

The video analyzer test falls into two general categories:

1. Testing tuner/i.f. circuits
2. Testing video and audio processing circuits

If you are testing tuner/i.f. circuits, apply the analyzer's radio frequency (rf) signals to the tuner input. To test audio and video processing circuits, feed the video analyzer's standard output signals to the corresponding Y/C, video, or audio input jacks. Use the troubleshooting guide shown in Fig. 7-1.

Analyzing the video tuning system

Cable-ready television tuners require extensive testing to ensure correct operation. Cable television systems shift channel frequencies as much as 2 MHz from standard television broadcast or conventional cable frequencies. Therefore, cable-ready tuners must perform a tuning search to locate these shifted carriers. This digitally controlled search occurs when a channel is selected. A cable-ready tuner may have problems tuning to either off-the-air or cable channels or have trouble tuning to specific channels.

With the analyzer you can see that it would be much easier to diagnose and troubleshoot television tuning systems if you had access to every television channel. Imagine having these channels with analyzing video test patterns and mono/stereo SAP audio test signals. The Sencore VG91 video analyzer has four rf functions: STD TV, STD cable, HRC cable, and ICC cable.

Use the STD TV position of the VG91's rf-i.f. signal to test single-channel or non-cable-ready tuning systems. Use the STD cable, HRC cable, or ICC cable position to duplicate the cable system that the tuner must receive or to test the tuning search function of the digital tuner. It is important to test several channels in each of the tuning bands to make sure all channels are being received.

On-board tuner repairs

For you to keep current customers and acquire new ones, you must keep your equipment and troubleshooting techniques up to date. In this day and age, you must do this in order to stay competitive.

One change in recent years is the introduction of the on-board tuners found in Thomson's newer RCA and GE television chassis. The tuner circuitry is manufactured right on the main circuit board with the rest of the television circuitry. Since the tuner cannot be replaced as a module, you have to troubleshoot it to isolate any defects and perform alignments as required. While this challenge is new and may be a little scary, it is in many ways no different from troubleshooting other discrete television circuits.

Knowing how the tuner functions and knowing how to troubleshoot and align the tuner are now essential to providing quality repairs. Let us now examine tuner circuits, especially those found in the RCA and GE chassis with on-board tuners. We will be showing simplified circuits of these RCA and GE sets; however, most other brands of tuners are similar in their operation.

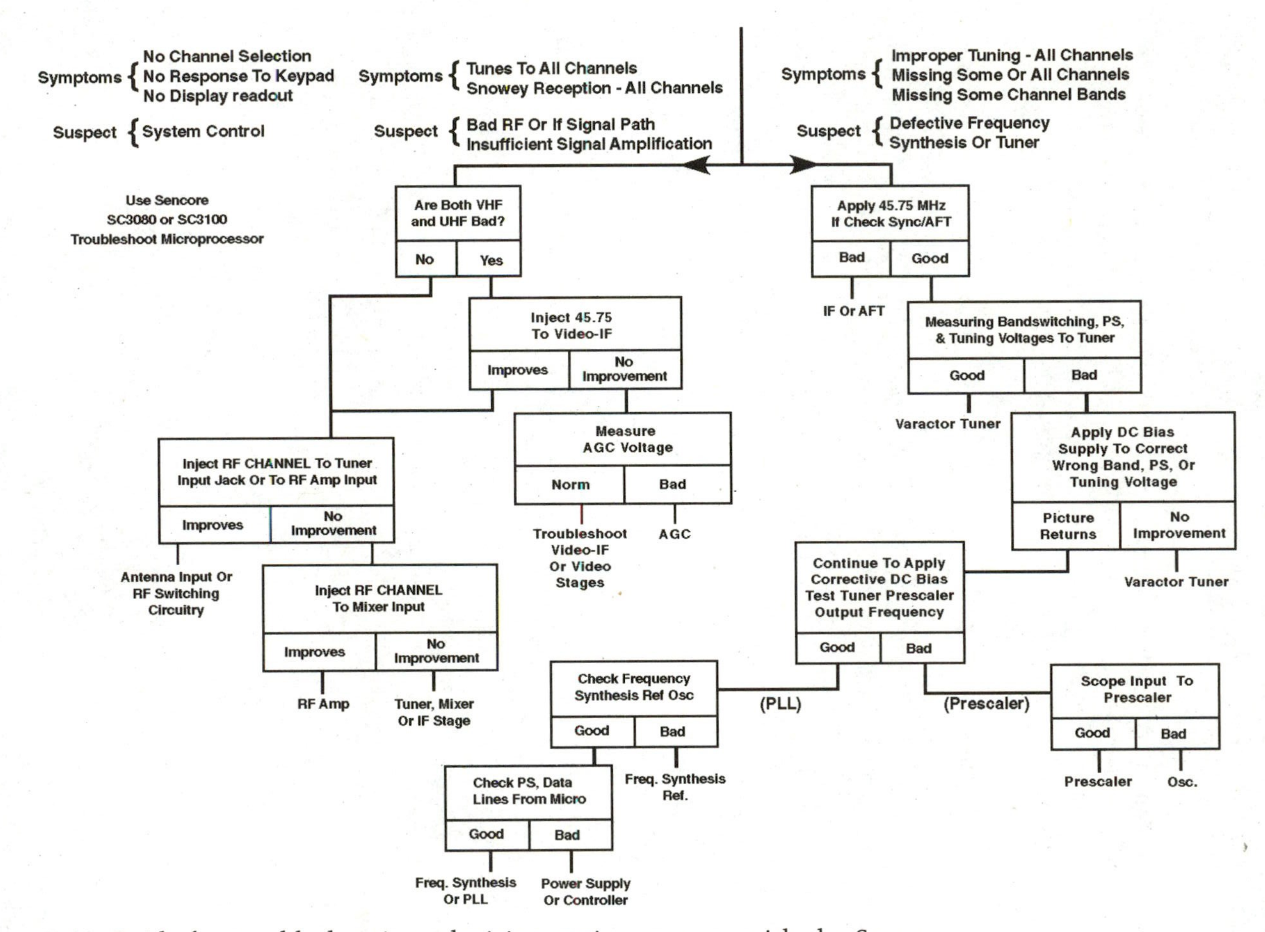

7-1 Guide for troubleshooting television tuning systems with the Sencore video generator. (*Courtesy of Sencore, Inc.*)

Tuner basics

The television tuner's job is to select a single rf television channel and amplify it while rejecting all other channels. It further must convert the selected channel to a lower i.f. carrier frequency (41.25 to 45.75 MHz) for more amplification and filtering.

The tuner is divided into VHF and UHF sections, as shown in the block diagram in Fig. 7-2. The VHF section processes the lower-frequency rf channels, while the UHF section handles the higher-frequency rf channels. Included in these sections are an rf amplifier, rf channel bandpass filters, oscillator, and mixer stage. Many television tuners include a preliminary i.f. amplifier and i.f. bandpass filter. Since the VHF and UHF sections have similar circuits, we will examine only those in the VHF section.

Rf amplifiers and automatic gain control

The rf amplifiers in the television tuner commonly are dual gate *N*-channel depletion-type MOSFETs. The transistors amplify the weak rf signal to improve the sensitivity of the receiver. The rf signal is input to one of the gates, and an automatic gain control (AGC) voltage is input to the other gate.

Rf amplifier FETs operate very much like vacuum tubes. When a negative voltage is applied to either gate with respect to the source, drain current is reduced. When a positive voltage is applied to either gate with respect to the source, drain current is increased.

During normal operation, 0.5- to 5-mV rf signal strength is applied to the input of the rf amplifier's transistor gate. The transistor increases and decreases conduc-

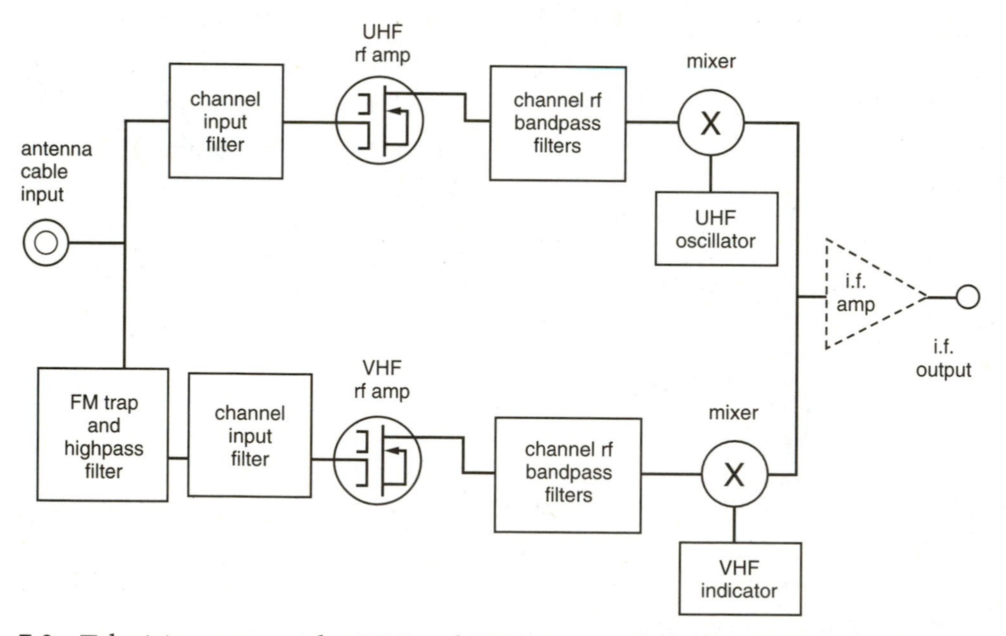

7-2 Television tuner with a VHF and UHF section. Each section includes channel filters, an rf amplifier, an oscillator, and a mixer. (*Courtesy of Sencore, Inc.*)

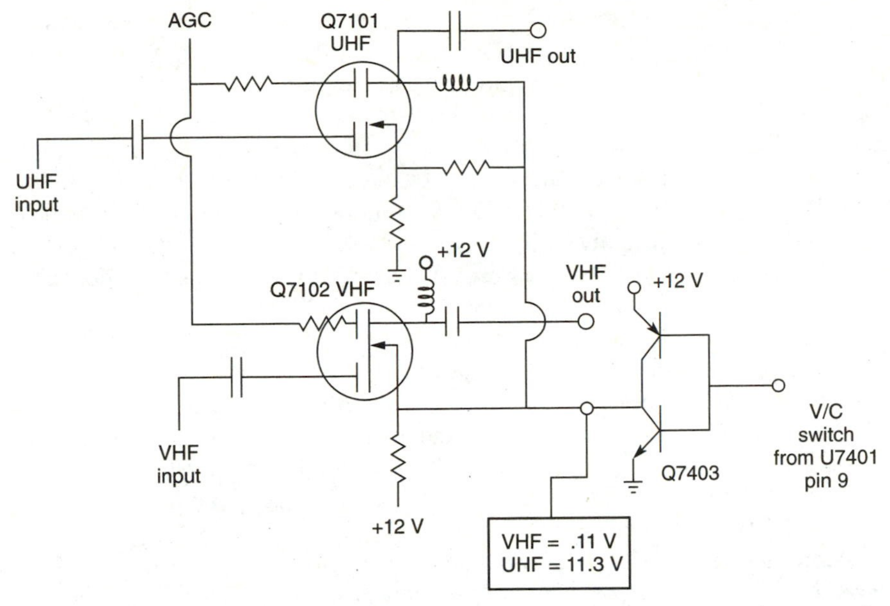

7-3 VHF and UHF amplifiers and band V/U switching in the Zenith chassis. (*Courtesy of Zenith Corporation.*)

tion with AGC control, and the rf signal is then amplified accordingly. The amplified signal is output onto the drain lead and coupled to the channel rf filters.

The AGC voltage applied to the other gate of the rf amplifier transistor controls the voltage gain of the transistor. A rising voltage increases the gain, while a decreasing voltage reduces the gain. At full gain, tuners with depletion FETs commonly have an AGC voltage ranging from 6 to 9 V. If the signal strength of the television rf signal to the tuner increases, the AGC voltage drops to lower the gain of the rf amplifier. This prevents interference produced by an overdriving signal to the mixer stage and i.f. amplifiers.

The VHF transistor amplifiers the lower-frequency rf channels, while the UHF transistor amplifies the higher-frequency rf channels, as seen in Fig. 7-3. For example, in the RCA chassis, transistor Q7102 amplifies channels 2 through 50, while transistor Q7101 amplifies channels 51 through 125.

Only one of the two rf transistors is active for any selected channel. The other transistor is turned off or disabled. The rf amplifier transistors are switched on or off with a voltage applied to the source and/or drain elements. For example, in Fig. 7-3, Q7102 is active when the band V/U switch voltage turns Q7403 on, connecting the source of Q7102 to ground. This removes the drain of Q7101, biasing it off. When the band V/U switch opens, voltage at the collector of Q7403 is applied to the source of Q7102, biasing it off while turning on Q7101.

Channel rf bandpass filters

The VHF and UHF sections of a television tuner commonly contain three rf channel bandpass filters. One bandpass filter is located before the rf amplifier, and two are located on the output of the rf amplifier before the mixer. Figure 7-4 shows the primary and secondary channel bandpass filters between the rf amplifier and mixer stage in the RCA chassis. A similar filter is found on the input of the rf amplifier. Each channel bandpass filter consists of inductors and capacitors that form a parallel resonant circuit. The filters are tuned so that the resonant frequency is that of the desired rf television channel. When the resonant frequency matches that of the channel to be selected, the LC filter passes the rf television channel of that frequency. The bandpass of the tuned circuits is roughly 6 MHz, passing the full rf channel signal information while filtering out other channels. In Fig. 7-4, the primary bandpass filter consists of L7111 and L7112 in parallel with capacitors V7119 and CR7108.

The bandpass filters are tuned to select different rf channels by changing the capacitance. A variable-capacitance diode or varactor, when reverse biased with applied dc voltages, forms a variable capacitor. For example, by applying a varying reversed-bias voltage from 1 to near 30 V, the capacitor may change from 200 down to 20 pF. This capacitance change tunes the LC bandpass filter through a range of television channel frequencies. As the capacitance decreases, the resonant frequency of the LC filter increases.

The capacitance range of a varactor diode tunes the channel bandpass filter through a range of frequencies covering a band of approximately 200 MHz. However, this is not sufficient to tune in all the channels that the VHF section of the tuner must select. To widen the frequency range of the channel bandpass filters, several inductors are placed in series within the LC resonant circuit. The inductors are switched in and out of the circuit by switching diodes.

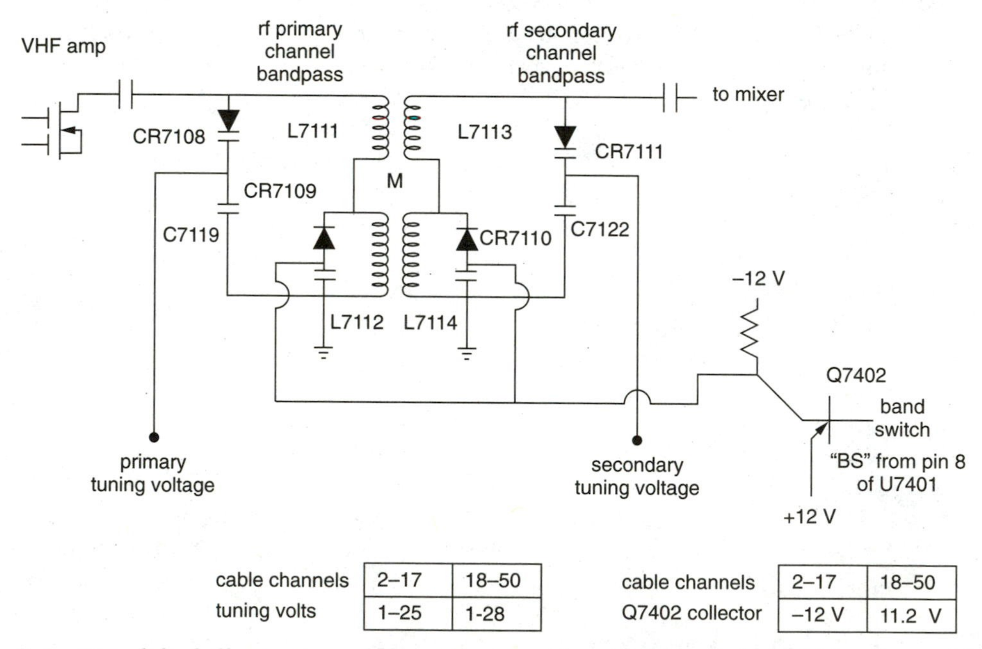

cable channels	2–17	18–50
tuning volts	1–25	1-28

cable channels	2–17	18–50
Q7402 collector	−12 V	11.2 V

7-4 Simplified illustration of the primary and secondary rf channel bandpass filters in the Zenith chassis. (*Courtesy of Zenith Corporation.*)

When a band-switching diode is biased on by an applied voltage, the inductor is shunted or removed from the LC circuit. When a switching diode is biased off, it is an open, and the inductor is part of the resonant circuit. The fewer inductors in series, the lower is the inductance of the LC circuit and the higher is the resonant frequency.

In the RCA chassis, CT7109 and CR7110 are the switching diodes. The diodes are biased off or open for cable channels 2 through 17. This permits the tuned circuit to tune from 55 to 170 MHz as the tuning voltage of from approximately 1 to 28 V is applied to the varactor diode. The switching diodes are switched on for cable channels 18 through 55. As the tuning voltage is again varied through its range, the tuned circuit resonates from 215 to 410 MHz.

The three LC channel bandpass filters of the VHF and UHF sections of the television tuner must have matched characteristics. This ensures that they will track or have the same resonance frequency and bandpass characteristics to the applied tuning and switching voltages. A slight shift in component characteristics in any of the filters reduces the gain or distorts the channel's 6-MHz-wide range of signals.

In addition to the three rf channel bandpass filters, the tuner commonly contains a high-pass and FM band filter. These filters are found on the tuner input before the first rf channel bandpass filter. The high-pass filter and FM trap block signals below channel 2 and in the FM band that may pass to the rf amplifier and mixer causing interference to the selected television signal.

Channel frequency synthesis

In the tuner's mixer stage, a carrier from the oscillator and the television rf signal are beat or heterodyned together. This converts the television channel down to an intermediate frequency (i.f.) containing all the channel's signal information. For example, channel 2 (55.25 MHz) is beat with an oscillator frequency of 101 MHz.

The difference between these frequencies is 45.75 MHz, the i.f. video carrier frequency. The i.f. circuits pass the channel's 6-MHz band of signals between the audio carrier at 41.25 MHz and the video carrier at 45.75 MHz.

The frequency of the tuner's oscillator determines which cable or off-the-air television channel is converted to the i.f. bandpass. To select cable channels, the oscillator must be adjusted to a frequency of precisely 45.75 MHz above the selected television channel's video carrier frequency. In the RCA chassis, the tuner's VHF oscillator is adjusted from 101 to 425 MHz to select cable channels 2 through 50.

In the RCA chassis, the VHF and UHF oscillators are part of IC U7301, which is shown in Fig. 7-5. Capacitors and inductors external to pins 9 and 11 form a resonant LC circuit to determine the frequency of the VHF oscillator. The LC circuit consists of L7304 and L7305 in parallel with capacitor C7314 and varactor CR7302.

The tuner's oscillator is part of a voltage-controlled frequency generator. To adjust the frequency of the LC oscillator, the capacitance of the LC circuit is varied with a varactor diode. Like the channel bandpass filters, a reversed-bias voltage applied to the varactor sets the LC circuit frequency 45.75 MHz above the desired television channel's video carrier. The voltage applied to the varactor diode tunes the LC circuit through a range of frequencies. In the RCA chassis, the oscillator ranges from 101 to 185 MHz for cable channels 2 through 17.

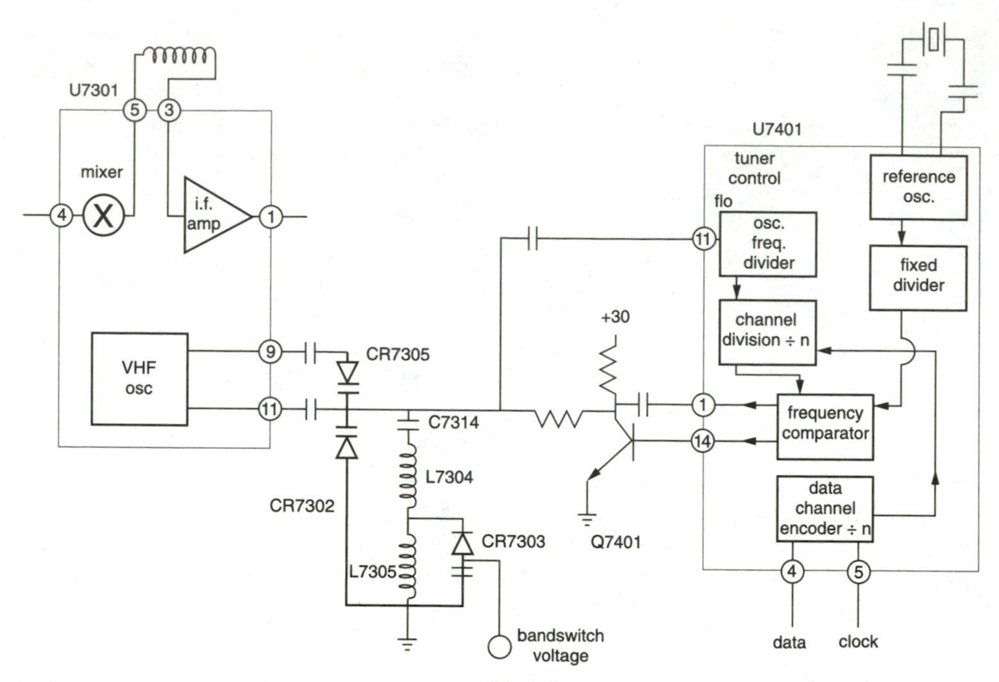

7-5 Frequency synthesis in the RCA television chassis. (*Courtesy of Thomson Electronics.*)

To widen the frequency range of the oscillator, an inductor is switched into the LC circuit in the same manner used in the bandpass filters. In Fig. 7-5, inductor L7305 is part of the LC circuit (switching diode open) as the frequency of the LC circuit ranges from 101 to 185 MHz to select cable channels 2 through 17. To permit the LC circuit to oscillate through a higher range, the diode is biased on, shunting L7305 and effectively removing it from the LC circuit. As the variable voltage is applied to the varactor diode, the LC oscillator's frequency ranges from 185 to 425 MHz to select channels 18 through 50.

The voltage-controlled LC oscillator must be set precisely to a frequency 45.75 MHz above the selected channel's video carrier frequency. Channel frequency-synthesis or phase-lock-loop (PLL) tuning provides a technique for precise control of the oscillator frequency.

Frequency synthesis consists of the tuner's oscillator, a crystal reference oscillator, frequency dividers, and a frequency comparator. The output of the frequency comparator is filtered to a dc voltage and applied to the voltage-controlled oscillator. In a modern television system, the frequency dividers are controlled by microprocessors for channel selection and automatic fine tuning (AFT) correction.

In a frequency-synthesis system, a sample of the tuner's oscillator from the LC circuit is fed to a frequency divider. Here, it is divided down, "prescaled," and then divided again by a divisor (N=channel divisor number) determined for the selected channel. The result is compared with the frequency of a reference oscillator divided down from a crystal. The frequency comparator digitally compares the frequencies and outputs a pulse-wave-modulated (PWM) signal. The PWM signal is filtered to an average dc voltage and fed to the LC oscillator to control its frequency. Any frequency error results in a variation of a PWM output filtered into a dc correction voltage and applied to the oscillator to correct the frequency.

In the RCA television chassis, the frequency dividers and comparison circuitry are part of U7401 (Fig. 7-6). Digital data are input to U7401 by the control microprocessor. The digital data tell U7401 what channel has been selected and provide information from a memory IC. U7401 selects the correct channel divisor (*n*) and outputs the proper UHF/VHF switching and bandswitching control voltages.

Checking out the electronic tuner

When all the tuner stages work properly, an rf channel's 6-MHz band of signals is properly amplified and converted to an i.f. signal. When a tuner stage fails or changes parameters, the output is altered. The output may be missing, noisy, or distorted on some or all channels. To isolate these defects, you need to use a step-by-step troubleshooting approach as you would with any television problem.

The first step in troubleshooting a tuner problem is to thoroughly performance test the tuner. When you are performance testing the tuner, keep in mind the operation of each of the tuner's major functional blocks that are shown in Fig. 7-6.

The first troubleshooting step is to evaluate the operation of the tuner at several rf channels with a known input level of a 1000-μV signal. The rf amplifier provides the proper signal gain to signals of 1000 to 2000 μV at its gate. At this input level, the AGC voltage should be 6–9 V.

This puts the rf amplifier FET at full gain. The proper input level is important because it indicates if the rf/i.f. amplifier has proper gain and if the bandpass filters are passing the signal without added signal losses. At this level, the CRT should show a noise-free picture on all channels. Applying too much or too little signal can cause misleading symptoms.

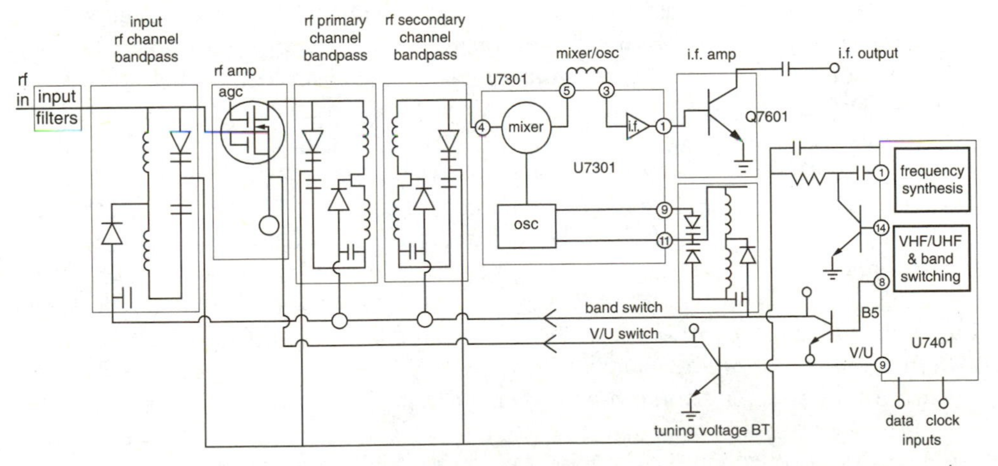

7-6 VHF section diagram of the RCA tuner. (*Courtesy of Thomson Electronics.*)

When checking the tuner, select rf channel signals that test the rf bandpass LC filters and oscillator LC stages through their tuning ranges. The rf channel bandpass filters before and after the rf amplifier pass the desired channel when resonant to the channel frequency. To set the resonant frequency, a tuning voltage is applied to the varactor diodes and bandswitching voltage to the switching diodes. The oscillator LC circuit is controlled in a similar manner but tuned 45.75 MHz above the channel carrier.

To test these LC circuits requires applying rf channels at the low, middle, and high ends of each tuning range. For example, in the VHF section of the RCA tuner, the rf channel bandpass tunes through two bands. Band one includes cable channels 2 through 17, and band two includes channels 18 through 50. You need to test these channels and several other channels in between. To test the UHF section of the tuner, you want to test at channels 51 and 125 and several other channels in between as well.

The results of the performance testing can give you important clues as to the possible problem. If all channels tune properly but are noisy, a gain problem, not a bandpass problem, likely exists. Look for an rf amplifier, AGC, or i.f. amplifier circuit as the likely culprit.

If the VHF channels come through nice and clear but not the UHF channels, a problem may exist with the VHF/UHF switching or in the UHF signal path section in between the input and mixer stages. Problems that cause a loss of channels in only one of the VHF or UHF bands may be due to a problem with the bandswitching voltage. Problems that affect channels at the low end of each tuning band in the VHF section may indicate a shift in a VHF bandpass filter component. Problems that affect the tuning of all channels are likely caused by circuits common to all channels, such as the tuner control circuits.

Testing key tuning and switching voltages

Since the tuner stages are controlled by dc voltages, you must make dc measurements to isolate defects. Of course, for most symptoms, first you will want to check power supply voltages and the 30- to 33-V tuning voltage.

Depending on the symptoms, you will want to measure switching and tuning voltages. Determining if these dc voltages are right or wrong for a particular channel requires an understanding of the normal control voltages. In the RCA tuner, as in other tuners, the voltage applied to the bandpass and oscillator LC circuits varies from approximately 1 to 28 V. While monitoring the tuning voltage test points, start at channel 2 and increment the television receiver upward in channels through a tuning band of channels. The tuning voltage should increment upward in voltage steps. To test the bandswitch and V/U switching voltages, measure circuit points corresponding to these control lines. The voltage should make distinct changes as you switch between the tuning bands and the VHF/UHF operation in the tuner. Refer to Table 7-1 for key tuning and switching voltages.

If the voltages are improper or do not change as the channels and bands are selected, the control line is either pulled to a low- or high-voltage condition because of

Table 7-1. Key tuning and switching voltages for the CTC175/176/177 chassis

Cable channels	Band 1, Ch. 2–17	Band 2, Ch. 18–50	Band 3, Ch. 51–125
Bandswitch	−12 V	11.4 V	11.4 V
V/U switch volts	0.11 V	0.11 V	11.3 V
Tuning voltages (Bt)	1–28 V	1–28 V	1–28 V

a problem component in the tuner or a problem in the control IC. Make voltage measurements and resistance checks or open the control line to isolate the problem.

On occasion, where a voltage is improper or missing, a dc power supply can be used to substitute the voltage. The Sencore TVA92 analyzer contains a dc bias power supply designed for this and other dc biasing uses. The bias supply, unlike conventional power supplies, can dominate the circuit point to either raise or lower the dc voltage. By substituting for a missing voltage or slightly shifting a tuning voltage, you can determine if the blocks of the tuner will work properly when the proper voltage is applied.

Tuner block signal injection information

When all the tuning and control voltages are normal but tuning is improper, a problem stage in the tuner is a likely consideration. You can further isolate the defective stage by substituting for the rf or i.f. signals at key points along the signal path of the tuner.

The input to the mixer stage is a good place to begin signal injection. The normal signal found at this test point is the rf television signal passed by the rf bandpass filters. To substitute at this test point, set the Sencore VG91 video generator to the same channel selected on the television receiver. Adjust the output level of the VG91 to approximately 1000–2000 μV. Connect the VG91's output to the input of the mixer (connect the ground to the tuner shield). Be sure to use the VG91's isolated test lead to prevent shorting out the dc voltage at the mixer input. If a good, clear video pattern is displayed on the CRT, the mixer, oscillator, and i.f. stages are good. Missing or poor video on the CRT indicates a problem in the mixer, oscillator, or i.f. stages.

To determine if the cause is a missing or improper oscillator frequency, feed the VG91's 45.75-MHz i.f. test signal into the mixer input. Adjust the output level of the VG91 from 1 to 5 mV. If a good, clear video pattern is displayed on the CRT, the mixer and i.f. stages are in good condition. You will need to troubleshoot the oscillator and its control circuits. If no improvement is seen, move your injection point to the i.f. amplifier input to further isolate the problem.

If injecting an rf channel signal into the mixer returns a good video display, you know the oscillator, mixer, and i.f. stages are operating properly. This indicates that the problem is before the mixer—in the input filters, rf bandpass filters, or rf amplifier. To isolate the problem, inject the VG91's rf television channel signal at points along the tuner's rf signal path from the input to the mixer input.

To isolate problems before the mixer, start by injecting the VG91's rf signal to the gate of the FET. Set the VG91 video generator to the same channel selected on the television receiver. Adjust the output level of the VG91 to approximately 1000 to 2000 μV.

Connect the VG91's isolated lead between the FET signal input gate and the tuner's ground. If good-quality video is produced on the CRT, the rf amplifier and circuits following it are good. The problem is now most likely in the input filter or bandpass filter. If missing or poor-quality video results, the rf amplifier is not working properly.

Zenith color television model SMS 1324

Let us now look at some tuner and remote control troubleshooting and alignment information for some late-model Zenith color television receivers. These data will cover block diagrams, IC pin outs, troubleshooting trees, and picture i.f./AFT adjustments. Refer to Fig. 7-7 for the complete block diagram information.

In Fig. 7-8 you will find the IC chip pin outs for the tuning control system and caption decoder. The information in Fig. 7-9 gives checks and hints for screen display and remote control problems, while Fig. 7-10 gives repair tips for channel control faults. Figure 7-11 presents many checks and tips for various picture problems. The IC tuning control chip pin outs are shown in Fig. 7-12.

Picture i.f./AFT adjustments (Zenith)

Note: The following precautions should be taken when aligning and adjusting this television receiver.

- Alignment requires an exacting procedure and should be undertaken only when necessary.
- An isolation transformer must be used to prevent electrical shock.
- The proper test equipment must be used to perform the alignment correctly. Use of equipment that does not meet these requirements may result in improper alignment and poor picture quality.
- Accurate test equipment is essential to obtain proper alignment for color television receivers.
- Use of excessive signal from a sweep generator can cause overloading of the receiver circuit. Overloading should be avoided to obtain a true response curve. Insertion of markers from the marker generator should not cause any distortion of the response curve.
- The ac power line voltage should be kept at 120 V while the alignment operation is being performed.
- Do not attempt to disconnect any components while the television receiver is in operation.
- Make sure the ac power cord is disconnected before replacing any part in the television receiver.

Test equipment requirements

- Digital volt/ohm meter
- Oscilloscope

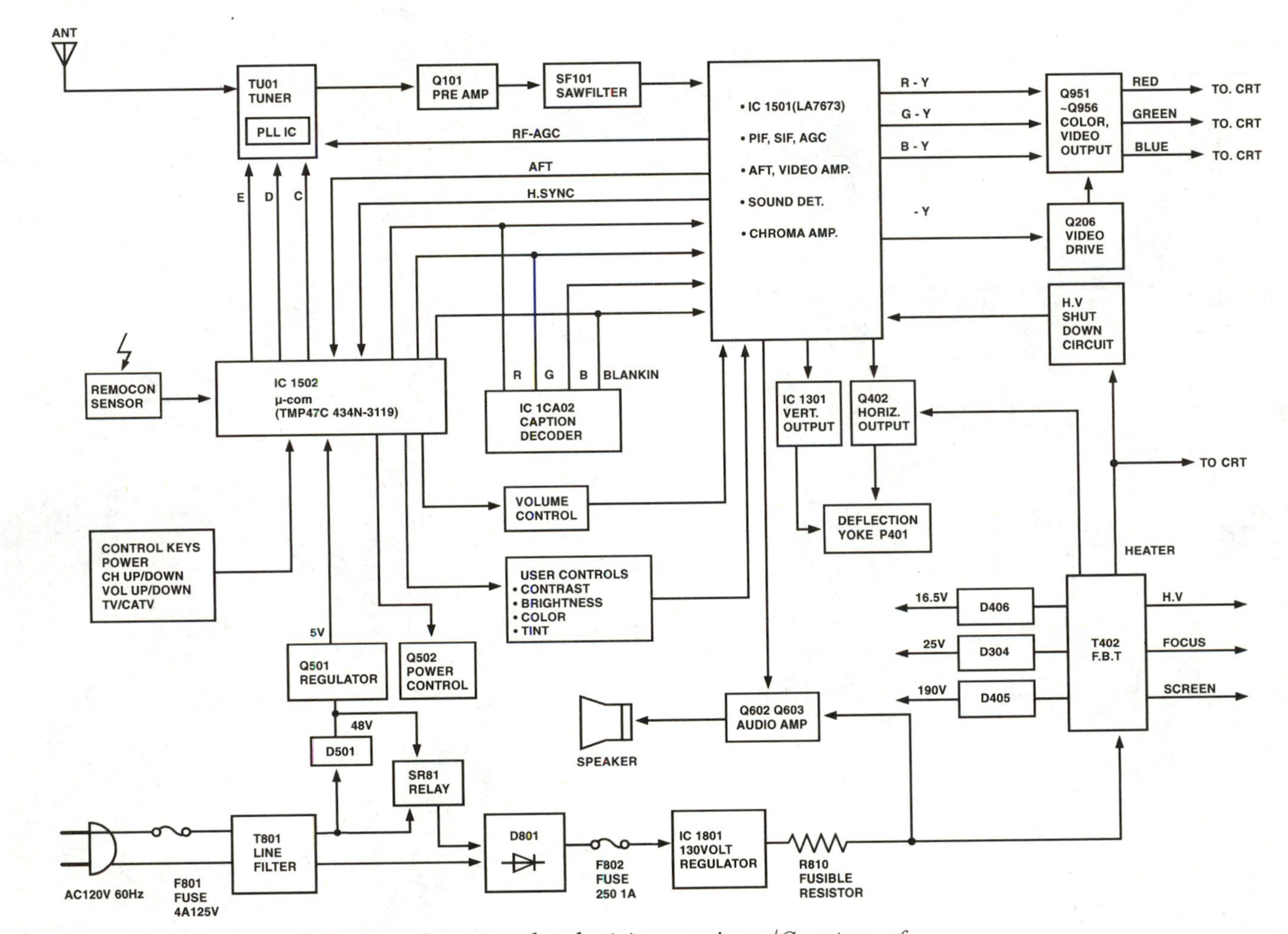

7-7 Block diagram of the complete Zenith television receiver. (*Courtesy of Zenith Corporation.*)

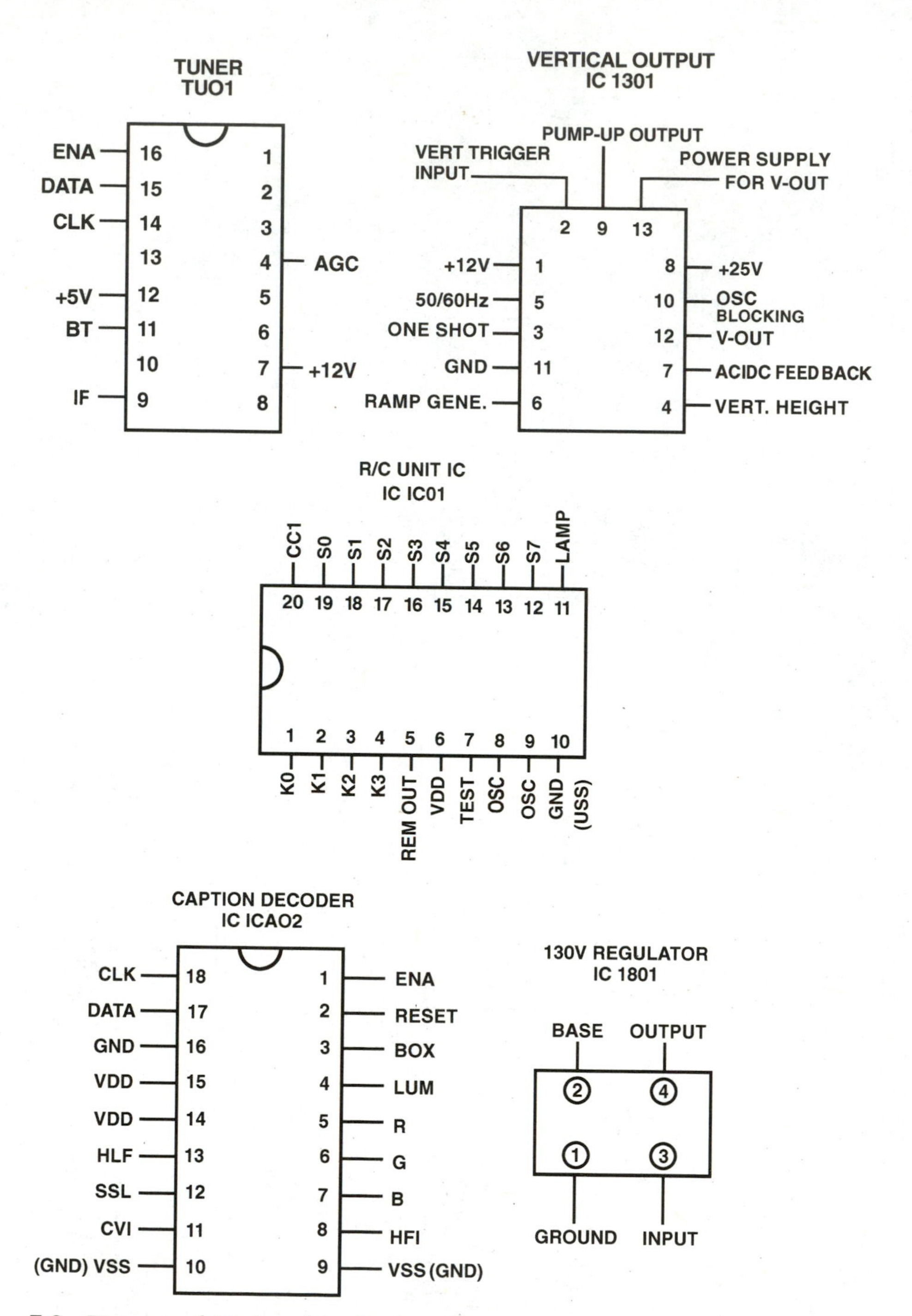

7-8 Pin-outs of ICs found in the Zenith tuner and control circuits. (*Courtesy of Zenith Corporation.*)

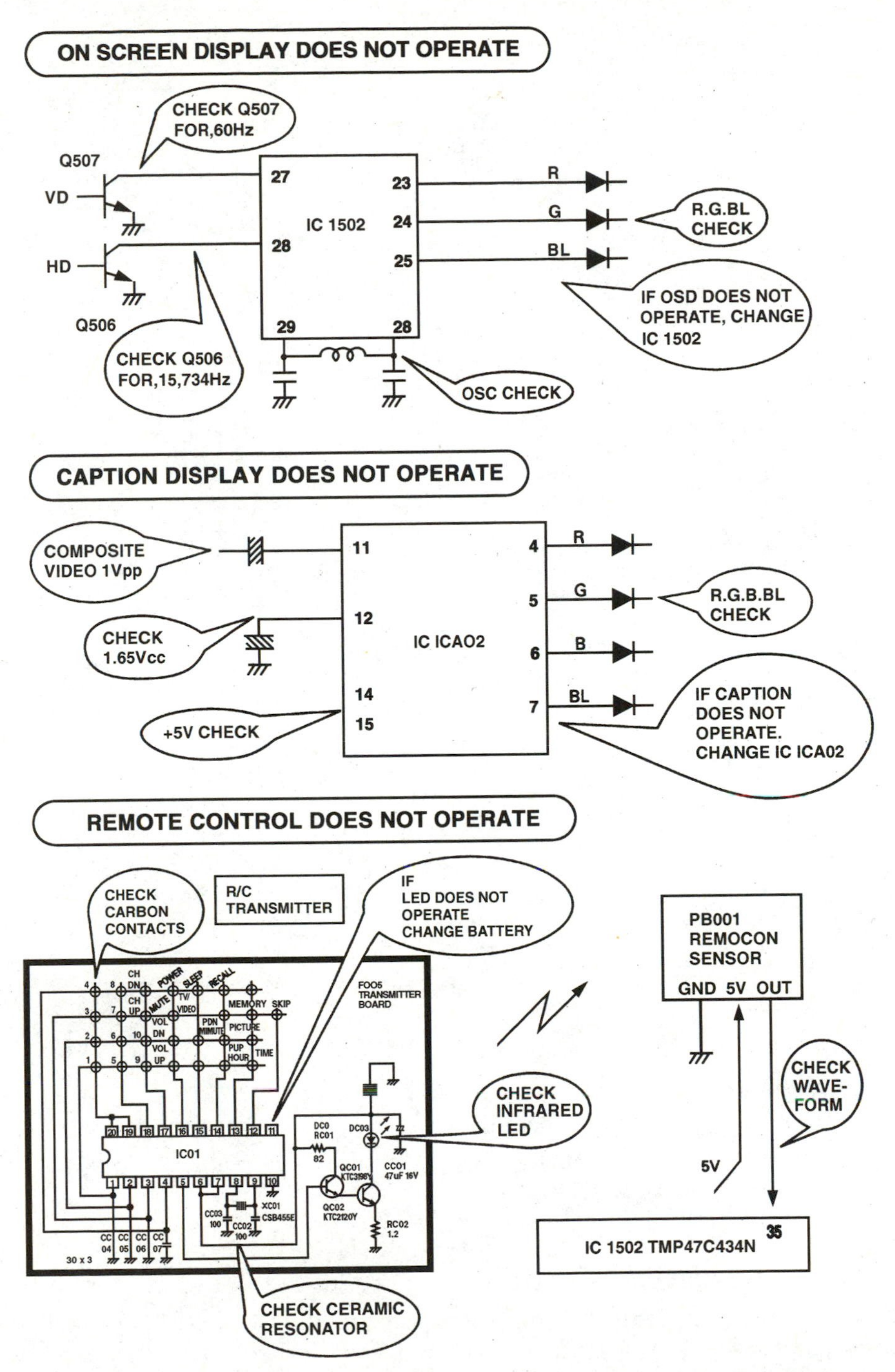

7-9 Troubleshooting information for problems in the screen display, caption display, and remote control operation. (*Courtesy of Zenith Corporation.*)

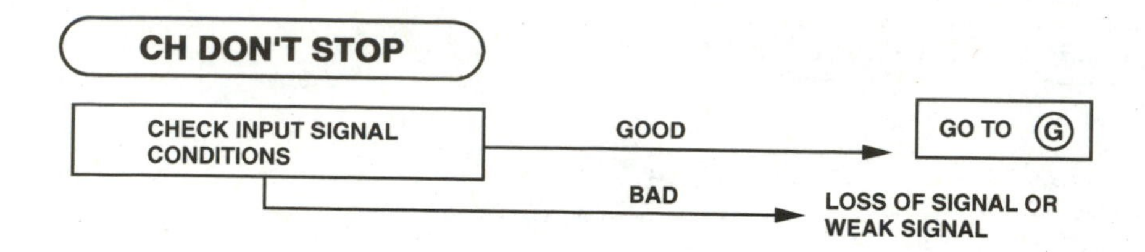

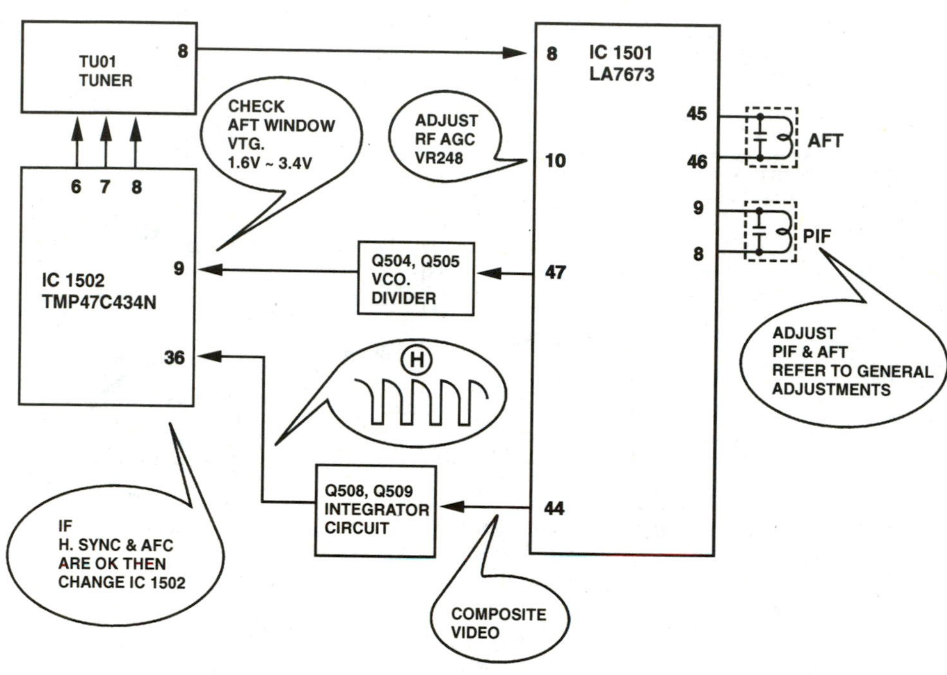

7-10 Troubleshooting hints when channel control will not stop. (*Courtesy of Zenith Corporation.*)

- Direct/low-capacity scope probe
- Color bar–dot–cross-hatch generator
- Postinjection sweep/marker generator
- Power supply–adjustable dc voltage
- Isolation transformer

Alignment information (Zenith)

Use the following steps to align this late-model Zenith color television chassis. For the proper test equipment setup, refer to the block diagram in Fig. 7-13.

1. Disconnect the tuner i.f. from the TP1 test point, and connect the equipment as shown in Fig. 7-13.
2. Set the sweep/marker generator for 30 V rms.

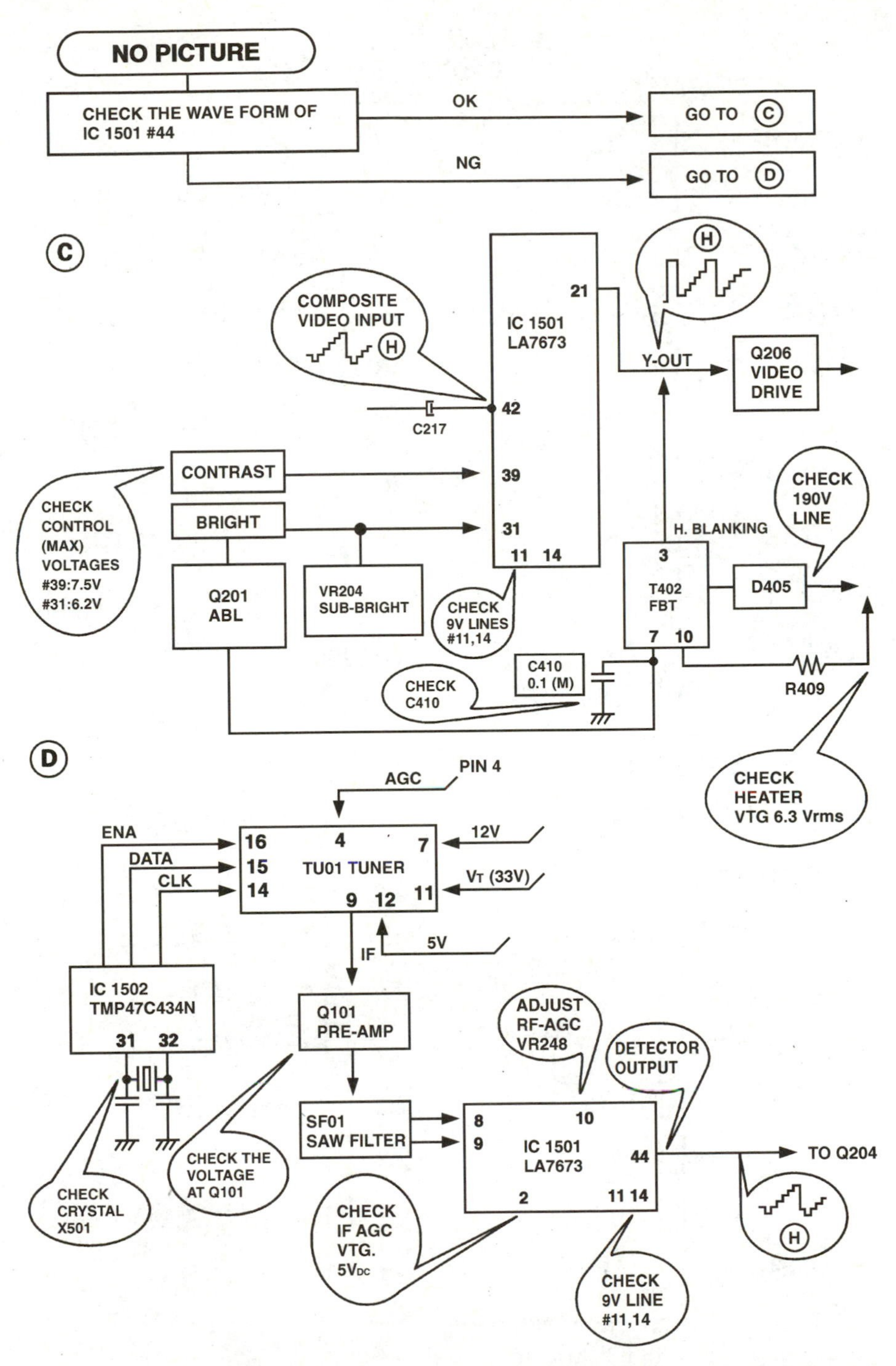

7-11 Hints and tips for "no picture" problems. (*Courtesy of Zenith Corporation.*)

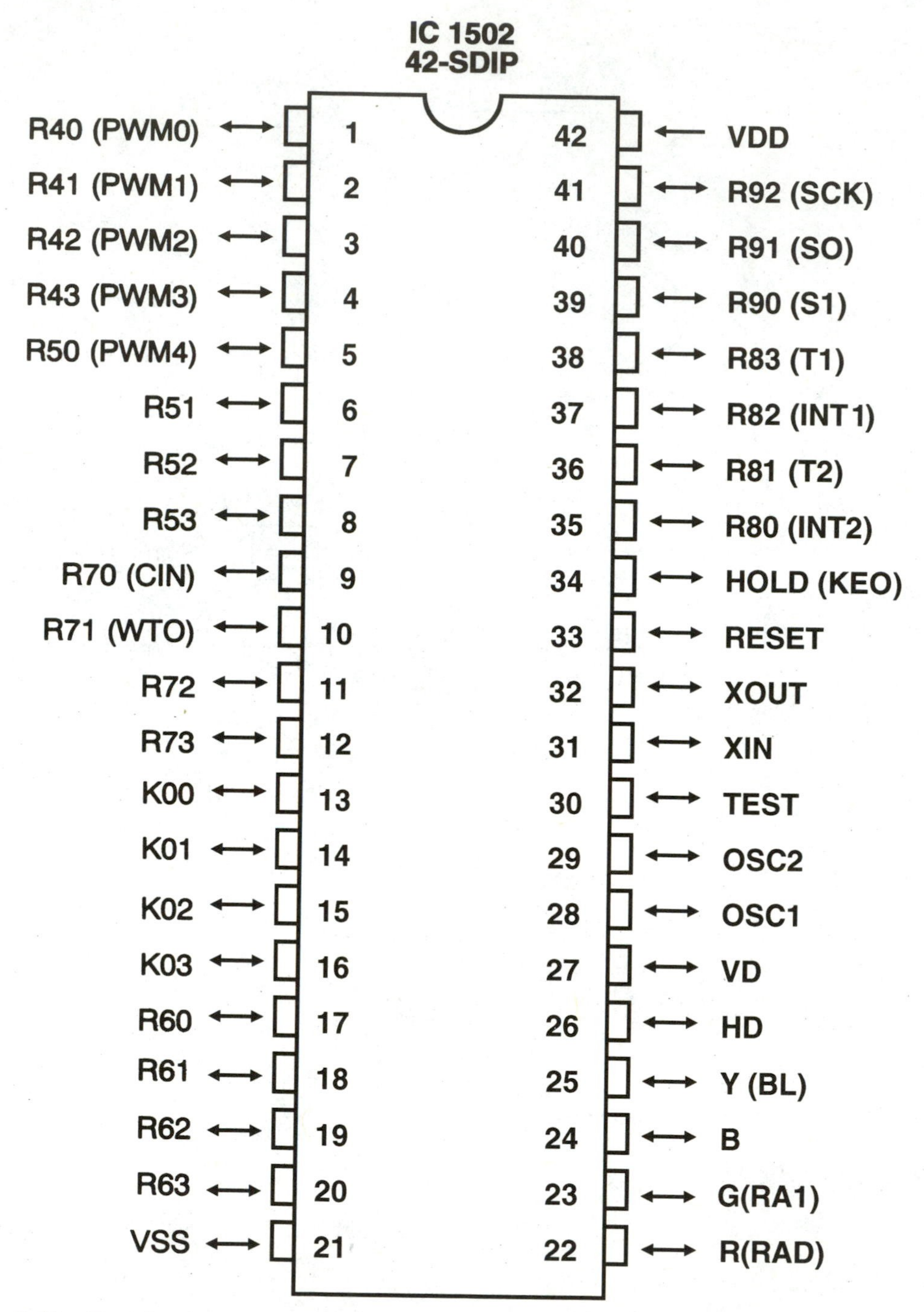

7-12 Zenith tuner control IC chip pin-outs. (*Courtesy of Zenith Corporation.*)

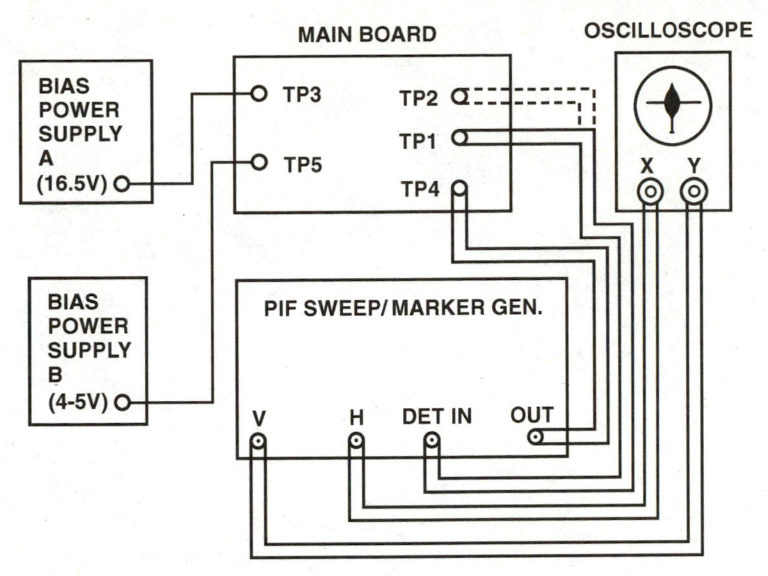

7-13 Equipment setup and connections for picture i.f. sweep alignment. (*Courtesy of Zenith Corporation.*)

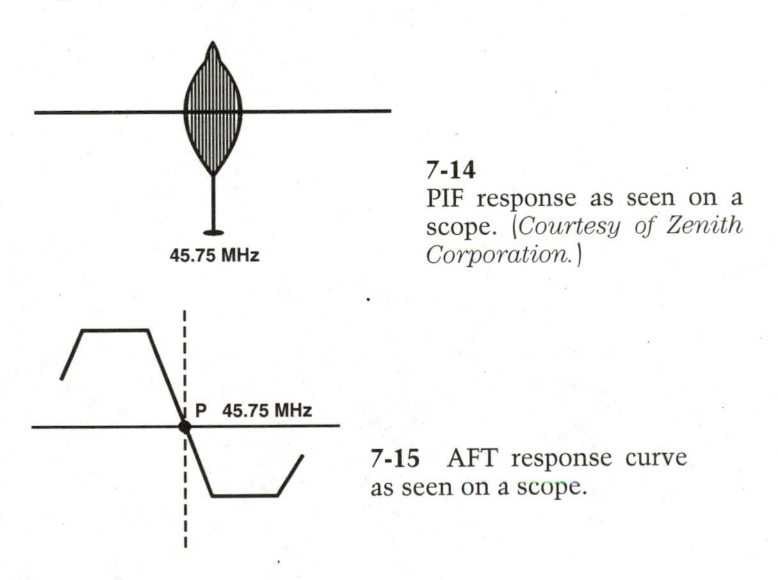

7-14 PIF response as seen on a scope. (*Courtesy of Zenith Corporation.*)

7-15 AFT response curve as seen on a scope.

3. Observe a 1-V peak to peak on scope by adjusting power supply dc bias (4 to 5 V).
4. Adjust PIF coil L205 for maximum display at 45.75 MHz on the scope, as shown in Fig. 7-14.
5. Connect the DET in cable to the TP1 test point.
6. Adjust AFT coil L207 for center display at 45.75 MHz on the scope, as shown in Fig. 7-15.
7. After completing the preceding steps, disconnect the equipment and adjust the AGC delay circuit.

8

CHAPTER

Analyzing video circuit operations

This chapter begins with an explanation of using video test patterns for analyzing color television video and chroma circuits. Then information is presented about how video signals and chroma signals relate to frequency. You will then find out how multiburst bar sweep patterns and chroma bar sweep patterns are used for testing video intermediate frequencies, comb filters, and luminance amplifier circuits. Next, the chapter looks at how to test circuit response by looking at the CRT picture and testing video response and troubleshooting with the oscilloscope.

The chapter continues with how to use dynamic test signals to isolate frequency-response problems in high-performance video stages. Next, a way to analyze chroma bandpass and tint circuits is presented. Then you will find out how to conduct performance tests, and techniques will be presented for aligning modern television sets with comb filters. Then we will see how to analyze color circuit performance by using industry standard color patterns. The chapter concludes with some useful information on direct-coupled (DC) circuits and ways to troubleshoot these "at times" very difficult circuit problems.

Bar sweep patterns

To accurately reproduce analog video scenes, the video i.f., comb filters, luminance amplifiers, color circuits, CRT driver, and video output stages must pass a wide range of video (luminance and chroma) frequencies. The Sencore VG91 video generator (Fig. 8-1) has special analyzing bar sweep video patterns to duplicate the full frequency range that video circuits must process. They are multiburst bar sweep, chroma bar sweep, and luma/chroma bar sweep. These three bar sweep patterns are available on the rf television channel, 45.75-MHz video i.f., standard video circuit block. This section will show you how video frequencies relate to picture quality and how to use these bar sweep video patterns.

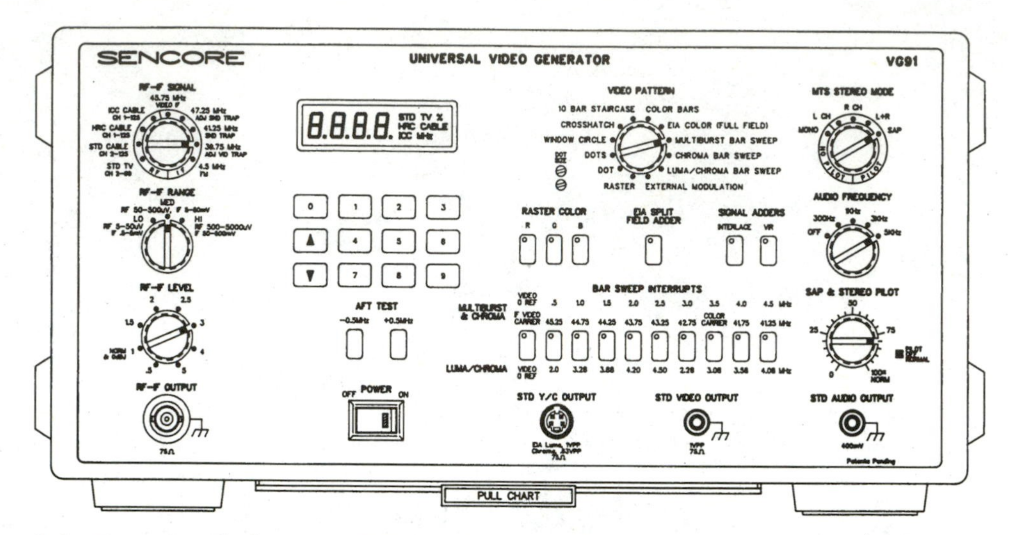

8-1 Front panel of Sencore VG91 universal video generator. (*Courtesy of Sencore, Inc.*)

Using the video analyzing patterns

Many problems in video systems affect the quality of the video that is produced on the CRT. Troubles such as smear, weak contrast, harsh edges, or poor picture detail are tough to track down. Often, these problems do not cause a noticeable difference in the dc bias or peak-to-peak signal voltages in the video circuits but only affect the overall bandwidth or waveshape of the composite video signal.

Problems that degrade the video picture may be related to problems in the video i.f., comb filter, luminance, or chroma circuits. These stages must pass wide bands of luminance and chroma frequencies to reproduce accurately each video scene viewed on the CRT.

Most sources of video patterns do not duplicate the full range of video or chroma signal frequencies. Color bar patterns contain only low-frequency signals. This is why they often do not show video-related performance problems. Video signals from DSS dishes, local television transmitters, and cable television systems change continuously in frequency content, making it difficult to isolate a video stage with poor frequency response from these various signal sources.

With the bar sweep video patterns you may judge the performance of video circuits by looking at the CRT screen or signal trace video problems to the defective circuit. To understand how this is possible, you need to have a better understanding of the bar sweep patterns and how different video and chroma signals relate to frequency.

How video signals relate to frequency

A video signal contains many different frequencies ranging from 0 to 4.2 MHz. Video circuits must be able to treat all the frequencies the same to produce a good picture. Let us now see how video frequencies relate to picture quality.

A television picture is made up of 525 horizontal scan lines, repeating at a rate of 15,734 Hz. Each line changes in amplitude to form the picture as the beam moves across the front of the screen. The video frequency depends on how often the amplitude changes during the scan line. You can calculate the frequency if you know how often the beam is interrupted as it moves across the screen.

To make these calculations, you need to consider the active scan time. This is the time between sync blanking and active video. To determine active video time, divide 1 by 15,734 to get 63.5 μs. Subtract approximately 11.1 μs for blanking and sync for 52.4 μs of active video.

Let us see what happens as we apply video to interrupt the beam, as illustrated in Fig. 8-2. We will start when the screen is completely white. The video signal is now interrupted only by horizontal blanking. Since the signal is at a fixed level for 84 percent of the time, it is more like a dc signal than an ac signal.

Next, let us see what happens when we change the picture by making the left half black and the right half white, as shown in Fig. 8-2. The video signal now becomes a square wave, forming one complete cycle during the active time of each horizontal scan line. Since one scan line takes approximately 52.4 μs and the video completes one full cycle, the video frequency is approximately 19 kHz. This is the lowest video frequency possible.

Now, let us double the interruption rate to form two white stripes of equal size. The time for each cycle is half as long as the first example. This results in twice the first frequency, or 38 kHz. Since we interrupt the screen more often, we can translate the interruptions into frequencies by multiplying the number of interruptions by 19 kHz or by taking the inverse of the scan time for one complete video cycle.

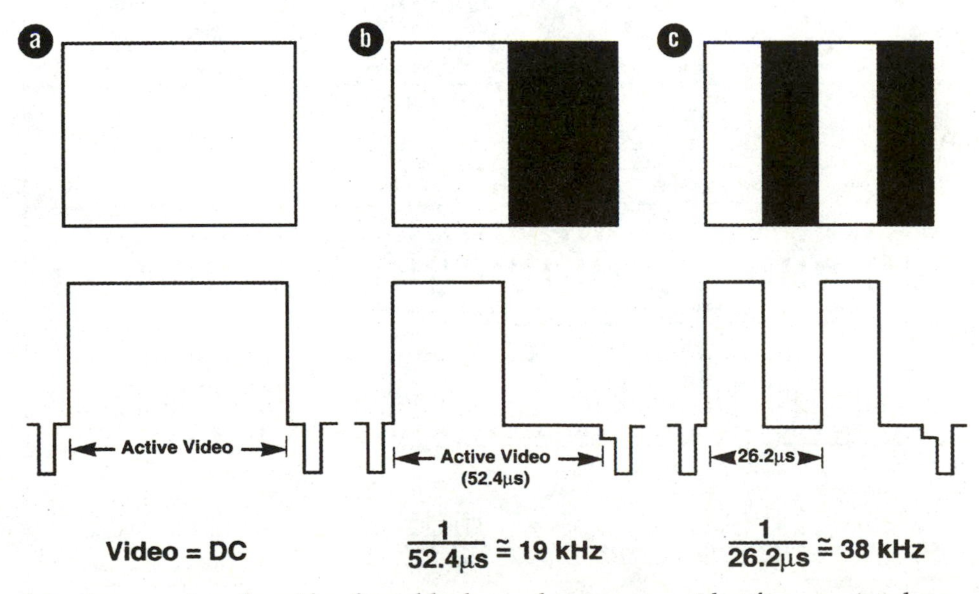

8-2 Interrupting the video from black to white creates video frequencies determined by the time for one cycle. (*Courtesy of Sencore, Inc.*)

Most pictures have a mixture of large and small objects instead of a row of identically sized objects. This causes the video signal to be a mixture of many frequencies. Each different sized picture element produces a signal of the same frequency as it would if it were repeated all the way across the screen. Large objects create low video frequencies and small objects create high frequencies. Figure 8-3 may put this into a better perspective.

The 10 soldiers across the screen are each one-twentieth the scan width of the screen. If we think of a soldier as a high level and the in-between space as a low level, the video is interrupted 10 times across the screen. The 10 soldiers then calculate to 10×19, or 190 kHz.

However, if a video circuit only amplified video frequencies from 0 to 190 kHz, we would not see the soldiers. We would only see their outline, and not a very clear outline at that. We need more bandwidth to see smaller details.

If the soldier's ties are one-fifth the width of the soldier, we need enough bandwidth to place 50 necktie-sized objects (plus equal blank spaces) on the screen. This requires 50 times 19 kHz, or approximately a 1-MHz bandwidth to provide 53 neckties across the screen.

Any picture element that is one-half the size of the necktie creates a video frequency twice that of the necktie, or 2 MHz. A 2-MHz signal represents one ob-

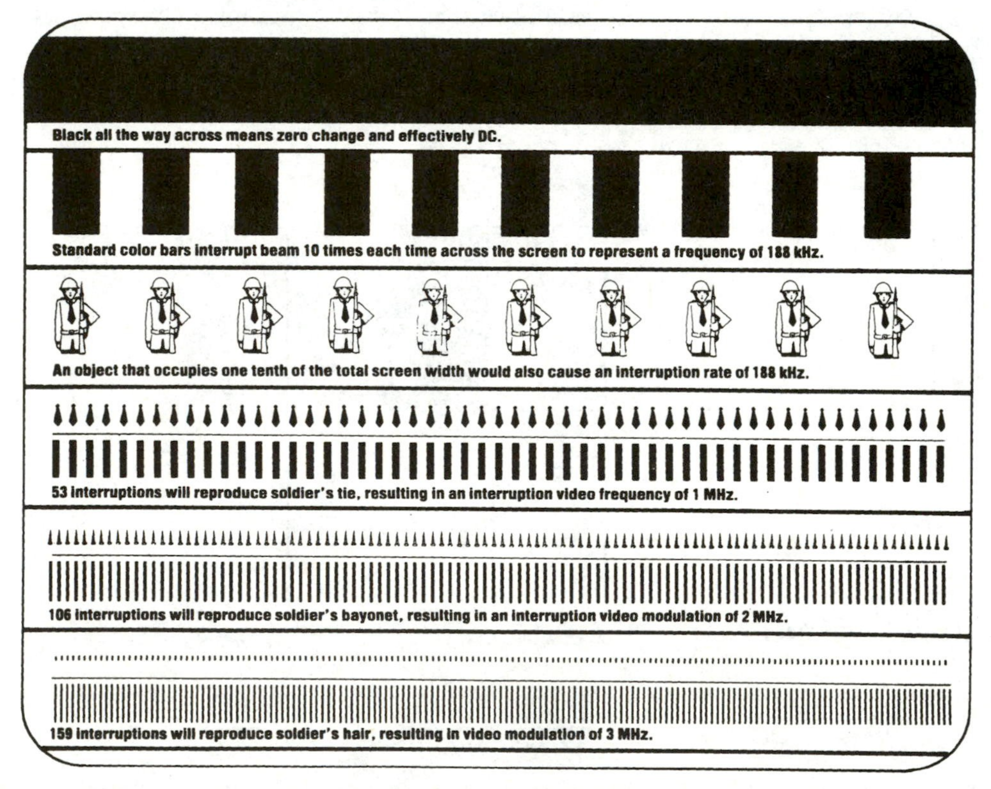

8-3 The presentation of a video picture comprising many video frequencies all mixed together. (*Courtesy of Sencore, Inc.*)

ject 1/106 the screen's width. This might be the bayonet on the end of each soldier's rifle.

Increasing the frequency to 3 MHz allows us to form objects 1/159th the width of the screen, such as the hair on the soldier's head. Since most television receivers have 3 MHz of video response, this is the smallest object they can show on the picture.

A monitor with a direct video input (which bypasses the tuner and i.f. circuits) may provide a response between 4 to 4.5 MHz. If so, 238 objects can be placed across the screen with spaces of the same size.

Relation of chroma to frequency

The color television frequencies included in the composite video signal add color to large and medium-sized objects. The color circuits must treat all color frequencies the same to produce a good color picture. Color frequencies relate to picture quality in a very similar manner as luminance frequencies.

During the horizontal scan lines, the color signals interrupt the conduction of the three color beams to add color to the objects of the picture. The color frequency relates to how often the color beam changes amplitude during the scan line or the size of the object being colored.

In the case of the EIA color bars shown in Fig. 8-4, the bars represent objects one-seventh the width of the screen. The frequency response needed to color objects this size is only approximately 68 kHz. (To solve for this, take the scan time of two bars, one complete signal cycle, which is 15 μs of scan time, and invert it.) To color smaller objects, a higher color frequency is required.

The response of the human eye to color changes with the size of the object. We cannot distinguish color in smaller objects. This makes it possible to reduce the range of color frequencies needed for a given video scene. In terms of a 23-in screen, object sizes ranging from 0.34 to 0.12 in correspond to a half cycle of video, or 1.5 MHz. The human eye is sensitive to objects in this size range only if they are comprised of orange or cyan. Objects 0.34 in or larger correspond to 0.5 MHz and require all primary colors for reproduction. The eye characteristics require only color frequencies ranging from 0 to 1.3 MHz to adequately reproduce color.

The color signals from a video camera are matrixed to produce R-Y and B-Y signals. These signals are increased in frequency and separated by 90 degrees with a balanced modulator centered at 3.58 MHz. This produces I and Q frequency color chroma signals that extend 1.3 MHz below and 0.5 MHz above 3.58 MHz. The chroma sideband signals are added to the luminance signals to form the composite video signals.

Most television receivers use only chroma frequencies 0.5 MHz above and below 3.58 MHz. This range of frequencies reproduces a satisfactory color picture. Some receivers increase the video i.f. bandpass by using a "wide-band" I color demodulator. Those receivers use the wider chroma frequencies that extend 1.3 MHz below 3.58 MHz, making it possible to color smaller objects.

A monitor with a direct video input or an S-video input also makes use of the increased range of color and luminance frequencies.

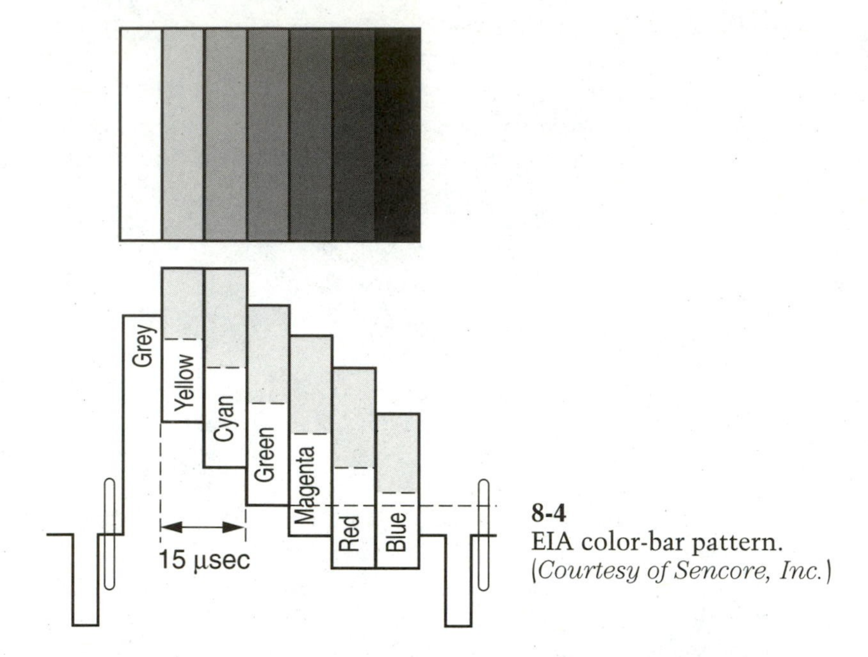

8-4
EIA color-bar pattern.
(Courtesy of Sencore, Inc.)

Using the Sencore VG-91 video generator

The multiburst bar sweep pattern shown in Fig. 8-5 provides 10 frequency bars (bursts) of video, each with a different test frequency. Any frequency-response roll-off or distortion in the video circuits will alter the amplitude or shape of the bars.

The multiburst bar sweep pattern consists of 10 frequency bars beginning with a solid white 0-MHz reference bar and increasing in 0.5-MHz steps to 4.5 MHz. The individual bars (frequencies) interrupt the video between black and white at rates between dc and 4.5 MHz. The bar frequencies sample the full range of luminance (Y) frequencies that video systems process.

Each of the multiburst bar sweep frequencies is generated as a square wave. A square wave needs correct phase and gain response to all its odd harmonics to retain its proper shape. The harmonics of the square waves for each bar frequency effectively fill in the gaps between the 0.5-MHz fundamental test frequencies. This ensures that problems between the fundamental test frequencies are not overlooked, as they might be with sine waves.

Use the multiburst bar sweep pattern to test and isolate frequency-response problems in any stage that must amplify the full range of luminance frequencies. This includes circuits such as video i.f.s, comb filters, and luminance amplifier circuits. The pattern also can be used to align video i.f. stages for the best overall frequency response and gain with minimum picture distortion.

Chroma bar sweep pattern

The chroma bar sweep pattern provides three chroma test bar frequencies. The chroma test frequencies are in the middle and edges of the color bandpass frequencies that the chroma stages must pass. Any response roll-off or distortion will change the amplitude or shape of these color bar frequencies.

The chroma bar sweep pattern consists of three chroma frequency bars at 3.08, 3.5, and 4.08 MHz, as shown in Fig. 8-5. All three color bars are phased as chroma and should appear at the color output of a properly aligned and working comb filter. The 3.08- and 4.08-MHz bars are interrupted with chroma changing at a 500-kHz rate. This produces chroma sideband frequencies at 3.08 and 4.08 MHz.

Chroma circuits with a poor response may result in no color, washed-out color, or color smearing. You can use the chroma bar sweep pattern to test and troubleshoot color problems in video i.f., comb filters, color bandpass circuits, or any chroma processing circuits.

Luma/chroma bar sweep pattern

The luma/chroma bar sweep pattern, shown in Fig. 8-5, provides luminance and chroma phase test frequency bars in a single video pattern. The combination of luminance and chroma signals in a single video pattern closely duplicates a composite video signal with a high luminance and chroma frequency content. The luminance and chroma test frequencies are specially chosen to dynamically test circuits in the band of frequencies where the luminance and chroma frequencies overlap. The frequencies are at the bandpass edges of the luminance and I color signals.

The luma/chroma bar sweep pattern includes six luminance-phased bars: a 0-MHz reference white and 2.0, 3.28, 3.88, 4.2, and 4.5 MHz. The luminance bars are identical to the multiburst bar sweep but differ in frequency. The 4.2-MHz frequency bar is the

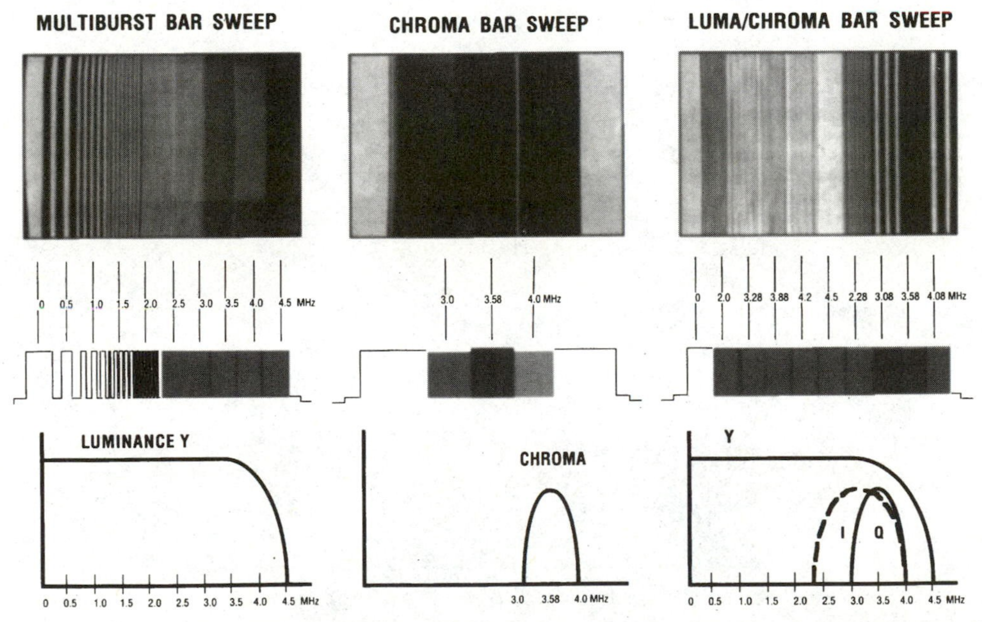

8-5 Bar sweep patterns test for bandpass response of video (luminance and chroma) stages. (*Courtesy of Sencore, Inc.*)

highest luminance frequency that can be transmitted. The 3.28- and 3.88-MHz bars occupy the same frequency that can be transmitted. The 3.28- and 3.88-MHz bars occupy the same frequency spectrum as the chroma frequencies for analyzing comb filter circuits.

Located to the right of the black-and-white luminance bars on the raster are four chroma-phased bars at 2.28, 3.08, 3.58, and 4.08 MHz. The chroma bars are the same as the chroma bar sweep pattern with the addition of a 2.28-MHz chroma frequency bar. This bar represents the highest I chroma sideband that may be transmitted in a composite video signal. These bars are color-phased and should appear on the color output of a properly operating comb filter.

Use the luma/chroma bar sweep pattern to analyze and align comb filters and to analyze other wideband luminance and chroma circuits, such as wideband I color demodulators or circuits fed by S-video (Y/C) inputs.

Identifying test frequencies with bar sweep interrupts

The bar sweep interrupt switches allow you to turn on and off individual bars of the multiburst bar sweep, chroma bar sweep, and luma/chroma bar sweep video patterns. This is useful for easily identifying a particular bar frequency.

> **Note:** Only the 3.0-, 3.5-, and 4.0-MHz bars can be activated in the chroma bar sweep pattern. Refer to Fig. 8-6.

The numbers shown above and below the bar sweep interrupt switches indicate the video, chroma, and equivalent i.f. frequencies produced by the video patterns. The top row of numbers indicates the composite video frequencies when the multiburst bar sweep and chroma bar sweep patterns are selected. The row just above the pushbuttons indicates the equivalent i.f. sideband frequencies. The values below the switches indicate the composite video frequencies produced when the luma/chroma bar sweep pattern is selected.

To identify frequencies, watch the CRT or oscilloscope as you turn the interrupt switches off and on. If the bar is not the one that you want to identify, move over a but-

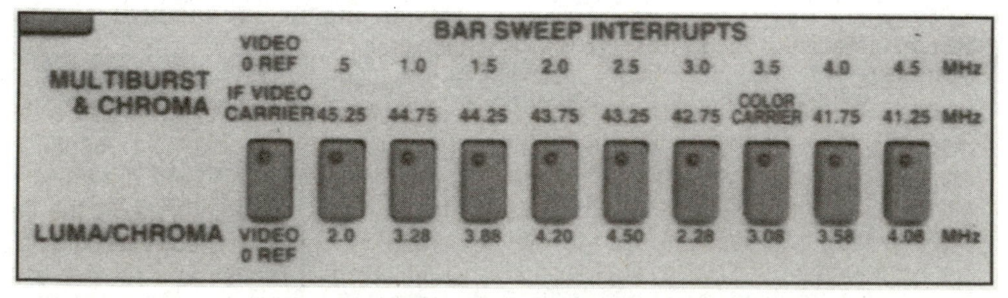

8-6 Bar sweep interrupt switches used to turn the individual bars of the bar sweep patterns on and off. Corresponding video and i.f. frequencies are indicated. (*Courtesy of Sencore, Inc.*)

ton or two until you are interrupting the one you want. Then look above or below the interrupt button and read the video frequency relating to the pattern you have selected.

You also may use the bar sweep interrupt switches to produce an all-white or all-black raster. When all the interrupt buttons are off, the full raster will be back in the multiburst or luma/chroma bar sweep patterns and a white raster will be produced when the chroma bar sweep pattern is selected.

Bar sweep patterns and the i.f. response curve

The i.f. response or gain to the range of i.f. frequencies is commonly shown with an i.f. response curve. The response curve offers a convenient way to indicate the relative amplification of the video frequencies within the bandpass of the video i.f. stages. Although the i.f. response between receivers varies, you may use the ideal response curve shown in Fig. 8-7 as a reference when performance testing or aligning the video i.f. stages using the VG91's bar sweep patterns.

The result of hetrodyning in the tuner reverses the frequency order of the luminance and chroma sideband frequencies in the video i.f. The video carrier is positioned at 45.75 to 41.75 MHz. The 1-MHz band of chroma sidebands ranges from 42.67 to 41.67 MHz.

From the ideal response curve you can see that video frequencies from 45 to 43 MHz receive about the same gain. Typically, the gains of the 45.75-MHz carrier and color carrier (42.17 MHz) are reduced to about 50 percent. The bandpass rolls off quickly to minimize gain to the audio carrier and to the adjacent channel signals.

The frequency bars of the VG91's bar sweep patterns can be used to performance test and align the video i.f. stages. The frequency bars in the bar sweep patterns are converted to i.f. sideband frequencies and receive gain relative to the i.f. response curve. The amplitude of each bar at the output of the video detector provides an indication of the i.f. frequency response curve.

Checking response by the CRT picture

The VG91 outputs each bar of the bar sweep patterns at approximately the same level. A video or chroma stage that has a limited frequency response will reduce the amplitude, distort the shape, or decrease the sharpness or brightness of one or several of the frequency bars. You may use the CRT to tell if each bar sweep frequency has made it to the CRT.

Looking at the CRT, you know that a particular luminance frequency of the multiburst or luma/chroma bar sweep patterns has passed through the video stages if the bar shows detail in the form of black-and-white vertical stripes. In addition, the background brightness (white level) of each bar on the CRT shows how the gain at one test frequency compares with others. Simply lower the brightness or contrast controls to see whether all bars have the normal brightness level. Restricted bars (reduced level) turn dark before others.

The shape of the video i.f. stages affects the squarewave bar sweep frequencies. The sharpness of the black-and-white stripes indicates how well the initial odd harmonics of each bar frequency are restricted by the video i.f. Any excessive gain to

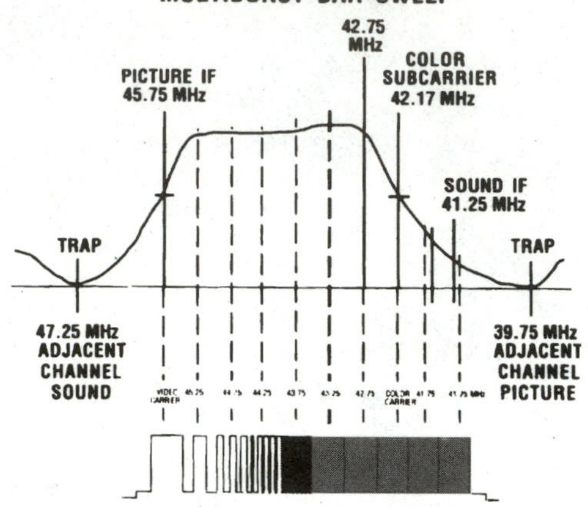

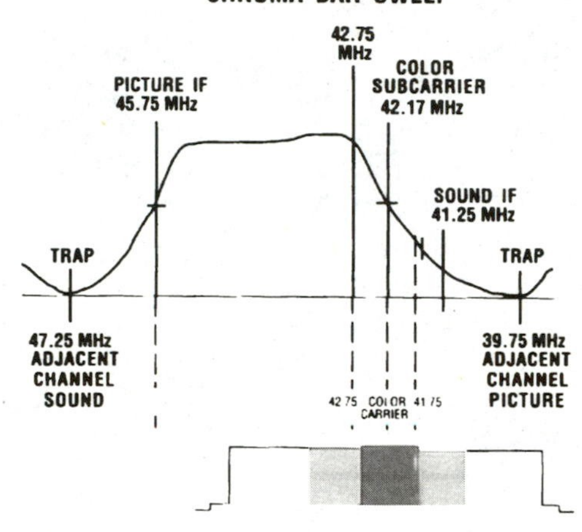

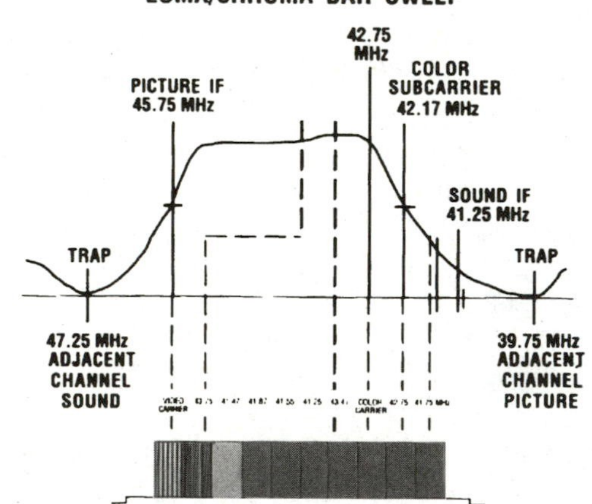

8-7
I.f. gain relative to individual bars of the bar sweep patterns. (*Courtesy of Sencore, Inc.*)

any of the lower-frequency bars produces harsh edges or ringing on the squarewave transition.

You can use a CRT to determine if a color-phased frequency bar of the chroma bar sweep or luma/chroma bar sweep patterns is passing through the video i.f. or chroma circuits. On a CRT, the 3.08- or 4.08-MHz frequency bar shows detailed color stripes if it is passing through the video i.f. stage as shown in Fig. 8-8. You know a luminance- or chroma-phased frequency bar has passed through the video stages if the CRT shows detail in the form of black-and-white or colored vertical stripes. A stage that is restricting the band of chroma signals will reduce the detail of the stripes. In addition, the amplitudes of the bars can be determined by reducing the color level control and observing when each bar turns gray.

A color television receiver should show video response to 3.0 MHz with an rf video signal applied to the antenna input. The response is limited by the i.f. response, which rolls off frequencies above 3 MHz. The 4.5-MHz frequency bar should be trapped out by the 4.5-MHz trap.

A monitor with a direct video input bypasses the bandpass restrictions of the video i.f. circuits. This extends the luminance frequency response as high as 4.0 or even 4.5 MHz. A direct video input provides a wider and more even gain of chroma frequencies as well.

A monitor with a Y/C input (S-video input) provides a wideband luminance and chroma input. The S-video input bypasses the video i.f. and comb filter circuits. Since the luminance and chroma are separate, the luminance response extends a full 4.5 MHz and the chroma a full 1.34 MHz. This produces clear luminance and chroma stripes on any of the bar sweep patterns.

In each case, the bar sweep patterns test every video circuit from the input right to the CRT. Simply look at the CRT to see whether the circuits show normal response.

Troubleshooting video circuits with the scope

The symptoms observed on the CRT during initial video performance checks with the bar sweep patterns point you to suspected circuits. Relate these symptoms

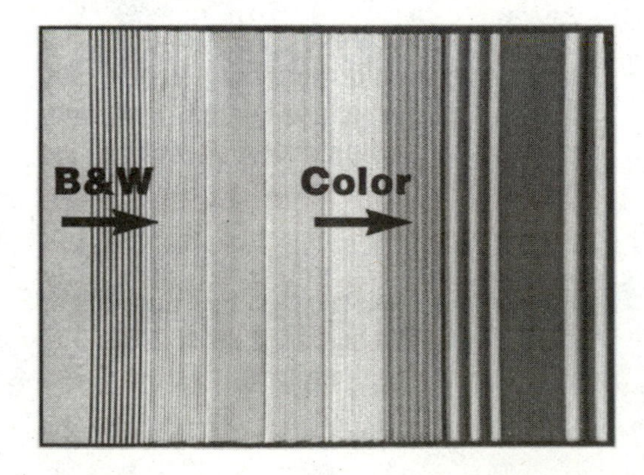

8-8 Determining if a color-phased frequency bar has passed through the video stages. (*Courtesy of Sencore, Inc.*)

to the particular luminance or chroma frequency bar(s), and use an oscilloscope to trace the defective video stage to isolate difficult video problems in the least amount of time.

The frequency bars of the bar sweep patterns repeat every horizontal scan line for easy interpretation on a scope. You will recall that the VG91 outputs the individual frequency bars of the bar sweep patterns at approximately the same level, so use the oscilloscope to analyze the relative level of each frequency bar. Judge the response of the video stage by comparing the amplitude of the frequency bars to the 0 Ref or 3.58-MHz (color carrier) bar. The amplitude of the individual frequency bars of the bar sweep patterns indicate how the video stage affects the range of luminance or chroma frequencies.

Using dynamic test signals

Video involves more than just putting a picture on a CRT. Luminance, chroma, setup, deflection, and audio circuits need to work simultaneously to produce proper output signals. For example, just because a television displays a good black-and-white image does not necessarily mean it will produce an acceptable color image.

The VG91 universal video generator's exclusive analyzing patterns let you identify video problems that are difficult to distinguish when using a conventional color bar signal generator. Analyzing the circuit stages using dynamic troubleshooting patterns can help you quickly isolate video defects just by looking at the CRT. In addition, video or deflection circuit adjustments and CRT setups are faster and more accurate, saving you precious troubleshooting time. Let us now see how to use the VG91 to identify and isolate specific video problems.

Isolating frequency-response problems

Video troubles such as smear, weak contrast, harsh edges, or poor picture detail are tough to track down. These problems often do not cause a noticeable difference in the dc bias or peak-to-peak signal voltages in the video circuits. They only affect the overall bandwidth (frequency response) of the composite video signal, causing poor video on the CRT. The problem could be in the video i.f., comb filter, video, or luminance circuits.

Most television video signals contain a mixture of large and small objects. The objects consist of a mixture of many low and high video (luminance) frequencies. A video circuit must have a wide frequency response (bandwidth) to pass frequencies from dc to 4.0 MHz to reproduce a crisp, detailed video picture.

Conventional color bar patterns contain only low-frequency luminance signals and do not duplicate the full video signal frequency range. This is why they often do not show video-related performance problems. Moreover, since video signals from local transmitters, DSS dishes, and cable systems continuously change in frequency content, it is nearly impossible to isolate a video frequency-response problem using these signal sources.

The VG91's multiburst bar sweep patterns produce a test signal to analyze the full frequency range processed by video circuits. The multiburst bar sweep patterns

(Fig. 8-9*a*) provide 10 video frequency test bars beginning with a solid-white 0-MHz reference bar and increasing in 0.5-MHz steps to 4.5 MHz. The individual bars (frequencies) interrupt the video between black and white. One cycle of the frequency in each bar causes a transition between black and white.

The frequency bars of the multiburst bar sweep patterns are an indication of picture quality. To test how well a stage is passing the full range of luminance frequencies, you apply this pattern to the input of the video stage. Then use the CRT or view the waveform on a scope to see if the frequency bars are passing through the video i.f. and luminance circuits. A video stage that has a limited frequency response will reduce the amplitude or distort the shape of one or several bars. On a CRT, you know a frequency has passed if the bar shows detail in the form of black-and-white vertical stripes. In addition, the brightness of each bar shows how the gain at one frequency compares with others. Frequency bars that are reduced in level will turn dark before the others. On a scope, the amplitudes of the individual bars will indicate the response of the video stages.

Analyzing chroma bandpass and tint circuits

Color frequencies relate to picture quality in a similar manner to video (luminance) frequencies. To reproduce color, chroma stages must pass a 1-MHz bandwidth of chroma signals between 3.0 and 4.0 MHz. Chroma circuits with poor response result in no color, washed-out color, or color smearing.

The chroma bar sweep pattern of the VG91 provides three chroma test bar frequencies at 3.08, 3.5, and 4.08 MHz. These chroma test frequencies are in the middle and edges of the color bandpass. All three color bars are phased as chroma and should appear at the color output of a properly aligned and working comb filter. The

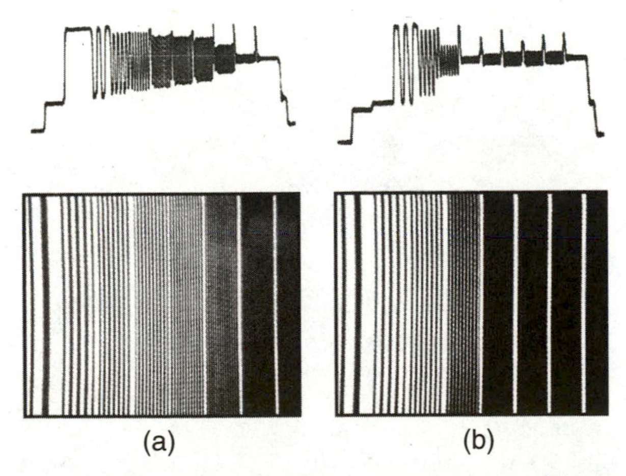

8-9 (*a*) Normal video frequency response. (*b*) Insufficient low-frequency video response resulting in reduced picture brightness and detail. (*Courtesy of Sencore, Inc.*)

3.08- and 4.08-MHz bars are interrupted with chroma changing at a 500-kHz rate. This produces chroma sideband frequencies at 3.08 and 4.08 MHz.

The middle (3.58-MHz) frequency bar is 75 percent color-saturated cyan that matches the phase and cyan bar of the EIA color bar pattern. This bar is not interrupted and should appear as a constant cyan color on the CRT screen. Cyan is the largest-amplitude color signal that a color circuit must pass without clipping or limiting. This is important when setting chroma levels such as the record chroma level in a VCR. The right and left sides of the chroma bar sweep provide a white reference level for special tests and alignment.

Use the chroma bar sweep pattern for testing and isolating color problems in video i.f., comb filters, color bandpass circuits, or any chroma processing circuits. Any frequency-response roll-off or distortion will change the amplitude or shape of the color frequency bars. On a CRT, you know the 3.08- or 4.08-MHz frequency bars have passed through the stage if the bar shows detailed color stripes. Reduced detail of the stripes indicates some reduction in signal amplitude. On a scope, the amplitudes of the individual bars will indicate the response of the stages to the color frequency bars. Note the bar pattern in Fig. 8-10.

Testing and aligning modern comb filters

The NTSC video system interleaves color and luminance information in the same bandpass (2.5 to 4.0 MHz). TV/video systems separate the luminance and chroma portions of the video signal using either inductor/capacitor filters or more modern comb filters.

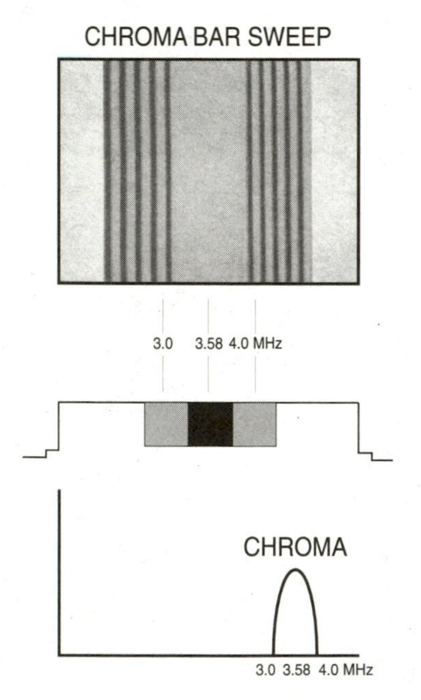

8-10
Chroma bar sweep pattern with three chroma test bar frequencies to troubleshoot color problems in video i.f., comb filter, and color bandpass circuits. (*Courtesy of Sencore, Inc.*)

The VG91's luma/chroma bar sweep pattern improves testing and simplifies alignment of comb filters. The luma/chroma bar sweep pattern differs from the multiburst bar sweep and chroma bar sweep patterns in two ways. First, the information for the luminance and chroma frequency-response test is combined into a single pattern. Second, the test frequencies are chosen to analyze the area of the bandpass where luminance and chroma signals are interleaved. Furthermore, the frequency range of the color bars is extended to test the lower color bandpass to 1.2 MHz.

The luma/chroma bar sweep pattern contains six luminance and four chroma frequency bars. The six luminance-phased bars include a 0-MHz reference white and 2.0-, 3.28-, 3.88-, 4.2-, and 4.5-MHz test frequencies. These should appear as black-and-white stripes on the left side of the CRT display. Located to the right of the black-and-white luminance bars are the chroma-phased bars (2.28, 3.08, 3.58, and 4.08 MHz). The 2.28-MHz chroma frequency bar represents the highest chroma sideband that may be transmitted in the composite video signal. As with the chroma bar sweep pattern, the 3.58-MHz bar is 75 percent color-saturated cyan. Note the pattern drawing shown in Fig. 8-11.

The frequency bars of the luma/chroma bar sweep pattern are output from the VG91 at approximately equal levels. If the luminance and chroma circuits have sufficient bandwidth to pass the frequency bars, they will show distinct vertical stripes on the CRT. When viewed on a scope, the amplitude of the frequency bars indicates the bandpass or response of the stage.

Analyzing color performance

With the VG91 you can analyze color television performance with industry-standard color patterns. Reproducing the transmitted color signals in their proper phase and amplitude depends on the proper operation of all the chroma circuits. Problems in the chroma amplifiers, 3.58-MHz oscillator, demodulator, matrix amplifiers, or any associated circuits can cause improper or missing colors. Service literature and manufacturer alignment procedures reference two industry-adopted color bar test patterns. These patterns provide color bars with different phases and amplitudes to test the chroma processing circuits.

The VG91's color bars pattern is improved over conventional color bar generators, ensuring proper performance with today's color circuits. The three main improvements that give the VG91 an edge are

1. A true color burst at the proper phase, amplitude, and location on the "back porch" of the horizontal sync.
2. The color information is phase locked to the horizontal sync for proper detection by comb filters.
3. The color information is at a 100 percent saturated level.

The EIA patterns produced by the VG91 can be used with a vectorscope and waveform monitor to analyze the relative amplitudes and phases of the color level (saturation).

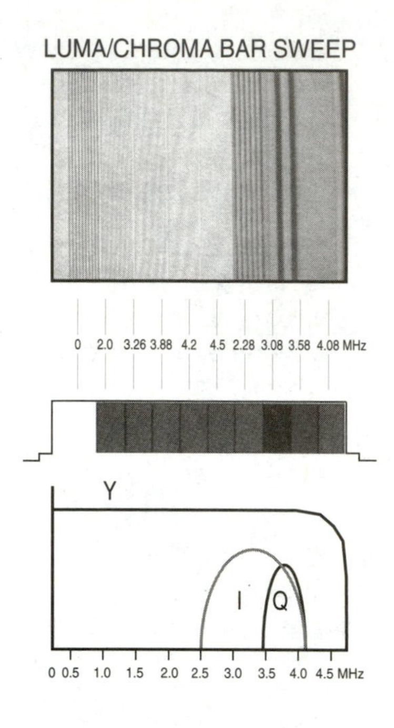

8-11
Luma/chroma bar sweep pattern. (*Courtesy of Sencore, Inc.*)

Servicing direct-coupled (DC) video circuits

Every day we are seeing more equipment that utilizes more direct-coupled (DC) solid-state circuit stages. This type of coupling is simple, and solid-state devices are easily adapted to this circuitry. Many service technicians first noted these direct-coupled stages in portable radio audio-output circuitry. Next came complete direct-coupled amplifiers for stereo audio systems. Now, with the advent of solid-state television, even more direct-coupled circuits are being used.

In the direct-coupled circuit shown in Fig. 8-12, resistor R3 serves as both the collector load for transistor Q1 and the bias resistor for transistor Q2. Resistor RL is the output load of the amplifier. If another stage were added, RL would serve the same function as R3, collector load for Q2 and bias supply for the added stage. Resistors R1 and R2 enhance circuit stability by providing a feedback path. Stability is of great importance in direct-coupled stages because temperature-caused bias variations in one stage will be amplified by all the following stages, resulting in some temperature instability problems. This may at times be the limiting factor for the number of stages that can be direct coupled.

Servicing direct-coupled stages requires a technique that is somewhat different from those used to troubleshoot R-C or transformer-coupled stages. A different approach is needed because each stage depends on the proceeding stage for bias. If several stages are direct coupled, a defect causing incorrect operation of one will affect the bias of the next and therefore all succeeding stages. Normal signal injection/tracing techniques cannot completely isolate the trouble.

The actual troubleshooting technique involves checking individual transistor element bias potentials. Start at the output stage of the circuit in question, and check

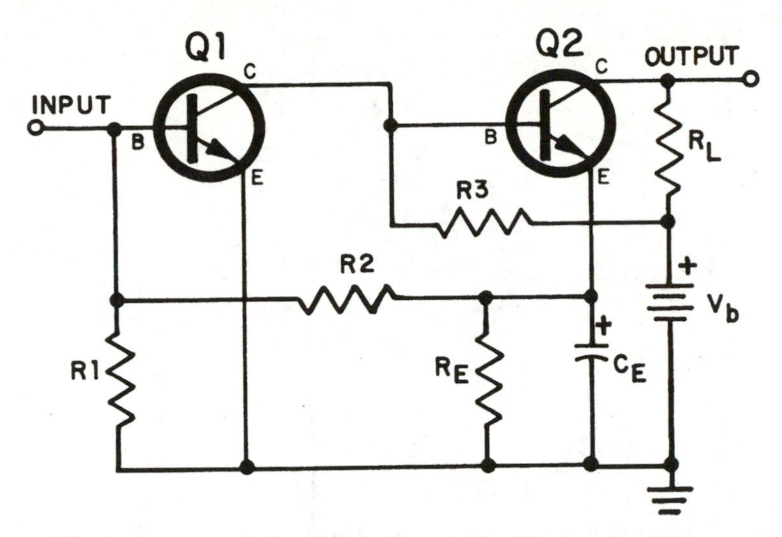

8-12 Direct-coupled solid-state circuit. (*Courtesy of Sencore, Inc.*)

back through each stage to the input of the circuitry involved. Each reading should be noted (written down, if necessary) and then compared with both the expected normal reading and the readings taken at other points in the circuit. The starting point for troubleshooting the circuitry shown in Fig. 8-12 would be to measure the voltage drop across RL to determine the operating conditions of transistor Q2. Little or no drop would indicate nonconduction, while heavy conduction would cause a large drop. In either case, it is necessary to measure the bias potentials on Q1 to determine whether the trouble is actually in the circuitry of Q2 or caused by incorrect bias supplied to Q2 as a result of Q1 circuit defects.

As an example, if tests indicate that Q2 is conducting heavily, the cause could be transistor Q2's emitter-to-collector leakage; however, the symptom would be similar if a defective Q1 supplied excessive bias to the base of Q2. The defect can be isolated to a particular stage by checking the bias potentials of all transistors in the circuit, starting at the output, and comparing these with the normal expected potentials and with each other. Different methods of direct coupling may be encountered. One such circuit is illustrated in Fig. 8-13. The servicing technique described above is also applicable to this method and to most other direct-coupled circuits.

Do not overlook the possibility of a defect in one stage supplying excessive bias to one or more succeeding stages, thereby causing other devices to fail. This is referred to as the *domino effect.* Some circuit designs have current-limiting built in to prevent subsequent *device* failure. In other designs, multiple device failure is quite possible. Remember, however, a logical analysis of all the direct-coupled bias potentials in a circuit will greatly simplify the troubleshooting procedure, regardless of the circuit configuration.

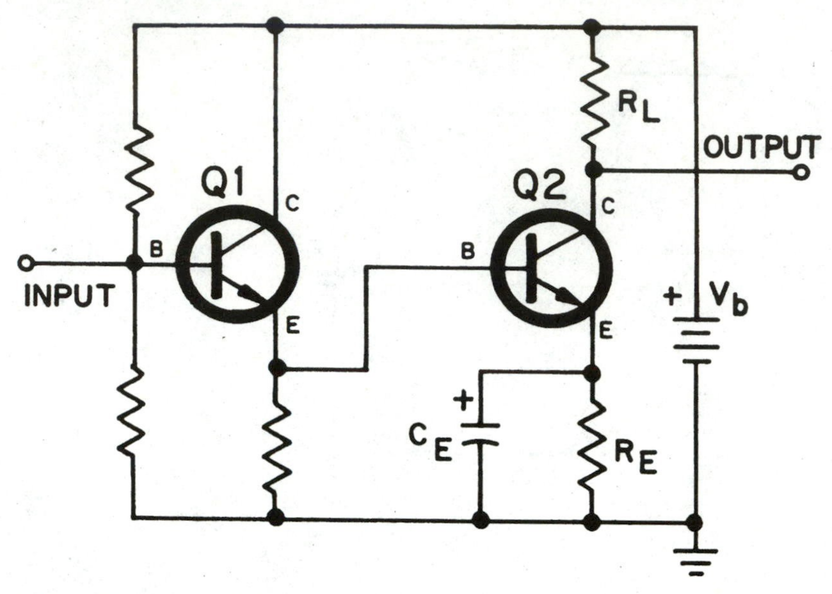

8-13 Direct-coupled transistor circuit. (*Courtesy of Sencore, Inc.*)

9 CHAPTER

Digital television video concepts

At this writing, AAVS by Sencore has just introduced the DSA309 digital studio video analyzer (Fig. 9-1). Designed for quality control monitoring in the television studio for performance testing and troubleshooting, the DSA309 digital studio analyzer is the first analyzer to include complete component and composite digital video monitoring and testing for both 525- and 625-line standards in one package. All the digital video signal tests are performed continuously and provide real-time, on-line measurements that operators can use to improve the quality of the final product. Television engineers can use these measurements to conduct system checks, do setups and adjustments, and ensure continuous, error-free operation of the video system.

Real-time measurements

Continuous real-time, on-line measurements of all key parameters permit live monitoring of the video for the following:

- Serial jitter
- Signal amplitude
- Color levels
- Nonrecommended value errors
- EDH errors
- Parity bit errors
- TRS errors
- Bit activity
- Reserved code errors

This chapter continues with digital and analog observations and the digital standard for component signal digitization. Other subjects covered include digital terrestrial television broadcasting, high-level encoding equipment, source coding and

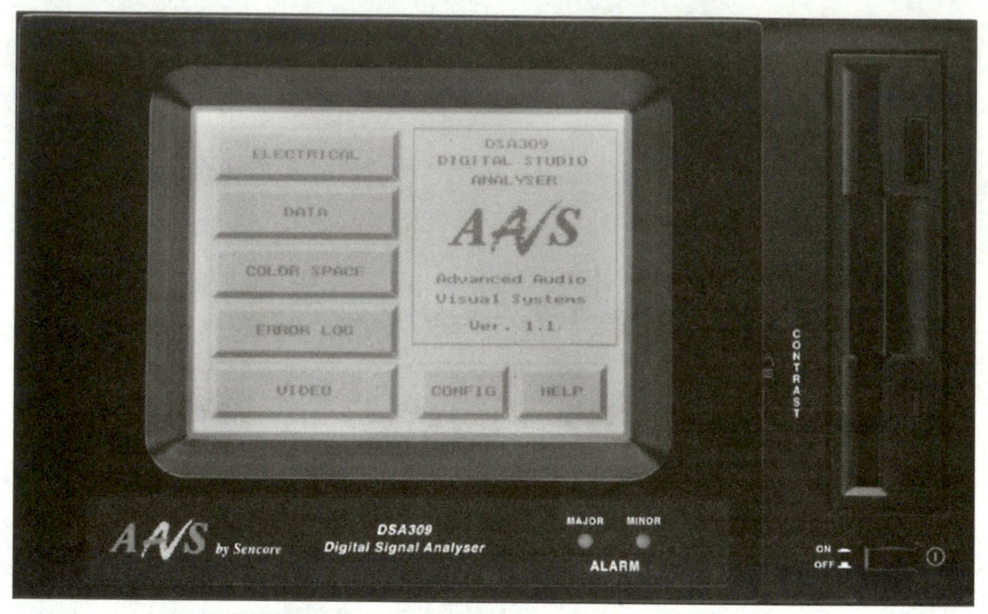

9-1 Sencore DSA309 digital studio video analyzer. (*Courtesy of AAVS, a Division of Sencore, Inc.*)

compression, MPEG-2 levels and profiles, video processing, sampling rates, colorimetry, film modes, antialias filtering, pixels, and square pixels.

The chapter continues with details of digital signal processing, component digital video information, and composite digital video theory.

Sencore and other test equipment companies are now developing instruments to help the video electronics technician troubleshoot these new digital video television receivers that will be coming into use in the near future.

Digital and analog observations

One of the problems with analog video is the need to maintain an accurate waveform throughout the distribution system. The analog signal progressively degrades as noise and cable attenuation accumulate, and the signal cannot be returned to its original integrity. However, digital video can be restored as long as the data can be recognized. The signal also can be manipulated for special effects in the digital domain without incurring any signal degradation.

The digital standard that will be discussed in this chapter is for component signal digitization, also referred to as D1, 4:2:2, and CCIR601. This standard defines the interface for digital television equipment using the Y, R-Y, and B-Y signals with provisions for 8- or 10-bit systems. The interface signals are transmitted on balanced conductor pairs in a parallel format.

The digitization process is performed on the component video signals Y, R-Y, and B-Y. The R-Y and B-Y signals that are digitized are actually gamma precorrected signals, Cr and Cb, which are 0.713 MHz (R-Y) and 0.564 MHz (B-Y), respectively. The sampling frequency for the luminance (Y) signal is 13.5 MHz, while the color difference signals (Cr and Cg) are each sampled at 6.75 MHz. If we define this in number of samples per line, we have 858 samples of luminance information, 720 during active time, and 429 samples of each color difference signal, 360 during active time. This information is then combined by multiplexing the signals in the following order: Cb, Y, Cr, Y, Cb. This gives us a total of 1716 (1440 active) samples per line, for an interface clock frequency of 27.0 MHz (Fig. 9-2).

Another advantage of digital video is that the horizontal and vertical blanking periods do not need to be digitized. Synchronization is accomplished by using a set of four restricted words at the end of active video (EAV) and at the start of active video (SAV). The first three words in each set make up a fixed preamble and are defined as 3FF 000 000h. The fourth word is used for odd/even field identification (F), active video/vertical blanking designation (V), EAV/SAV designation (H), and four protection bits that depend on the state of the F, V, and H bits.

Since the horizontal and vertical blanking periods are not digitized, ancillary data such as digitized stereo audio can be inserted into this period. The only requirement for ancillary data is that the information be preceded by a three-word header defined as 000 3FF 3FFh.

The signal characteristics of D1 video are defined as 11 signals (10 data bits and clock) transmitted via balanced signal pairs with compatible levels as determined by the Electronic Components Laboratory (ECL) (Fig. 9-3) to permit the use of ECL parts for either or both ends. The clock is a 18.5±3 ns clock, and the peak-to-peak

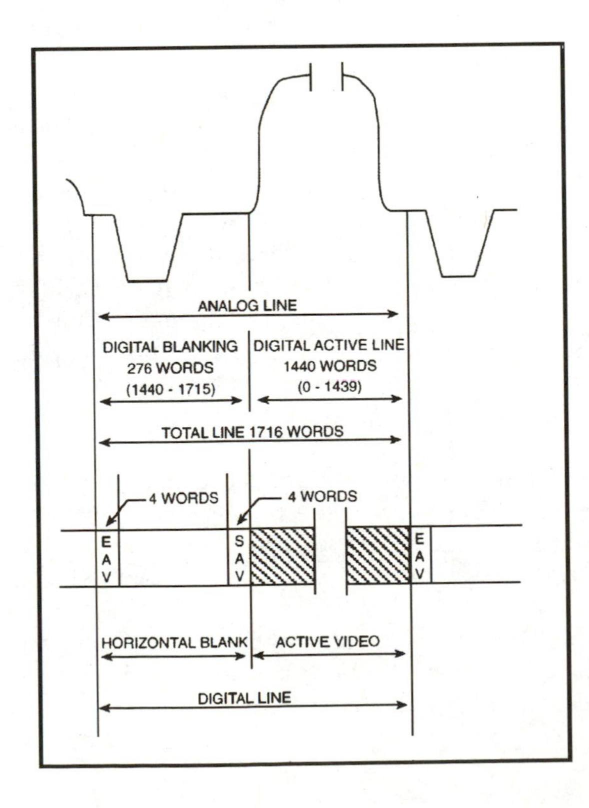

9-2
Timing relation. (*Courtesy of AAVS, a Division of Sencore, Inc.*)

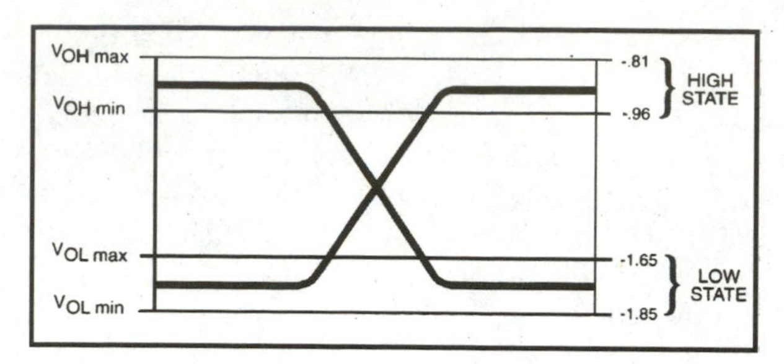

9-3 ECL levels. (*Courtesy of AAVS, a Division of Sencore, Inc.*)

jitter between positive edges must be within 3 ns of the average time of the rising edges computed over at least one field. Data transitions occur on the negative edge of the clock and should be captured on the positive edge, as shown in Fig. 9-4.

And then there is the serialization test of the parallel video, which allows the transfer of data to be carried on a single coaxial cable rather than on 24 conductors. This makes the routing of the video signals much easier and allows some of the existing cabling to be used. The serial data rate is 270 Mbits/s (10 bits at 27 MHz). At this rate, equalization may be required at the receiving end of the signal to correct for typical cable losses.

Digital formats are rapidly becoming the preferred method for the transfer of video information in many applications and will require adherence to many standards for both transmission and testing.

Digital system overview

The *Digital Television Standard* describes a system designed to send high-quality video and audio and ancillary data over a single 6-MHz channel. The system can deliver reliably about 19 Mb/s of throughput in a 6-MHz terrestrial broadcasting channel and about 38 Mbps of throughput in a 6-MHz cable television channel. This means that encoding a video source whose resolution can be as high as five times that of conventional television (NTSC) resolution requires a bit rate reduction by a factor of 50 or higher. To achieve this bit rate reduction, the system is designed to be efficient in utilizing available channel capacity by exploiting complex video and audio compression technology.

The objective is to maximize the information passed through the data channel by minimizing the amount of data required to represent the video image sequence and its associated audio. The objective is to represent the video, audio, and data sources with as few bits as possible while preserving the level of quality required for the given application.

Although the rf/transmission subsystems described are designed specifically for terrestrial and cable applications, the objective is that the video, audio, and service multiplex/transport subsystems be useful in other applications.

System block diagram

A basic block diagram representation of the system is shown in Figs. 9-5 and 9-6. This representation is based on one adopted by the International Telecommunication Union, Radiocommunication Sector. According to this model, the digital television system can be seen to consist of three subsystems:

1. Source coding and compression
2. Service multiplex and transport
3. rf/transmission.

Source coding and compression refers to the bit rate reduction methods, also known as *data compression*, appropriate for application to the video, audio, and ancillary digital data streams. The term *ancillary data* includes control data, conditional access control data, and data associated with the program audio and video services, such as closed captioning.

Ancillary data also can refer to independent program services. The purpose of the coder is to minimize the number of bits needed to represent the audio and video information. The digital television system employs the MPEG-2 video stream syntax

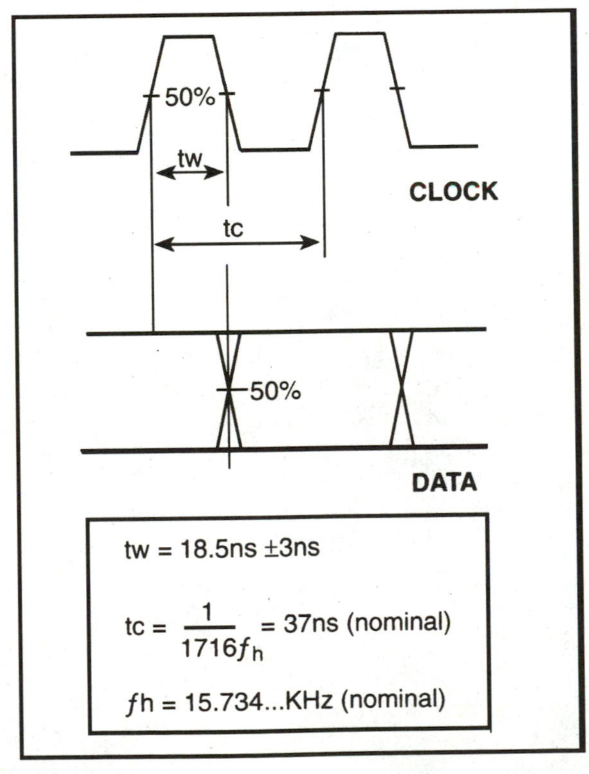

9-4 Clock-to-data timing. (*Courtesy of AAVS, a Division of Sencore, Inc.*)

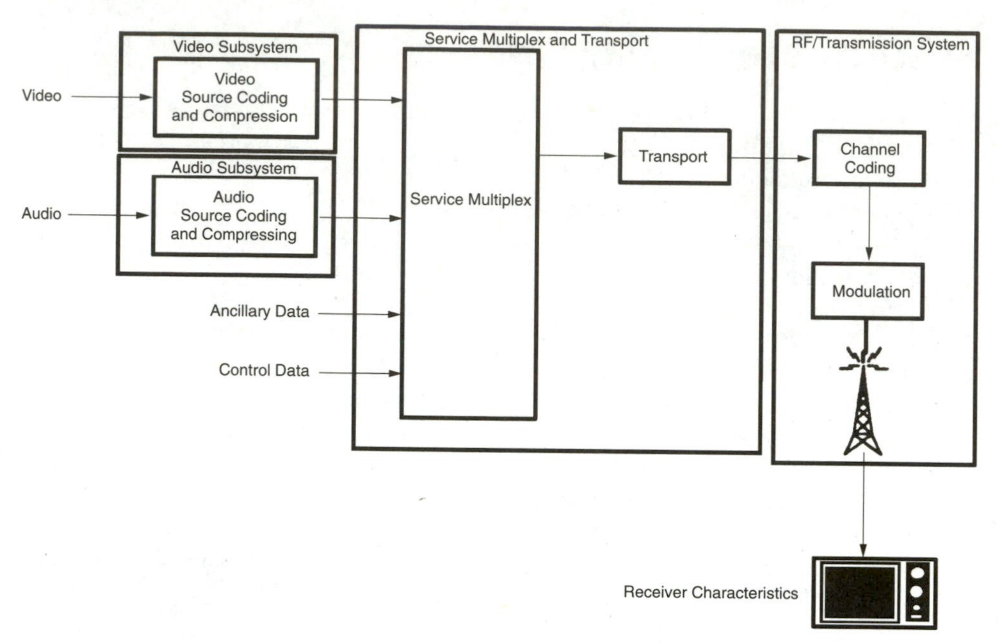

9-5 ITU-R digital terrestrial television broadcasting model. (*Courtesy of AAVS, a Division of Sencore, Inc.*)

for the coding of video and the *Digital Audio Compression Standard* (AC-3) for the coding of audio.

Service multiplex and transport refers to the means of dividing the digital data stream into "packets" of information, the means of uniquely identifying each packet or packet type, and the appropriate methods of multiplexing video data stream packets, audio data stream packets, and ancillary data stream packets into a single data stream. In developing the transport mechanism, interoperability among digital media, such as terrestrial broadcasting, cable distribution, satellite distribution, recording, and computer interfaces, was a prime consideration. The digital television system employs the MPEG-2 transport stream syntax for the packetization and multiplexing of video, audio, and data signals for digital broadcasting systems. The MPEG-2 transport stream syntax was developed for applications where channel bandwidth or recording media capacity is limited and the requirement for an efficient transport mechanism is paramount. It also was designed to facilitate interoperability with the ATM transport mechanism.

Rf/transmission refers to channel coding and modulation. The channel coder takes the data bit stream and adds additional information that can be used by the receiver to reconstruct the data from the received signal, which, due to transmission impairments, may not accurately represent the transmitted signal. The modulation (or physical layer) uses the digital data stream information to modulate the transmitted signal. The modulation subsystem offers two modes: a terrestrial broadcast mode (8 VSB) and a high-data-rate mode (16 VSB).

Illustrated in Fig. 9-7 is a high-level view of the encoding equipment. This block diagram is not intended to be complete but is used to illustrate the relationship of various clock frequencies within the encoder. There are two domains within the encoder where a set of frequencies are related, the source-coding domain and the channel-coding domain.

The source-coding domain, represented schematically by the video, audio, and transport encoders, uses a family of frequencies based on a 27-MHz clock (f27 MHz). This clock is used to generate a 42-bit sample of the frequency that is partitioned into two parts defined by the MPEG-2 specifications. These are the 33-bit program_clock_reference_base and the 9-bit program_clock_reference_extension. The former is equivalent to a sample of a 90-kHz clock that is locked in frequency to the 27-MHz clock and is used by the audio and video source encoders when encoding the presentation time stamp (PTS) and the decoded time stamp (DTS). The audio and video sampling clocks, f_a and f_v, respectively, must be frequency locked to the 27-MHz clock. This can be expressed as the requirement that there exist two pairs of integers (n_a) and (n_v, m_v) such that

$$f_a = (n_a \backslash m_a) \times 27 \text{ MHz}$$

$$f_v = (n_v \backslash m_v) \times 27 \text{ MHz}$$

The channel-coding domain is represented by the FEC/sync insertion subsystem and the VSB modulator. The relevant frequencies in this domain are the VSB symbol frequency (f_{sym}) and the frequency of the transport stream (f_{TP}), which is the frequency of transmission of the encoded transport stream. These frequencies must be locked, having the relationship

$$f_{TP} = 2 \times (188 \backslash 208)(312 \backslash 313) f_{\text{sym}}$$

The signals in the two domains are not required to be frequency locked to each other and in many implementations will operate asynchronously. In such systems,

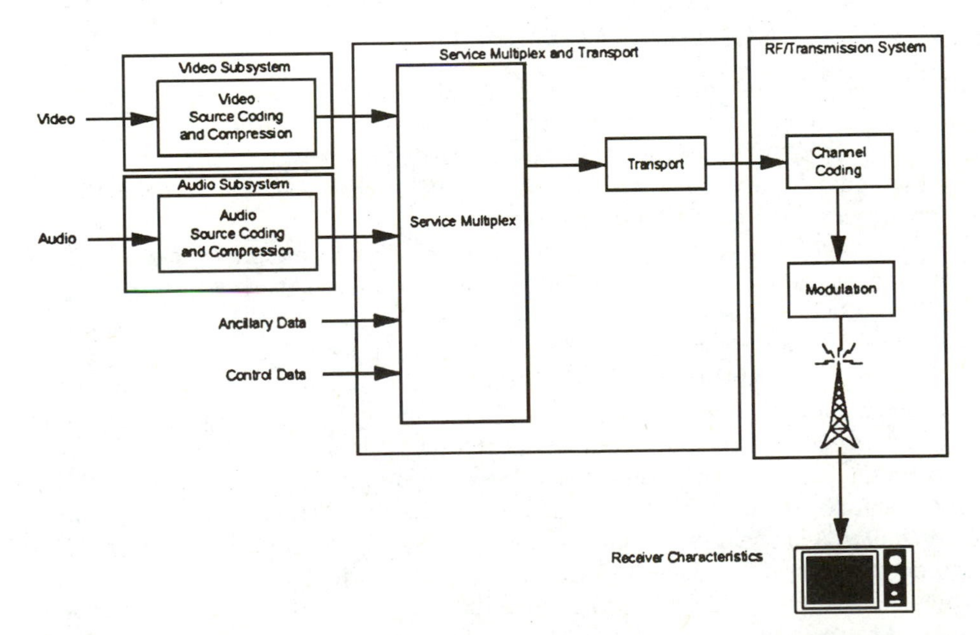

9-6 High-level view of encoding equipment. (*Courtesy of AAVS, a Division of Sencore, Inc.*)

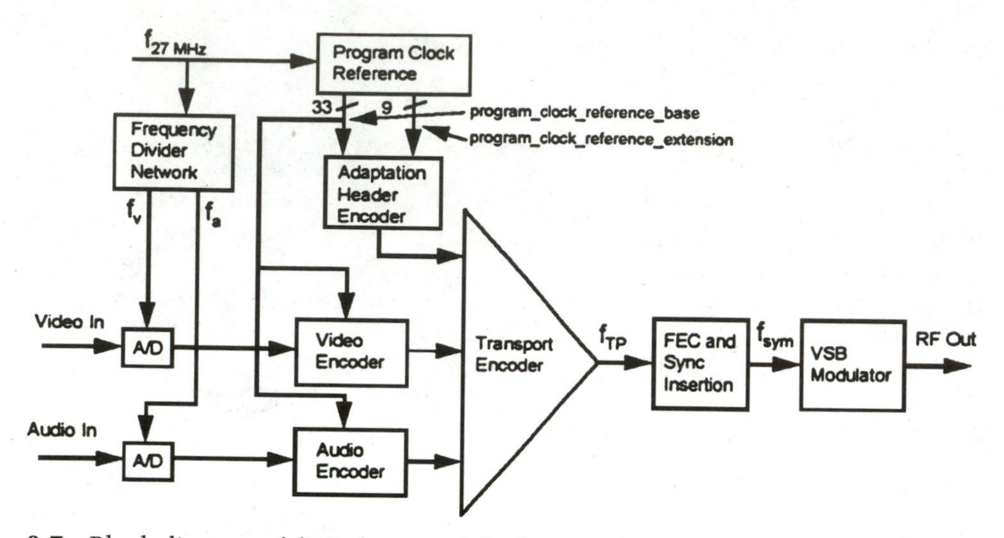

9-7 Block diagram of digital terrestrial television broadcasting model. (*Courtesy of AAVS, a Division of Sencore, Inc.*)

the frequency drift can necessitate the occasional insertion or deletion of a null packet from within the transport stream, thereby accommodating the frequency disparity.

Overview of video compression and decompression

The need for compression in a digital high-definition television (HDTV) system is apparent from the fact that the bit rate required to represent an HDTV signal in uncompressed digital form is about 1 Gb/s, and the bit rate that can reliably be transmitted within a standard 6-MHz television channel is about 20 Mb/s. This implies a need for about a 50:1 or greater compression ratio.

The *Digital Television Standard* specifies video compression using a combination of compression techniques, and for reasons of compatibility, these compression algorithms have been selected to conform to the specifications of MPEG-2, which is a flexible internationally accepted collection of compression algorithms.

The purpose of this tutorial exposition is to identify the significant processing stages in video compression and decompression, giving a clear explanation of what each processing step accomplishes but without including all the details that would be needed to actually implement a real system. The necessary details in every case are specified in the normative part of the standard documentation, which shall in all cases represent the most complete and accurate description of video compression. Because the video coding system includes a specific subset of the MPEG-2 toolkit of algorithmic elements, another purpose of this tutorial is to clarify the relationship between this system and the more general MPEG-2 collection of algorithms.

MPEG-2 levels and profiles

The MPEG-2 specifications are organized into a system of profiles and levels so that applications can ensure interoperability by using equipment and processing that adhere to a common set of coding tools and parameters. The *Digital Television Standard* is based on the MPEG-2 *main profile.* The main profile includes three types of frames for prediction (I-frames, P-frames, and B-frames) and an organization of luminance and chrominance samples (designated 4:2:0) within the frame. The main profile does not include a scalable algorithm, where *scalability* implies that a subset of the compressed data can be decoded without decoding the entire data stream. The *high level* includes formats with up to 1152 active lines and up to 1920 samples per active line and for the main profile is limited to a compressed data rate of no more than 80 Mb/s. The parameters specified by the *Digital Television Standard* represent specific choices within these constraints.

Compatibility with MPEG-2

The video compression system does not include algorithmic elements that fall outside the specifications for MPEG-2 main profile. Thus video decoders that conform to the MPEG-2 can be expected to decode bit streams produced in accordance with the *Digital Television Standard.* Note that it is not necessarily the case that all video decoders that are based on the *Digital Television Standard* will be able to properly decode all video bit streams that comply with MPEG-2.

The video compression system takes in an analog video source signal and outputs a compressed digital signal that contains information that can be decoded to produce an approximate version of the original image sequence. The goal is for the reconstructed approximation to be imperceptibly different from the original for most viewers, for most images, for most of the time. In order to approach such fidelity, the algorithms are flexible, allowing for frequent adaptive changes in the algorithm depending on scene content, history of the processing, estimates of image complexity, and perceptibility of distortions introduced by the compression.

Figure 9-8 shows the overall flow of signals in the ATV system. Note that analog signals presented to the system are digitized and sent to the encoder for compression, and the compressed data then are transmitted over a communications channel. On being received, the possibly error-corrupted compressed signal is decompressed in the decoder and reconstructed for display.

Video preprocessing

Video preprocessing converts the analog input signals to digital samples in the form needed for the subsequent compression. The analog input signals are red (R), green (G), and blue (B) signals.

Video compression formats

In Table 9-1, "Vertical lines" refers to the number of active lines in the picture. "Pixels" refers to the number of pixels during the active line. "Aspect ratio" refers to the picture aspect ratio. "Picture rate" refers to the number of frames or fields per

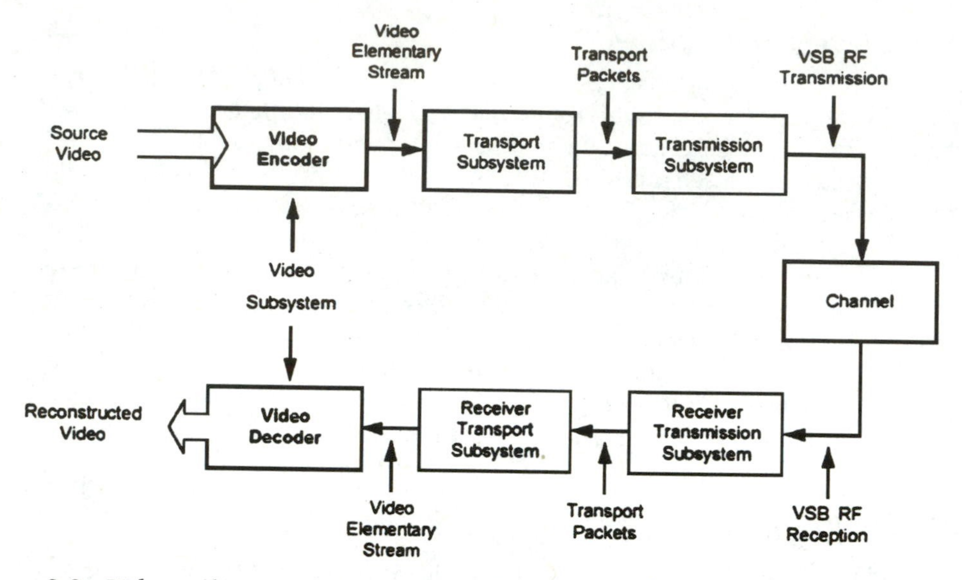

9-8 Video coding in relation to the ATV system. (*Courtesy of AAVS, a Division of Sencore, Inc.*)

second. In the values for picture rate, "P" refers to progressive scanning, and "I" refers to interlaced scanning. Note that both 60.00-Hz and 59.94-Hz (60×1000/1001) picture rates are allowed. Dual rates are also allowed at the picture rates of 30 and 24 Hz.

Possible video inputs

While not required by the *Digital Television Standard,* there are certain television production standards, shown in Table 9-2, that define video formats that relate to compression formats specified by the standard. The compression formats may be derived from one or more appropriate video input formats. It may be anticipated that additional video production standards will be developed in the future that extend the number of possible input formats.

Sampling rates

For the 1080-line format, with 1125 lines per frame and 2200 total samples per line, the sampling frequency will be 74.25 MHz for the 30.00 frames per second (fps) frame rate. For the 720-line format, with 750 total lines per frame and 1650 to-

Table 9-1. Compression formats in the *Digital Television Standard*

Vertical lines	Pixels	Aspect ratio	Picture rate
1080	1920	16:9	60I, 30P, 24P
720	1280	16:9	60P, 30P, 24P
480	704	16:9 and 4:3	60P, 60I, 30P, 24P
480	640	4:3	60P, 60I, 30P, 24P

Table 9-2. Standardized video input formats

Video standard	Active lines	Active samples per line
SMPTE 274M	1080	1920
SMPTE S17.392	720	1280
ITU-R BT.601-4	483	720

tal samples per line, the sampling frequency will be 74.25 MHz for the 60.00 fps frame rate. For the 480-line format using 704 pixels, with 525 total lines per frame and 858 total samples per line, the sampling frequency will be 13.5 MHz for the 59.94-Hz field rate. Note that both 59.94 and 60.00 fps are acceptable as the frame or field rate for the system.

For the 480-line format, there may be 704 or 640 pixels in the active line. If the input is based on ITU-R BT.601-4, it will have 483 active lines with 720 pixels in the active line. Only 480 of the 483 active lines are used for encoding, and only 704 of the 720 pixels are used for encoding. The first 8 and the late 8 pixels are dropped. The 480-line, 640-pixel picture format is not related to any current video production format. It does correspond to the IBM VGA graphics format and may be used with ITU-R BT.601-4 sources by using appropriate resampling techniques.

Colorimetry

For the purposes of the *Digital Television Standard, colorimetry* means the combination of color primaries, transfer characteristics, and matrix coefficients. Video inputs conforming to SMPTE 274M and S17.392 have the same colorimetry; in this chapter they will be referred to as SMPTE 274M colorimetry. Note that SMPT 274M colorimetry is the same as ITU-R BT.709 (1990) colorimetry.

In general bit streams, technical engineers should understand that many receivers will likely display all inputs, regardless of colorimetry, according to default SMPTE 274M colorimetry. Some receivers also may include circuitry to properly display SMPTE 170M colorimetry (color primaries, transfer characteristics, and matrix coefficients have the value 0X06). It is believed that few receivers will display properly the other colorimetry combinations.

Precision of samples

Samples are typically obtained using analog-to-digital converter circuits with 8-bit precision. After processing, the various luminance and chrominance samples typically will be represented using 8 bits per sample of each component.

Source-adaptive processing

The image sequences that constitute the source signal can vary in spatial resolution (480, 720, or 1080 lines), in temporal resolution (60, 30, or 24 fps), and in scanning format (2:1 interlaced or progressive scan). The video compression system accommodates the difference in source material to maximize the efficiency of compression.

Film mode

When a large number of pixels do not change from one frame in the image sequence to the next, a video encoder may automatically recognize that the input was film with an underlying frame rate less than 60 fps.

In the case of 24-fps film material that is sent at 60 Hz using a 3:2 pull-down operation, the processor may detect the sequence of three nearly identical pictures followed by two nearly identical pictures and only encode the 24 unique pictures per second that existed in the original film sequence. When 24-fps film is detected by observation of the 3:2 pull-down pattern, the input signal is converted back to a progressively scanned sequence of 24 fps prior to compression. This avoids sending redundant information and allows the encoder to provide an improved quality of compression. The encoder indicates to the decoder that the film mode is active.

In the case of 30-fps film material that is sent at 60 Hz, the processor may detect the sequences of two nearly identical pictures followed by two nearly identical pictures. In this case, the input signal is converted back to a progressively scanned sequence of 30 fps.

Color component separation and processing

The input video source to the ATV video compression system is in the form of RGB components matrixed into luminance (Y) and chrominance (Cb and Cr) components using a linear transformation (3×3 matrix, specified in the standard). The luminance component represents the intensity, or black-and-white picture, while the chrominance component contains color information. The original RGB components are highly correlated with each other; the resulting Y, Cb, and Cr signals have less correlation and are thus easier to code efficiently. The luminance and chrominance components correspond to functioning of the biologic vision system; that is, the human visual system responds differently to the luminance and chrominance components.

The coding process also may take advantage of the differences in the ways that humans perceive luminance and chrominance. In the Y, Cb, and Cr color space, most of the high frequencies are concentrated in the Y components; the human visual system is less sensitive to high frequencies in the chrominance components than to high frequencies in the luminance component. To exploit these characteristics, the chrominance components are low-passed filtered in the ATV video compression system and subsampled by a factor of 2 along both the horizontal and vertical dimensions, producing chrominance components that are one-fourth the spatial resolution of the luminance components.

Antialias filtering

The Y, Cb, and Cr components are applied to appropriate low-pass filters that shape the frequency response of each of the three components. Prior to horizontal and vertical subsampling of the two chrominance components, they may be processed by half-band filters in order to prevent aliasing.

Number of lines encoded

The video coding system requires that the coded picture area has a number of lines that is a multiple of 32 for an interlaced format and a multiple of 16 for a noninterlaced format. This means that for encoding the 1080-line format, a coder must actually deal with 1088 lines (1088=32×34). The extra 8 lines are in effect "dummy" data that simplify the implementation. The extra 8 lines are always the last 8 lines of the encoder image. These dummy lines do not carry useful information but add little to the data required for transmission.

Representation of picture data

Digital television uses a digital representation of the image data, which allows the data to be processed using computer-like digital processing. The process of digitization involves sampling of the analog television signals and their components and representing each sample with a digital code.

Pixels

The analog video signals are sampled in a sequence that corresponds to the scanning raster of the television format, that is, from left to right within a line and in lines from top to bottom. The collection of samples, in a single frame or in a single field for interlaced images, is treated together as if all corresponded to a single point in time (in case of the film modes, they do in fact correspond to a single time or exposure interval). The individual samples of image data are referred to as *picture elements,* or *pixels,* or *pels.* A single frame or field can be thought of as a rectangular array of pixels.

Square pixels

When the ratio of active pixels per line to active lines per frame is the same as the display aspect ratio, which is 16:9, the format is said to have *square pixels.* The term refers to the spacing of samples and does not refer to the shape of the pixel, which might ideally be a point with zero area from a mathematical sampling point of view.

Spatial relationship between luminance and chrominance samples

As described previously, the chrominance component samples are subsampled by a factor of 2 in both horizontal and vertical directions. This means that the chrominance samples are spaced twice as far apart as the luminance samples.

Digital signal processing theory

Before going into the theory of digital video signals, let us review the basics of digital signal processing. The theory of converting an analog waveform to a digital bit stream involves sampling a waveform at regular intervals and recording the signal amplitude at each sample point as a digital integer value. By reversing the process, the original waveform is restored from the stream of samples. The process of digitization is illustrated in Fig. 9-9.

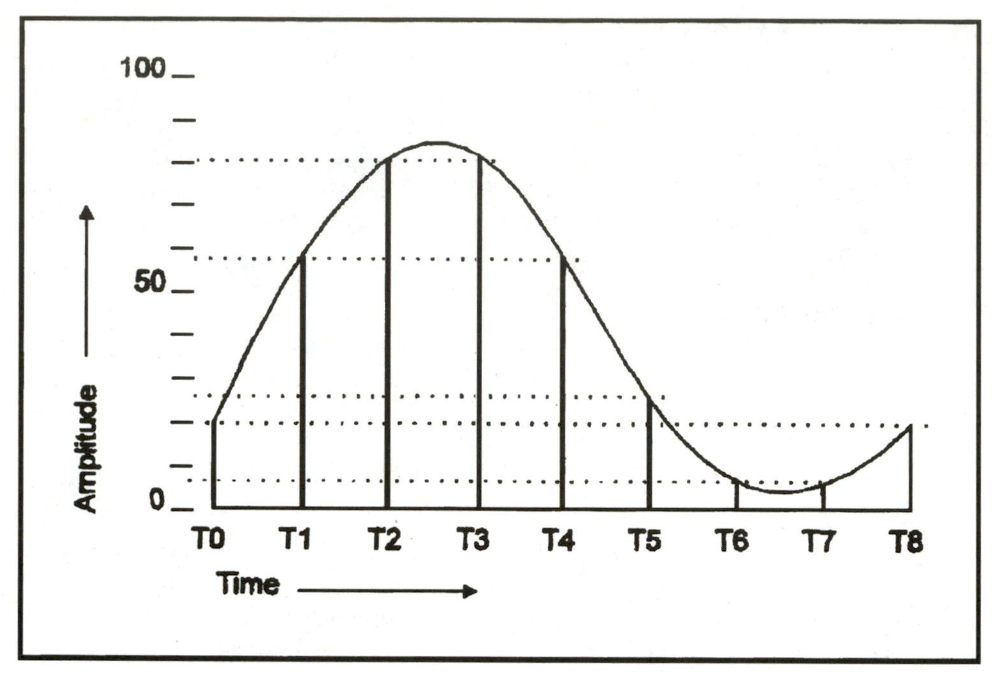

9-9 The process of digitization. (*Courtesy of AAVS, a Division of Sencore, Inc.*)

This waveform is being sampled at an interval that is about one-eighth its period. The digitizer would sample the amplitude at T_0 and return the value of 20. The next sample (at T_1) would return 58. T_2 would have a value of 90, as would T_3. T_4, T_5, T_6, T_7, and T_8 would have values of 58, 25, 8, 8, and 20, respectively.

If we know the sample frequency and the values for each point, we can reconstruct the waveform by producing a voltage proportional to each value at the corresponding time. This is illustrated in Fig. 9-10.

The output of the digital-to-analog converter (DAC) is shown by the heavy "stairstep" line. This signal resembles the original waveform, but with a bad case of the "jiggles." After electronically integrating and filtering the output of the DAC, you will again see the familiar original waveform, shown by the dashed line.

A couple of pitfalls can be deduced from this simple explanation. The sample rate must be high enough to make steps that can reasonably follow the original waveform. The higher the frequency of the waveform being sampled, the higher is the required sample rate. If the frequency of the waveform approaches the sample rate, we have too few samples to effectively reconstruct the waveform. The waveform shown in Fig. 9-11 was sampled at a frequency that was too low.

Figure 9-11 shows a waveform (solid line) that was sampled at a rate that was too low. When converted back to analog, the DAC would reconstruct a lower-frequency waveform than the original. This is shown as the dashed line. This is known as *aliasing*. To avoid aliasing, the Nyquist theorem states that the sample rate must be at least twice the highest frequency to be sampled. We must therefore pass the incoming signal through a low-pass filter that prevents frequencies above the Nyquist limit from reaching the analog-to-digital converter and introducing aliasing into our signal.

The second possible pitfall in digitization concerns the size of the "steps" in the digitization process. Depending on the precision required, we can divide the total range of amplitude to be sampled into a set number of steps. These are called the *quantization levels.* Figure 9-12 illustrates quantization.

Here we are digitizing a signal that goes from 0 to a value X linearly. We will divide this total range of levels into eight discrete digital values. The goal is to express the levels of the straight-line ramp digitally. At time T_1, the closest digital value that can express the level of our line is 2. At time T_2, 3 is the nearest digital value. At T_3, 5 is very close. If we use these values and record them, reconverting to analog would yield the curve shown by the dotted line. We have introduced errors by quantizing in steps that were too large. We cannot express the actual values accurately at each sample time.

An additional consideration in the choice of the number of quantization levels concerns the available signal-to-noise ratio (SNR). Since each additional bit doubles the available number of levels, each additional bit increases the SNR by about 6 dB. Thus 8-bit quantization has a theoretically possible SNR of about 12 dB lower than 10-bit quantization. The number of quantization levels must be chosen to provide signal-to-noise performance for the required application.

Thus we see that during the digitization process, a high sample rate and a large number of quantization levels are desirable. However, these two parameters obviously have a direct effect on the resulting data rate. Doubling the sample rate or the quantization levels doubles the data rate. In real-world situations, increased data rates are always accompanied by increased cost. We therefore choose the lowest sample rates and minimum number of quantization levels that will allow adequate fidelity of the reproduced signal.

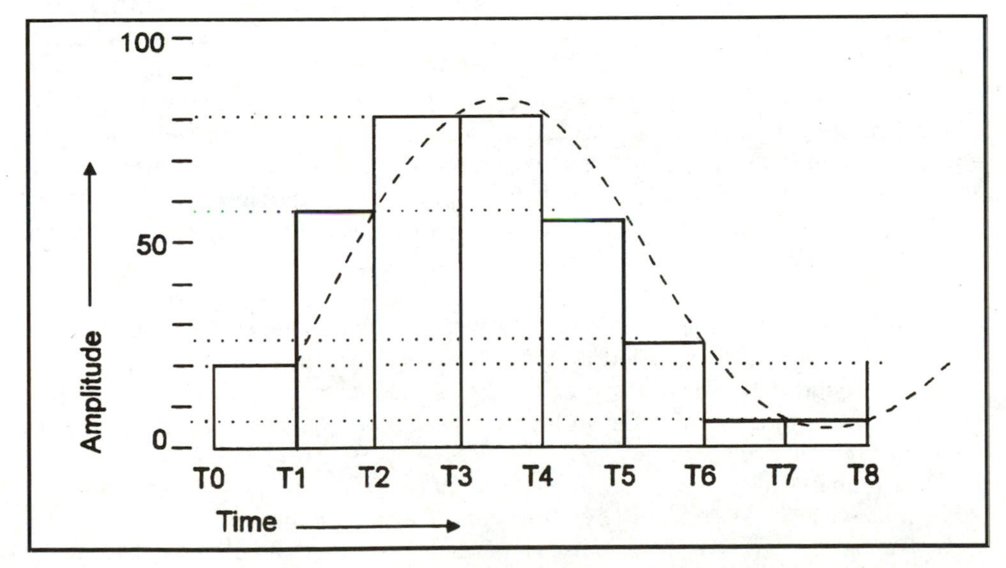

9-10 Waveform produced by a proportional voltage value at a corresponding time. (*Courtesy of AAVS, a Division of Sencore, Inc.*)

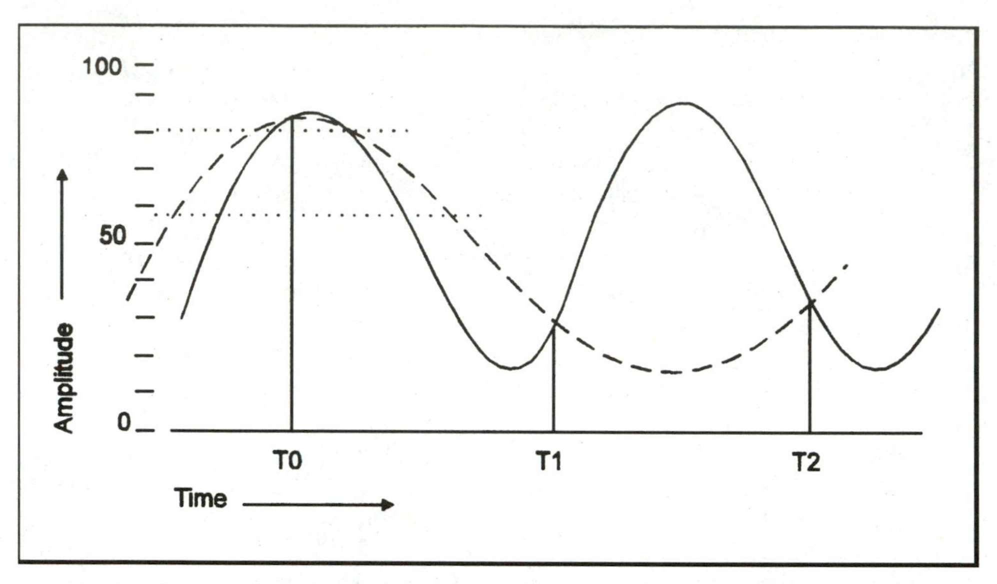

9-11 Waveform sampled at too low a frequency. (*Courtesy of AAVS, a Division of Sencore, Inc.*)

Component digital video theory

Applying the theories of digital signal processing to component analog video signals has yielded the component digital video signal standard. We will begin by studying the data structure of the signal and the methods chosen to represent the signal with digital values, converting the component analog video signal to component digital video.

To digitize the analog video components, a sample rate of 13.5 MHz was chosen to sample the Y signal. One important factor in the choice of sample rate was the desire for easy interchange between 525-line systems and 625-line systems.

Even if the analog luminance signal is considered to have maximum bandwidth of 6 MHz, the sample rate of 13.5 MHz represents a sample rate well above the lower limits defined by Nyquist. The chrominance difference signals are sampled at the lower rate of 6.75 MHz (one-half the luminance sample rate). This is acceptable because their bandwidth can be restricted to one-half the luminance bandwidth without serious loss of detail.

The three components are quantized to 8 bits (256 levels) or 10 bits (1024 levels). Linear quantization is used, meaning that each quantizing step is of equal size. The original specifications called for 8-bit quantization, but it was felt to be inadequate. A second optional standard of 10 bits was implemented. The tendency is to use 10 bits, but much equipment in the field uses 8-bit quantization, and new 8-bit equipment is still produced.

The human eye typically can discern luminance differences of about 1 percent. Since a television system has an available contrast ratio of about 100:1, it would appear that the 256 levels available with 8 bits would be more than adequate to express all the luminance levels necessary. Generally, this is true, but when extremely low noise sources are used, as is the case with electronically generated color backgrounds, some "banding" will sometimes be observed in 8-bit systems. This is especially true when these signals are passed through various digital effects equipment where mathematical manipulations have been performed on the signal.

It is possible to exchange signals between 8- and 10-bit equipment, but care must be exercised to avoid distortions that can occur if the 2 least significant bits are merely eliminated. The magnitude of the signal does not change in conversions from 8 to 10 bits or vice versa because the least significant bits (LSBs) represent fractional values below the radix point. However, simply shorting the LSBs can introduce distortion.

These distortions can be avoided by using a process called *dynamic rounding* that introduces a random or pseudorandom element into the rounding process by which the fractional values are rounded to 8 bits. This process is not used universally, however, and manufacturers should be asked about the method used for 10- to 8-bit conversion in their equipment when required.

Fully saturated color bars with 100 percent white will produce the analog color difference component signal waveforms shown in Fig. 9-13. It is apparent from an examination of the waveforms in the upper part of the figure that the two color difference signals have peak-to-peak excursions that are greater than that of the luminance signal. In fact, the R-Y component has a peak-to-peak value 1.402 times that of luminance. The maximum values occur on the red and cyan bars. The B-Y component has a peak-to-peak value 1.772 times that of luminance. The peaks occur at the blue and yellow bars. In addition, the values for R-Y and B-Y include both positive and negative values.

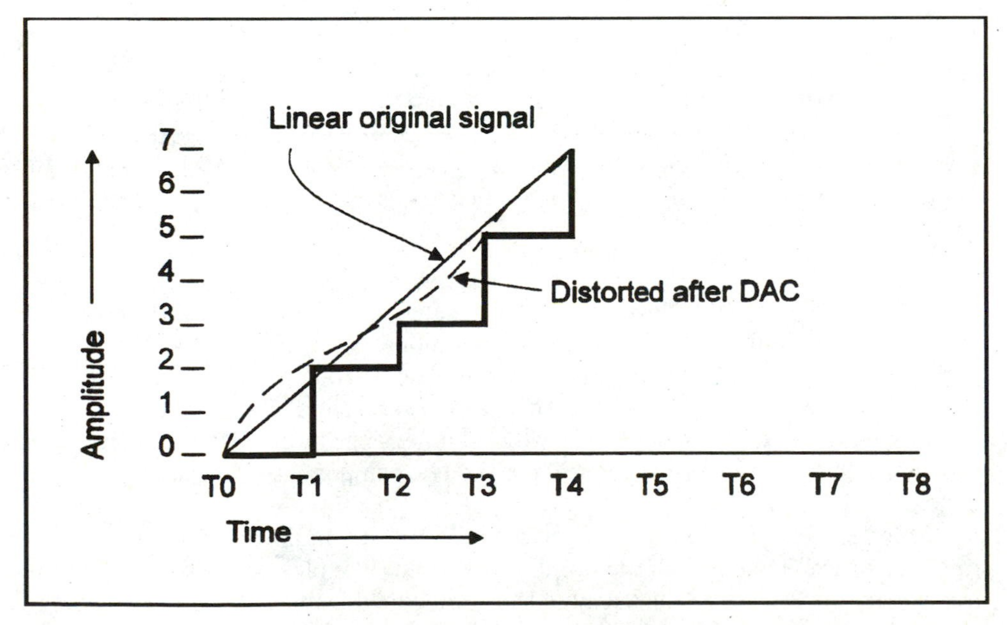

9-12 Illustration of quantization. (*Courtesy of AAVS, a Division of Sencore, Inc.*)

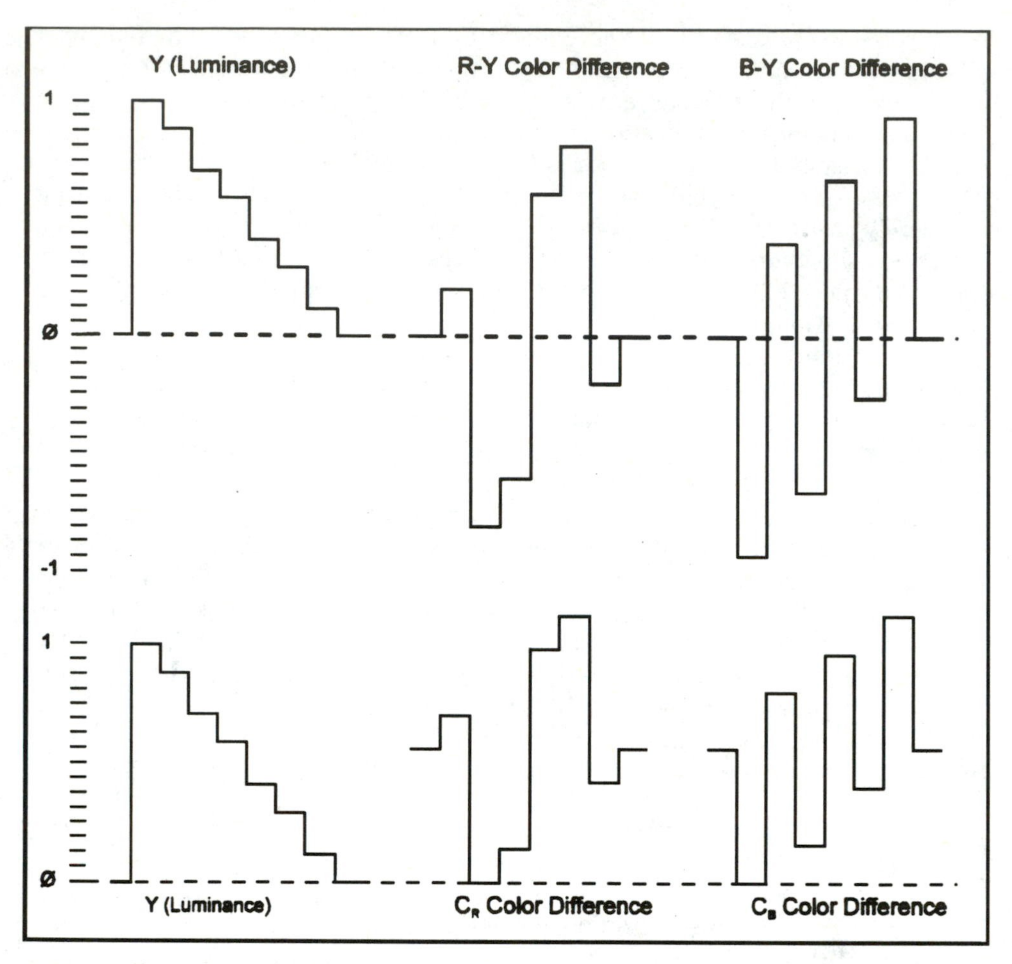

9-13 Fully saturated color bars with 100 percent white producing the analog difference component signal waveform. (*Courtesy of AAVS, a Division of Sencore. Inc.*)

In order to take advantage of the full dynamic range available, and in order to avoid negative numbers, the R-Y and B-Y color difference signals are modified before digitization. They are scaled by multiplying R-Y by 0.713 and B-Y by 0.564. In addition, a dc offset is introduced so that their zero levels is at digital 128 (8 bits) or 512 (10 bits). In this way, the negative swings can be reproduced digitally without the need for negative numbers. After scaling, the two components are called Cr and Cb instead of R-Y and B-Y.

Since the chroma components are sampled at one-half the luminance sample rate, there are only one-half the number of chroma samples per television lines as luma samples. Figure 9-14 illustrates the principle of sampling an image.

In Fig. 9-14 we see that on line 1, sample zero has a Y sample, a Cr sample, and a Cb sample. Sample one has only a Y sample, whereas sample two has all three components sampled. Thus odd-numbered samples are luminance-only samples, and even-numbered samples contain all three samples.

There are a total of 858 luminance samples per horizontal line (864 in the 625-line system). The active picture contains 720 samples. This is the same in both 525- and 625-line systems. This makes up a total of 1440 samples during the active picture area of each line.

In order to avoid having to use three transmission paths, the three components, Y, Cr, and Cb, are time-multiplexed by clocking the data out of buffers at twice the normal luminance rate. The luminance samples are interleaved with the chrominance samples. Figure 9-15 shows the timing relationship between the various samples and the analog television lines.

The synchronizing pulses are not digitized in the component digital video system. Instead, a special set of digital words, called the *timing reference signal* (TRS), is used to signal the sync time to each piece of the equipment. There are two types of TRSs, one called *start of active video* (SAV) and the other called *end of active video* (EAV). These signals provide all the synchronizing information needed to reconstruct and synchronize the image at the receiver. Note that Fig. 9-16 shows the meaning of each bit of the *XY* or *XYZ* word.

The most significant bit is always set to 1. The next bit, the F bit, indicates whether we are in the first scan field (the F bit is set to 0) or the second field (F=1). The third bit, the V bit, is set to 1 during vertical interval and to 0 for the rest of the time. The fourth bit, the H bit, indicates where we are in the horizontal line. It is set to 0 for SAV and 1 for EAV.

The 4 least significant bits (of the 8-bit word) make up a *Hamming code,* which allows us to detect errors in the TRS and correct one errored bit. In a 10-bit signal, the 2 least significant bits are normally set to 0. All the necessary sync information must be carried by the 8 most significant bits to ensure compatibility with 8-bit equipment.

Since the analog sync pulse information is not needed, the data space between EVA and SAV is available to carry other information. Any type of data can be inserted into this space and are called *ancillary data.* The most common type of ancillary

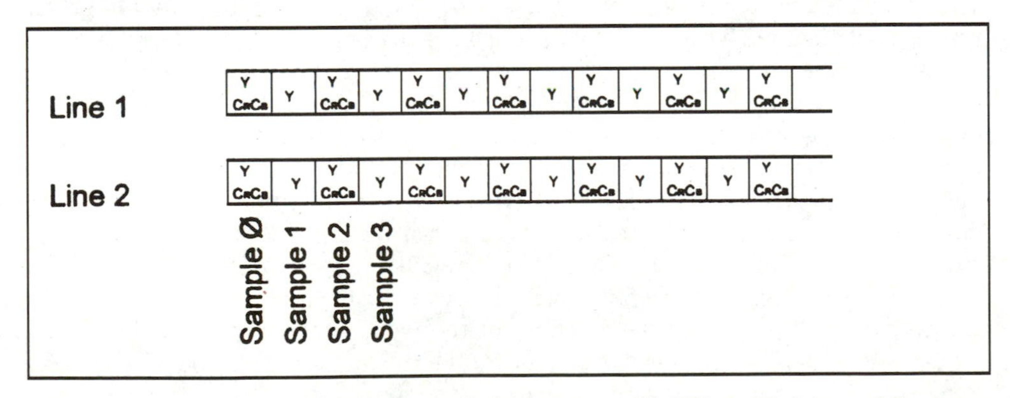

9-14 Principle of sampling an image. (*Courtesy of AAVS, a Division of Sencore, Inc.*)

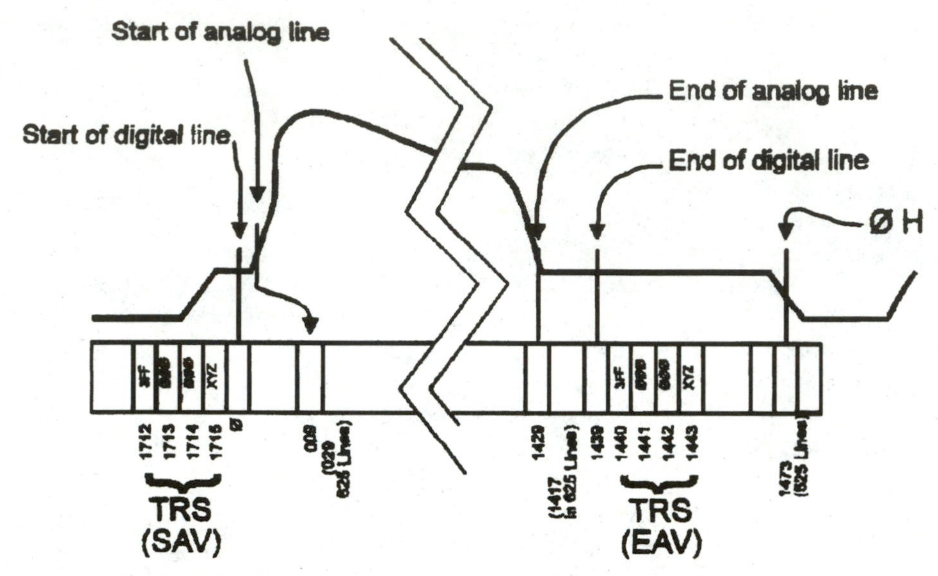

9-15 The timing relationship between the various samples and the analog television line. (*Courtesy of AAVS, a Division of Sencore, Inc.*)

data to be inserted are digital audio. It is increasingly common to use AES/EBU digital audio in this application.

Ancillary data between EAV and SAV are called *horizontal ancillary data* (HANC). There are 268 words available for ancillary data use in the 525-line system. Space for 282 words is available in the 625-line system. In addition, a large part of the vertical interval is available for ancillary data. These data are called *vertical ancillary data* (VANC). During the vertical interval, the active video portion (1440 words in both 525- and 625-line systems) of the line is available for ancillary data, since there is no picture information.

The total available ancillary data space in 625-line systems is slightly greater. There are 176,250 words of HANC data and 70,560 words of VANC data, for a theoretical total of 246,810 words. Figure 9-17 shows the structure of ancillary data.

A data header consists of a three-word sequence and begins the ancillary data block. This data header signals the equipment that the data that follow are ancillary data. Following the data header is the data ID word that identifies the type of ancillary data that are to follow. The next word is used either as a data block number to identify a sequence of data blocks that make up a part of a whole or as a secondary data ID to further identify the data block. The last word before the user data is the data count, which indicates how much data are in the user data block.

User data, a maximum of 525 words, follow the data count and end with a one-word checksum to protect the integrity of the data. There may be multiple data blocks after the EAV reference signal. Several forms of data have specific ID codes assigned, such as digital audio (one or several channels), timecode, and the error detection and handling (EDH) signal. In addition, any other type of data that the user wishes to send via the digital data stream can be embedded.

The digital values made up of all ones and zeros in the 8 most significant bits are reserved for preambles and thus are not to be used for any other purpose. It is important that all generating equipment avoid generating these codes, since their presence in active video can cause serious difficulties in the receiving equipment.

Composite digital video theory

After the component digital standards were established, a second standard was established to allow the interchange of digitized composite video. In this standard, composite NTSC- or PAL-encoded video is digitized directly. This allows some of the same advantages of lossless dubbing and other digital video characteristics at what was initially a much lower cost. In addition, a subsequent tape format associated with this standard used much smaller cassettes, which allowed an integrated camera/recorder that was very desirable for electronic news gathering applications.

The cost difference has decreased with the advance in technology, and it appears that this standard will be a "temporary" one, particularly with the advent of advanced (HDTV) television, which will transmit in the component format. In addition, new component tape formats also use narrower tape widths and smaller cassettes than the original (D1) component format. This has made component integrated camera/recorders possible. Thus encoded formats (composite digital video) eventually will disappear. The composite standard currently enjoys some popularity, particularly among broadcasters in the United States, who work in an NTSC world, and in small postproduction houses, where most of the equipment is analog NTSC.

The composite digital signal is produced by sampling a baseband composite NTSC signal at a frequency that is equal to 4 times the frequency of the chroma subcarrier and is phase locked to the B-Y axis. This means that sampling is done at 4×3.58 MHz=14.31818 MHz, or nominally, 14.3 MHz. This sample rate produces exactly 910 samples per line.

The samples produced by this method are at the same position in each line, since the phase of the subcarrier has a constant relationship to the phase of the horizontal sync pulse. The sampling is thus orthogonal, which means that samples on two different horizontal scan lines having the same sample number will be exactly

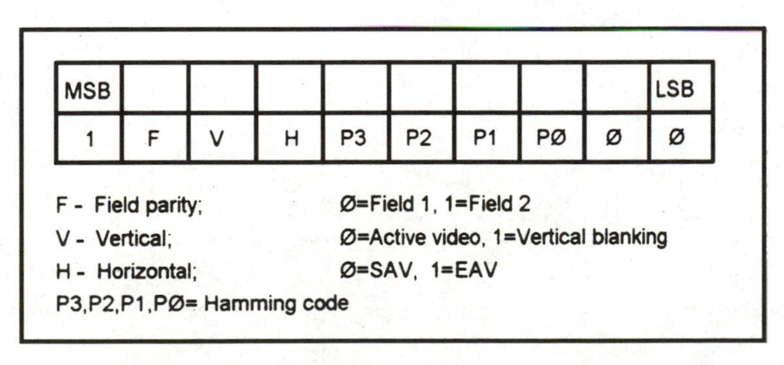

9-16 The meaning of each bit of the *XY* or *XYZ* word. (*Courtesy of AAVS, a Division of Sencore, Inc.*)

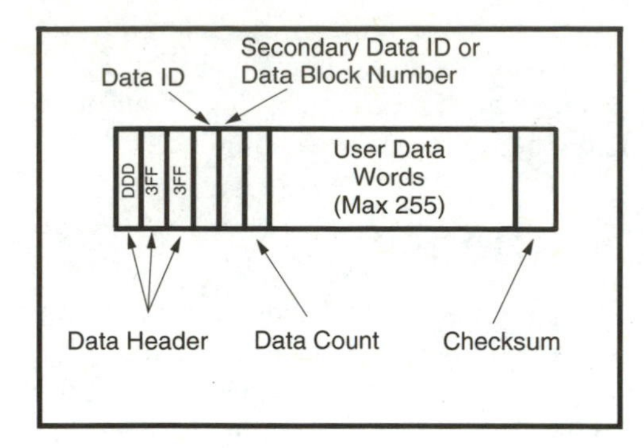

9-17 Ancillary structure data. *(Courtesy of AAVS, a Division of Sencore, Inc.)*

aligned vertically in the image structure. Sample zero occurs immediately prior to the beginning of the active video line, and sample 909 occurs at the end of the horizontal sync pulse, following the color burst.

In the PAL standard, sampling is done in phase with burst at 4 times the subcarrier, or 4×4.43, or nominally, 17.7 MHz. The samples in the PAL system do not fall in the same position in each line, due to the fact that the subcarrier does not have the same phase relationship to horizontal sync for each line. This produces a situation where the sample position moves in relationship to sync, producing a noninteger number of samples per line. All lines are considered to have 1135 samples except two; they have 1137 samples.

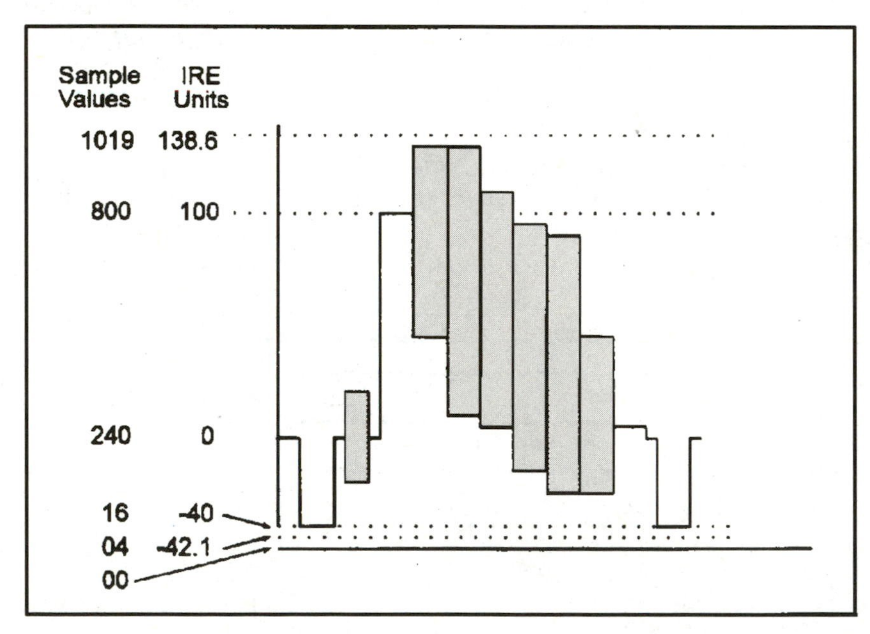

9-18 Levels that are produced by the analog-to-digital converter. *(Courtesy of AAVS, a Division of Sencore, Inc.)*

In contrast to component digital video, the entire line is digitized, including the sync pulse and burst. In the NTSC system, the sample values for burst will be the same for each line, since the sampling is done in phase with the I and Q axes. Thus samples will always be taken at the same position in the burst sine wave. This allows the digital-to-analog converters to calculate and adjust the position of the horizontal sync pulse so that the desirable zero-phase relationship between burst and sync can always be maintained. Levels that are produced by the analog-to-digital converter are defined as shown in Fig. 9-18.

Ancillary data may be inserted in the horizontal sync tip as shown in Fig. 9-18. The preamble, or data flag, for ancillary data in composite digital video consists of only one digital word. With the exception of the data flag, the format of the ancillary data is the same as that of ancillary data in component digital video.

Space for a maximum of only 55 words is available in each horizontal sync pulse. In addition, 21 words can be placed on each equalizing pulse and 376 on each vertical sync pulse. Adding the total capacity per frame reveals that there is a total of 32,901 words of ancillary data space available (55×507 horizontal pulses+21 words×24 equalizing pulses+376 words×12 vertical pulses). This represents a modest data rate that must be allocated carefully, especially if four channels of embedded audio are desired.

10

CHAPTER

Large-screen projection receivers

This chapter discusses several popular rear-projection color television receivers. The chapter reviews the overall projection television system and presents block operational diagrams. Then the special circuits found in these large screen television sets are discussed. In addition, special features such as SPIP, DPIP, and picture-in-a-picture (PIP) are discussed. Then some picture tube replacement cautions and adjustment information are presented. The chapter concludes with circuit information and projection television troubleshooting information and strategies.

GE/RCA projection television chassis

The CTC168 and CTC169 color television chassis are the latest additions to the RCA and GE unitized chassis. The CTC169 chassis is used in both direct-view and projection television receivers.

New chassis features

A number of new features are available in these chassis and are as follows:

- Pix-in-pix
- Digital comb filter
- White stretch and black stretch circuits
- Channel labeling
- Auto VCR input select
- Commercial skip
- Universal remote control

The pix-in-pix module is not as highly featured as the CHIP module used in the CTC140 chassis. The module provides picture in picture and picture swap but does not have the zoom feature of CPIP. The receivers that have the digital comb filter also have the white stretch and black stretch circuits. The comb filter removes chroma noise from the luminance signal. The white and black stretch improve the perceived contrast in the picture. The white stretch circuit operates in the low-brightness scenes and affects the white portion of the picture content. The black stretch circuit operates in high-brightness scenes and pulls the low-level portion of the picture to black.

Channel labeling allows the user to add up to 4 characters per channel and a total of 40 labels. The remote hand unit is used to create or erase labels. The commercial skip feature places a timer display on the screen. The channel can be changed while the timer is displayed. When the timer reaches zero, the channel is switched back to the original station. Each time the channel skip button is pressed, the timer is increased in 30-s steps. The timer can be set for a maximum of 4 minutes. Pressing the clear button on the hand unit cancels the channel skip command.

The auto VCR input select feature automatically changes the channel to the selected video input when the VCR1 button on the hand unit is pressed. The selected channel can be any of the video inputs or any rf channel. The VCR channel setup is located on the setup menu. The universal remote control hand unit will control the television, two compatible VCRs, and a cable converter box.

New circuit areas

These chassis have a number of new circuits. Most of the changes are refinements of circuits used in previous chassis, but some are new circuits that perform the same functions in different ways. These new areas are as follows:

- *Switching regulator.* The use of a switching regulator is not new. However, the regulator used in the CTC168 and CTC169 chassis provides both the standby and run supplies for the chassis. There is no standby transformer.
- *Vertical.* The vertical circuit is similar to a switching power supply. A horizontal rate signal is used to provide the power for the vertical deflection. The scan is controlled by an SCR.
- *Tuner.* The tuner has changed to include band switching and tuning voltage generation. The inputs to the tuner are the B+ supply and clock and data inputs from the system control microprocessor.
- *System control microcomputer.* The functions of the system control microcomputer and the AIU have been combined into one integrated circuit. The system control microcomputer has a communications bus to send commands to the tuner, the digital audio circuit, and digital comb filter, and the EEPROM. It also provides control signals to color processing IC U1001.
- *Digital audio.* The MTS decoder is now a digital IC with fewer adjustments than in previous designs.
- *Digital comb filter.* The digital comb filter removes chroma noise from the luminance signal.

Functional operation

The RCA/GE projection television chassis can be divided into five major circuit areas. Refer to the block diagram shown in Fig. 10-1. These major circuit functions are:

- Power supply
- System control/tuning

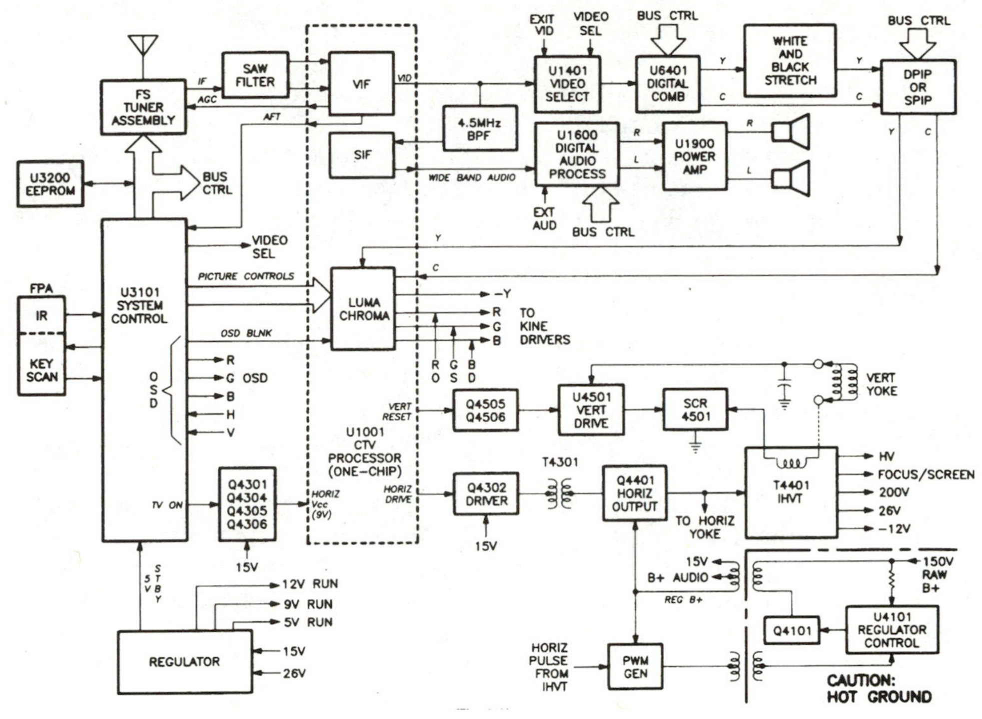

10-1 RCA CTC168 television receiver. (*Courtesy of Thomson Electronics Corporation.*)

- Signal processing
- Audio selection
- Deflection circuits

Power supply operation.
The power supply is a chopper-type switching regulator that supplies both the standby and run B+ supplies. Q4101 is controlled by the regulator IC to provide the regulator B+ 15-V supply and B+ for the audio output amplifier. The 15-V supply is routed to a regulator to develop the 5-V standby power. The 5-V standby power supply keeps the system control microcomputer operational whenever ac power is applied. The 5-V standby supply is also routed to the front panel assembly for the remote receiver and the keyboard scan circuits.

In the standby mode, the audio circuit is off, so there is no load on the audio B+ line. The regulator B+ also has no load, since the horizontal deflection circuit is not operating. The 15-V supply is routed to the horizontal driver stage to provide power to the driver stage. The regulator uses a different regulation system mode than is used in the run mode.

The chassis is turned on by the system control microcomputer. The microcomputer biases a switching transistor "on" to supply power to the horizontal section of U1001. When the horizontal circuit begins to operate, the scan-derived 26-V supply is generated. The 26-V supply is used by the regulator as a bias supply for the run power supplies. If the horizontal circuit fails to operate, none of the run B+ supplies will be present.

When the chassis is operational, the regulator circuit uses a pulse-width-modulated (PWM) feedback signal to control the regulator B+ supply. The PWM generator samples the level of the regulator B+ supply and uses horizontal rate pulses to develop the control signal. The control signal is transformer coupled to the regulator control IC. The regulator control circuit is referenced to hot ground, while the remainder of the chassis is referenced to cold ground.

System control.

The CTC168 and CTC169 chassis contain one microcomputer (U3103) to perform the system control functions. U3101 contains a serial communications bus that transmits commands and data to peripheral circuits within the chassis. Devices under bus control are the tuner, digital audio IC, digital comb filter, pix-in-pix module, and an optional EEPROM.

The system control microprocessor also receives infrared (ir) remote commands from the ir receiver and scans the keyboard on the front panel assembly (FPA). On-screen display characters are generated by U3101 and applied to the CRT driver stages to produce characters on the screen. U3101 receives horizontal and vertical signals from the chassis for OSD timing.

U3101 generates pulse-width-modulated (PWM) signals that are filtered to provide dc control voltages for the picture controls such as color, tint, brightness, sharpness, and contrast. Video selection control lines from U3101 are applied to the video input selection circuit to select between the television tuner or an external video input for viewing on the screen.

Signal processing.

The i.f. signal from the tuner is processed in the CTV processor U1001 (one chip) to produce the composite video signal that is applied to the video selection stage. The rf AGC voltage is developed in the one chip to control the gain of the tuner. The AFT voltage is also developed in U1001 and applied to the system control microprocessor for use during tuning operations.

The video input selection stage selects between the internal tuner video source and the external video inputs on the back of the set. The selected video source is applied to either an analog or digital comb filter.

Sets with digital comb filters also contain the white and black stretch circuits, sometimes referred to as *automatic gamma correction.* The white stretch stage boosts the whites in low average picture level (APL) scenes, while the black stretch circuit stretches the blacks in high APL scenes to enhance the contrast between black and white picture content.

Some sets contain a DPIP or SPIP module to create the pix-in-pix feature. The DPIP module only produces the small pix on the screen, while the SPIP module contains multiple features such as big pix freeze, zoom, and multichannel mode.

Audio processing.

The 4.5-MHz bandpass filter removes the audio i.f. signal from the video signal for application to the sound i.f. stages within U1001. The wide-band audio output of the one chip is applied to the digital audio processing IC U1600. This IC is used for MTS/SAP decoding, audio source selection, volume/mute/tone/balance control, and expanded stereo operation. The IC receives its operational commands through the serial communications bus from the system control microprocessor U3101.

The right and left audio signals from U1600 are applied to the audio power amplifier to drive the speakers. Direct-view sets contain either 1- or 5-W per channel power amplifiers, while projection sets contain 10-W per channel amplifiers.

Deflection circuits.

The horizontal deflection circuit is similar to that in many recent chassis. Chip U1001 supplies the horizontal driver signal. The horizontal drive pulse is routed to the base of the horizontal driver transistor. The output of the horizontal driver is transformer coupled to the base of the horizontal output transistor. The horizontal output transistor drives both the primary of the high-voltage (HV) transformer and the horizontal yoke. U1001 also contains the x-ray protection circuit.

The vertical circuit uses an output from U1001 to reset the ramp generator. The ramp generator provides a reference ramp to the vertical drive IC. The vertical drive IC generates the gate drive signals for the SCR. A winding on the HV transformer provides the power for the vertical scan.

B+ regulator overview

The B+ regulator provides both the standby and run power supplies. The regulator is a switched-mode supply that operates whenever ac power is applied. The supplies developed by the secondary of the transformer are all switched off in the standby mode to reduce the power consumption when the set is off. Refer to the regulator block diagram in Fig. 10-2.

The regulator control IC has two different regulation modes. In the standby mode, a sample from the hot ground side of the regulator provides a dc input to the error amplifier in the regulator control IC. In the run mode, a pulse-width-modulated (PWM) signal is transformer to the regulator control IC. The PWM signal is developed from the regulator B+ supply and a pulse from the HV transformer. The output of the regulator control IC drives the chopper output transistor. By controlling how long the output transistor is conducting, the amount of energy transferred to the secondary is controlled.

Standby operation

The 120-V ac input is rectified to become the 150-V raw B+. The raw B+ is routed via the output transformer to the collector of the chopper output transistor. The raw B+ is also passed through a dropping resistor to provide the startup supply for the regulator control IC. In the standby mode, the regulator control IC develops a series of output pulses. The pulses are produced at a 20-kHz rate but are not continuous. The internal reference voltage is switched from 2.5 to 2.25 V to provide a reference voltage for the error amplifier. The reference voltage is 2.5 V when the regulator is producing output pulses and 2.25 V when production of the output pulses

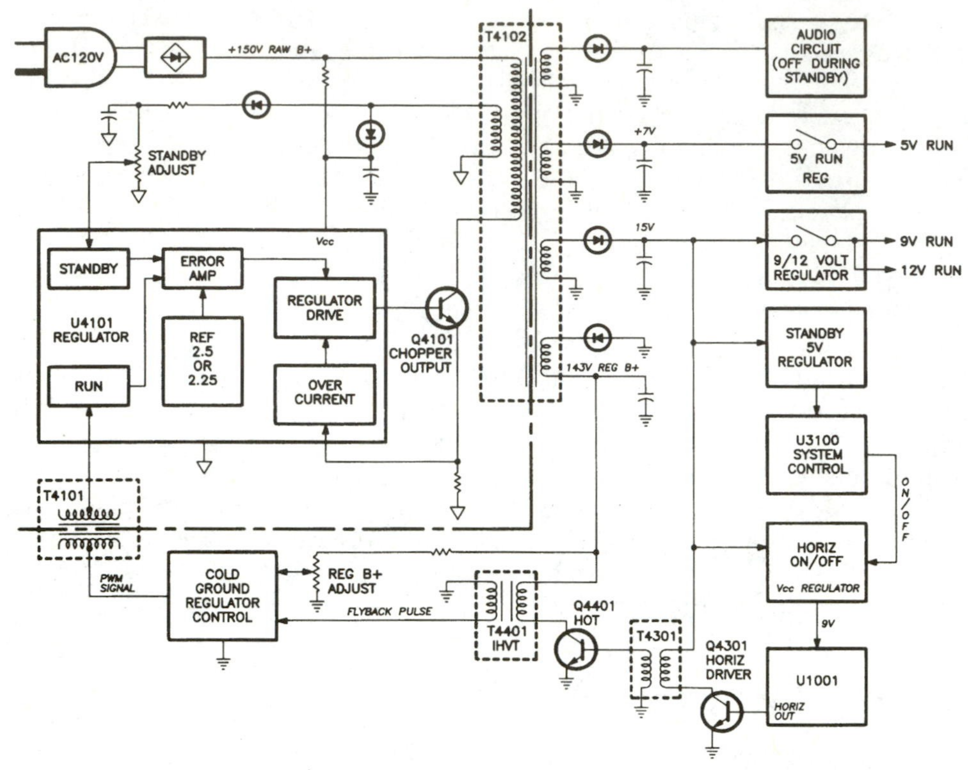

10-2 B+ regulator circuit. (*Courtesy of Thomson Electronics Corporation.*)

is inhibited. The switching occurs when the regulator output reaches the minimum duty cycle. No pulses are generated until the feedback voltage drops below 2.25 V.

The regulator IC turns on when the input voltage at the Vcc input rises above 10 V. The IC remains on until the voltage drops below 7.5 V. The raw B+ input is only used to start the IC. Once the output pulses are generated, a winding in the output transformer is used to supply both the power to the regulator IC and the feedback signal to the error amplifier. The pulses from the run winding are rectified and filtered before being routed to the IC as a dc voltage. The other outputs of the switching regulator are present, but the circuits are disabled to reduce the power consumption during standby. The regulator B+ supply is present, but since the horizontal circuit is not operating in the standby mode, the load on the supply is minimal. In the standby mode, the regulator B+ supply is approximately 140 V. The supply provides 143 V when the television set is on.

On/off operation

The 5-V standby supply is routed to the microcomputer when the set is off. The system control microcomputer must remain operational in order to detect when a front panel key is pressed or when a remote command is received. Power is also routed to the horizontal driver and horizontal output stages. When the horizontal circuit is operational, the scan-derived system provides the power to the remainder of the chassis.

When the power on button is pressed, the system control microcomputer detects the key press and sends the television on signal to an on/off switch transistor. When the on/off switch is turned on, the B+ regulator for the horizontal portion of U1001 is enabled and develops a horizontal drive pulse. Once the horizontal circuit is operational, the scan-derived supplies are used to turn on the 5-V run regulator and the 9/12-V run regulator, thus supplying power to the remainder of the chassis. When the set is turned off, the power to the horizontal oscillator is removed, causing the horizontal circuit to cease operation.

Run mode

In the run mode, the load on the switching power supply increases. The chopper output transistor is turned on longer to provide the additional power to the secondary supplies. The regulator changes modes of operation to handle the increased load. The internal reference voltage of the regulator IC is switched from 2.25 to 2.5 V. The on time of the pulse-width-modulated signal varies with the regulator B+ level in order to change the duty cycle of the chopper output transistor. During the run mode, the chopper runs at a horizontal rate.

The regulator control IC has an overcurrent shutdown circuit. If the cause of the overcurrent condition remains, the overcurrent circuit again disables the regulator output voltage.

Troubleshooting overview

When troubleshooting the B+ regulator circuit, it is necessary to remember that there are two distinct modes of operation. The switching regulator is used to provide both the standby and run supplies. If the standby operation is incorrect, the run mode cannot be checked. If the cold ground regulator control circuit fails to operate, the regulator tries to provide the proper outputs using the standby control circuit; however, the regulator cannot respond quickly enough. The set will operate, but the picture size will change depending on the scene content.

Most problems in the switching regulator circuit will result in a "dead set" symptom. If the fuse is blown, suspect a shorted chopper output transistor. If the standby supplies are present, a "dead set" symptom could still be caused by either the on/off circuit and system control microcomputer or the horizontal deflection circuit.

B+ regulator standby circuit operation

The switching regulator circuit is used to provide the standby supplies, taking the place of the standby transformer. Refer to the circuit in Fig. 10-3. In order to reduce the power consumption of the set in the standby mode, the switching regulator has two different modes of operation. In the standby mode, the output transistor is turned on for a series of pulses and then remains off. The power demand in the standby mode is small, so this "short on time" is sufficient to provide all the standby power supply

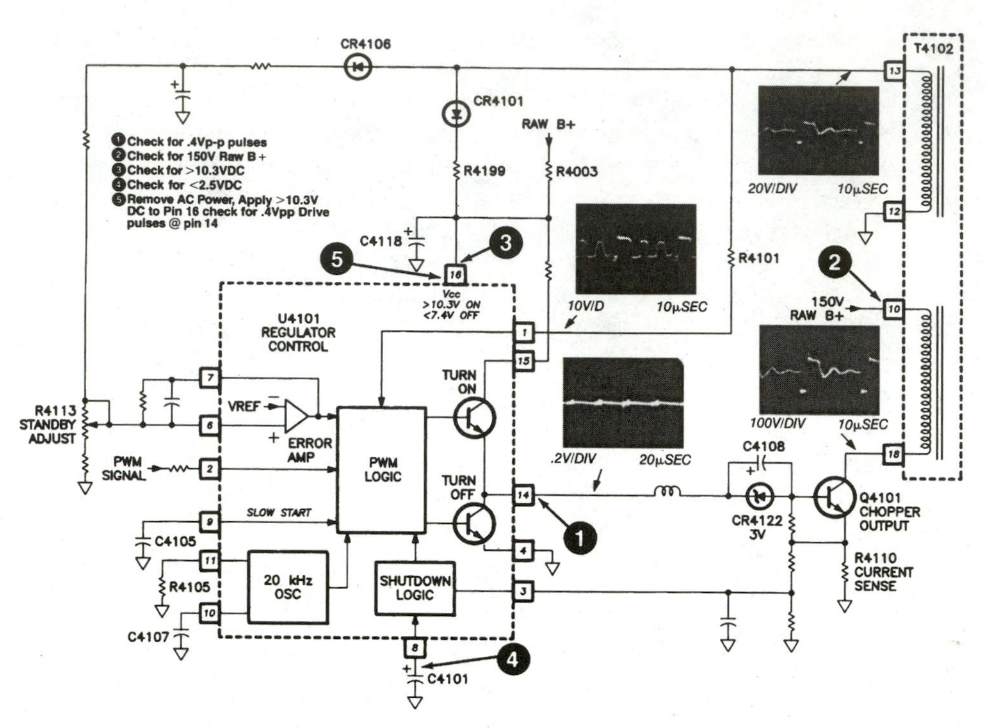

10-3 B+ regulator standby circuit. (*Courtesy of Thomson Electronics Corporation.*)

consumption. The regulator uses a feedback winding in the output transformer to provide the B+ for the regulator control IC and the feedback signal for the error amplifier. This circuit does not provide tight regulation of the standby power supplies, but this is not critical. In the run mode, the regulator control IC provides a horizontal-rate pulse-width-modulated drive signal. The cold ground control circuit enables the regulator to control the regulator B+ much more closely than in the standby mode.

The feedback signal is compared with a fixed reference signal. The difference between the feedback signal and the reference signal enables the logic circuit to determine if the output voltage is too high or too low.

The voltage from the secondary winding is also used by the logic circuit in U4101 to determine when to turn the output transistor on. The output transistor is not turned on until the transfer of energy in the transformer is complete. The logic circuit checks the input at pin 1 when changing from the standby mode to the run mode and vice versa.

Troubleshooting strategies

Let us now look at some ways to troubleshoot this power supply and B+ regulator circuits when various problems develop.

"Dead set" symptom

A dead set can be caused by any of the five major circuit areas. For example, in any receiver with the 10-W per channel audio system, a malfunction in the audio can cause the system control microcomputer to cycle on and off. Malfunctions in the digital comb filter can cause the microcomputer to lock up, again giving a "dead set" symptom. Problems in the microcomputer will give a "dead set" symptom, and any faults in the horizontal deflection circuit will produce a similar result. Finally, most problems in the chopper power supply also produce a "dead set" symptom. A quick check of the chopper power supply operation involves checking the standby 15-V supply. If this supply is present, the power supply is probably operating normally, and the problem is located elsewhere on the chassis.

Standby supply voltage missing

If the standby supply voltages are missing, first check the raw B+ input, line fuses, and surge resistor.

Raw B+ voltage missing.

If the raw B+ is missing at the collector of the chopper output transistor, the line fuse or the surge resistor is probably open. The usual cause is a shorted chopper transistor. If the chopper transistor is open, it is necessary to check the current sense resistor before restoring power to the set. In addition, it is necessary to check the resistance of the secondaries of T4102 to try to determine the cause of the failure. The resistance to ground from the cathode of all the secondary supplies except the 15-V line should be greater than 100 kΩ. The resistance from the cathode of CR4118 to ground should be greater than 4kΩ. If any resistance measurement is less than this, correct the problem on that supply before restoring power to the set. If the chopper output transistor is shorted, the IC also could be damaged. After the transistor is replaced, measure the voltage at U4101 pin 16. If the voltage is less than 2.5 V, the IC is shorted. If the voltage bounces between 7 and 10 V, suspect overcurrent shutdown.

Raw B+ is present.

If the raw B+ is present at the collector of the chopper output device but the standby supplies are missing, the problem could be overcurrent shutdown, overvoltage shutdown, or a malfunction in the regulator control IC and associated components. Measure the dc voltage to hot ground at pin 8 of U4101. If the voltage is greater than 2.5 V, the IC is in the overcurrent shutdown mode. If the voltage at pin 16 is greater than 10.5 V and less than 15 V, the IC should begin operation. An overcurrent condition could exist that does not cause the IC to latch off. Using an oscilloscope, measure the peak-to-peak voltage at U4102 pin 3. The voltage is less than 0.6 V during normal operation. If the voltage is between 0.6 and 0.9 V, the regulator cycles on and off. If the voltage is greater than 0.9 V, the IC latches off until ac power is removed.

Overcurrent shutdown

An overcurrent shutdown condition is normally caused by a shorted or leaky component on one of the secondary supplies. For example, a shorted horizontal output transistor or S-shaping capacitor causes an overcurrent condition. Check the resistance of the secondaries of the transformer to determine which supply is defective.

Overvoltage shutdown

U4101 enters an overvoltage shutdown mode if the input Vcc rises above 15.7 V. The IC stays off until ac power is removed. Overvoltage conditions can be caused by an open in the standby adjust voltage input to the error amplifier or a spike on the Vcc line due to insufficient filtering. Remove the ac power, and discharge C4118. Verify the components in the feedback input to the error amplifier. Restore ac power, and check for normal operation. An overvoltage condition can occur if the cold side B+ regulator is regulating at too high a voltage or if C4118 becomes leaky.

No output from U4101

If there is no output from U4101 and there is no overcurrent or overvoltage condition, use an external dc power supply to confirm operation of the IC. Remove ac power, and connect a dc supply to pin 16 of U4101. Monitor the output of the IC at pin 14. Raise the dc input above 10.5 V, and check for an output waveform. If the output is present, check for base drive directly on the base of the chopper transistor. If the output pulses are missing at pin 14 of U4101, check for the 20-kHz clock signal at pins 10 and 11. If the signal is missing, suspect C4107, R4105, or U4101.

If the IC operates normally when using an external dc supply but does not operate when ac power is applied, check the voltage at pin 3 for an overcurrent shutdown condition. The IC latches off if the peak voltage input at pin 3 rises above 0.9 V. Measure the Vcc input. The voltage at pin 16 must go above 10.5 V before the IC will begin operation. Once the IC turns on, the voltage at pin 16 can drop to as low as 7.5 V before the IC shuts off. If the input voltage is too low, suspect a leaky C4118. If the Vcc input is greater than 10.5 V and there is no overcurrent condition but the IC still does not produce an output, suspect a defective U4101.

Standby power supply operation

The standby power supply circuit (Fig. 10-4) provides the power supplies required by the chassis to detect when the power on command is received and to enable the chassis to begin operation. The system control microcomputer is on whenever ac power is applied. The system control microcomputer scans the front panel keys to detect when the power button has been pressed. The system control microcomputer also receives the input from the ir receiver to determine when the remote on command was issued. The regulator B+ supply is present to enable the horizontal circuit to begin operation quickly. The remainder of the chassis receives power after the horizontal circuit begins operation.

When the set is turned on, the horizontal output transistor is turned on by the horizontal driver. The current through the HV transformer causes energy to be transferred to the secondary windings. One winding is used to provide both the 26- and 200-V supplies. The 26-V supply is used as a control signal to turn on the 9- and 12-V run power supplies. When the 12-V run supply is present, the 5-V run regulator is also turned on.

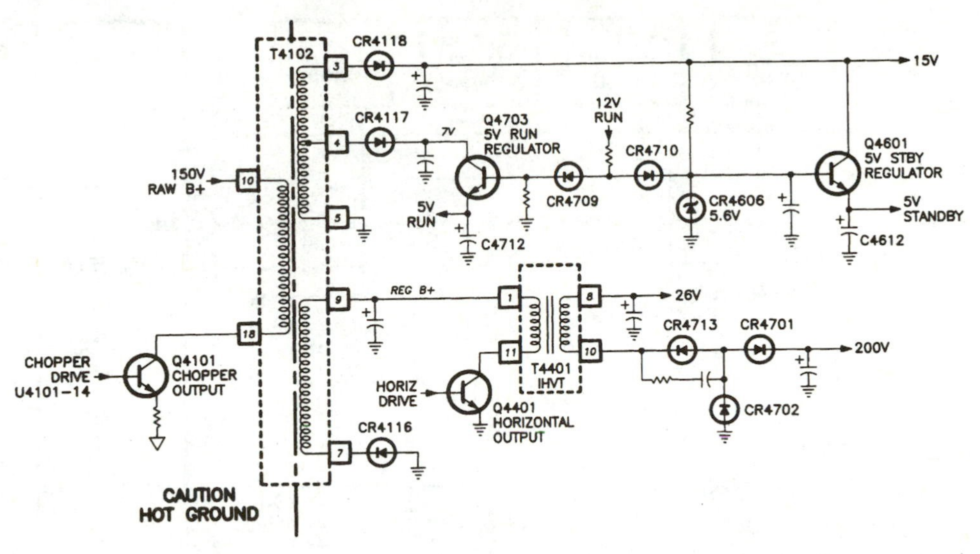

10-4 Standby power supply circuit. (*Courtesy of Thomson Electronics Corporation.*)

Communications bus

Many of the major stages in the CTC168 chassis receive their operational commands via a three-wire serial communications bus from the system control microprocessor U3101. Figure 10-5 shows the devices that are connected to the bus. Although the devices are tied to the same bus, there are three bus formats (IM, IIC, and ATE) implemented at different times over the bus. Note that not all devices use all three lines of the bus. The enable line is used only in the IM bus format.

Clock operation

The clock synchronizes the transfer of data between devices. Information on the data line is read or transferred at the appropriate edge or level (depending on bus format) of the clock pulse. The clock line is not a continuous signal but is active only during data transmissions. When no buttons are pressed and the set is on, the clock line should pulse low about four times every second to request stereo/SAP presence information from the digital audio IC U1600. The activity should vary when a local or remote function is executed. Similar activity can be seen on the data and enable lines.

Data line

The data line contains the command or status information sent between the various devices on the bus, while the rest of the signals are used for timing and directing the data transfer. The viewable activity on this line is very similar to the activity seen on the clock line, as described above.

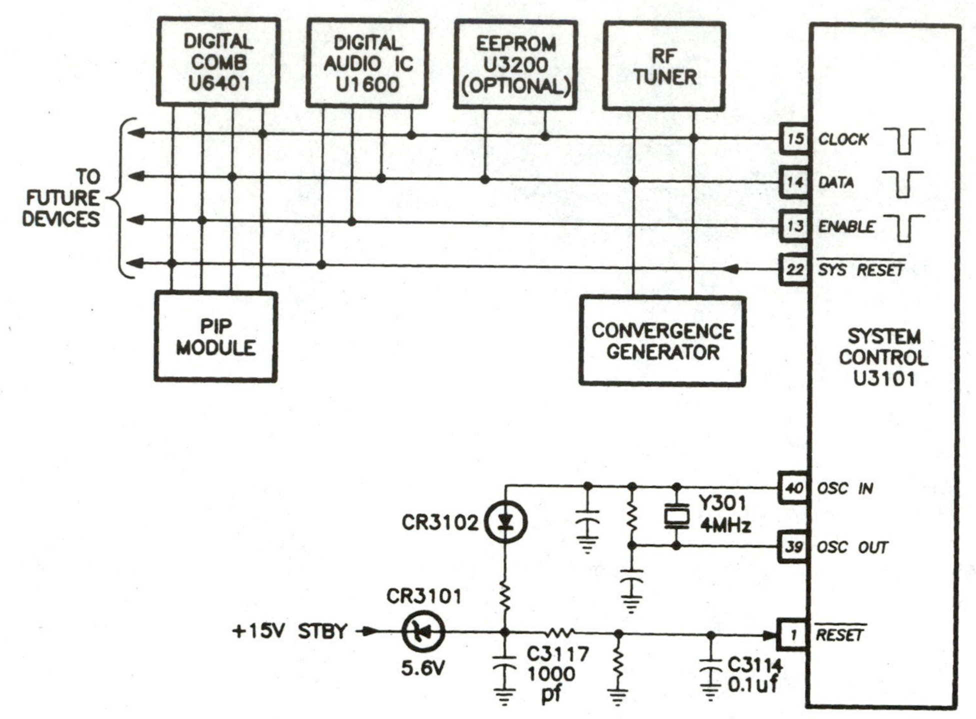

10-5 RCA system control serial bus and reset circuit. (*Courtesy of Thomson Electronics Corporation.*)

Enable (IM bus)

The IM bus uses this line to distinguish between address and data communications on the bus. With multiple devices on the bus, individual addresses are assigned to each device. Before commands can be sent to a specific bus device, the system control microprocessor sends the address of the desired device to inform it that it is going to be receiving a command. A low on the enable line signifies that device address information is being transmitted on the data line, while a high signifies command data. The viewable activity is about the same as the clock and data lines.

Bus troubleshooting tip

The system control microprocessor requests stereo presence status from the digital audio IC U1600 about four times per second over the serial bus. Check for this activity on the bus when the set is on and no buttons are pressed. The best way to verify proper communications bus activity is to check for 0- to 5-V peak-to-peak signals on the clock, data, and enable lines. Intermediate voltages (between 1 and 4 V) signify a defect on the bus.

System reset

This line is used to reset all peripheral devices when the set is first turned on. Do not confuse this reset with the reset input at pin 1 of U3101. The system reset line is low when the set is off. When the set is first turned on, the system reset line stays low for about 1 s and goes high until the set is turned off. The peripheral devices are ready to receive bus communications after the system reset line goes high.

EEPROM information

All these sets feature an electronically erasable programmable read-only memory (EEPROM) chip (U3200). The EEPROM is a *nonvolatile* memory device, which means that it retains its stored information even when power is lost. It can be altered to store new information, unlike read-only memory (ROM) devices, whose data are permanent.

The EEPROM stores the following information:

- *Channel scan list.* Sets with EEPROMs installed will be programmed with channels 2 through 13 and channel 91 at the factory. The set will not automatically enter autoprogram mode when turned on after an extended power loss. Autoprogramming must be initiated by the setup menu. Sets without the EEPROM will automatically enter autoprogramming mode when turned on after an extended loss of ac power. The autoprogramming must be initiated by the setup menu.
- *Alphanumeric channel labels.* The four-character alphanumeric label assigned to a channel by the customer is also stored in the EEPROM. This feature is implemented only in sets with an EEPROM to prevent loss of these data.
- *Customer convergence adjustment (projection sets).* Projection sets allow the customer to adjust center convergence from the remote using the setup menu. The EEPROM stores a numeric value assigned to the convergence setting for the blue and red CRTs.
- *VCR channel setting.* The VCR channel feature automatically selects a preprogrammed channel on the television whenever the remote VCR button is pressed. If your VCR is connected to input 1 on the back of the set, the customer programs channel 91 into the setup menu. The television will automatically select channel 91 when the remote VCR button is pressed. The channel selected is stored in the EEPROM.
- *Rf switch option and menu format.* The EEPROM is programmed at the factory to tell the system control microprocessor whether an rf switch is not installed in the set. If the rf switch is not installed, the system control microprocessor will delete the antenna selection feature from the setup menu.
- *Oscillator and power loss memory retention.* The system control microprocessor U3101 is reset at pin 1 on ac power application. The 4-MHz oscillator at pins 40 and 39 is stopped by the reset circuit during power loss to prevent discharging of the 5-V standby supply. The longer the 5-V supply is maintained, the longer the stored memory information within U3101 can be retained. Memory retention is guaranteed for at least 10 s of power loss but will probably last a few minutes.
- *System control functions.* As we go over these system control functions, refer to Fig. 10-6, which shows an overview of the pins and functions of system control microprocessor chip U3101.

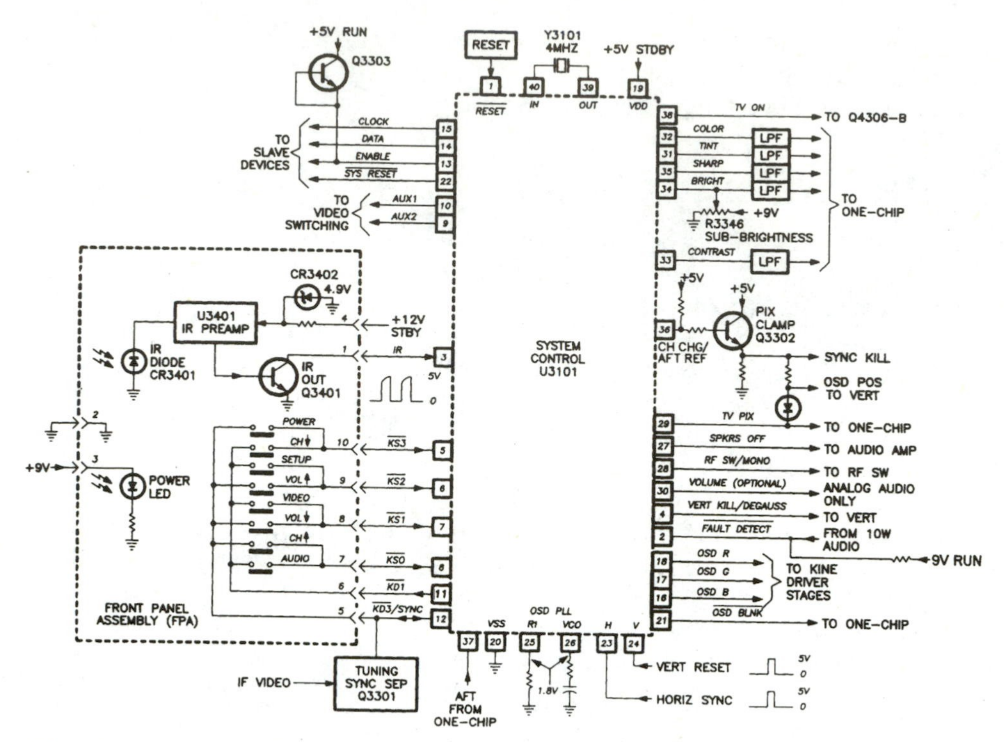

10-6 System control functions circuit. (*Courtesy of Thomson Electronics Corporation.*)

Keyboard decoding

With no keys pressed, the KS0 through KS3 lines are pulled high by resistors on the main chassis. KD1 and KD3 are set low while no key is pressed. When a key is closed (pressed), a low from either the KD1 or KD3 is applied to one of the KS lines. When a low is detected at one of the KS lines, KD1 and KD3 are toggled in sequence to decode which key has been pressed.

If the power, volume up or down, and the channel buttons are held, KD1 will keep scanning, but KD3 will remain low after the key is decoded. If channel down, setup, video, and audio keys are held, both KD1 and KD3 continue to scan after the key is decoded.

Tuning sync and AFT

The KD3 line is shared by the tuning sync input. When the set is on, tuned to an active channel, and no key is being pressed, positive horizontal pulses enter the microprocessor at pin 12. These pulses are used during tuning to detect active channels. When pressing the channel up or down keys, the system control microprocessor interrupts the key scan sequence for about 10 ms to check for sync pulses at pin 12.

Ir receiver

The FPA also houses the infrared (ir) remote receiver. CR3401 receives the ir signal, which is amplified and demodulated by U3401. The demodulated ir output is buffered and inverted by Q3401 to produce the 5-V peak-to-peak square wave for system control microprocessor pin 3. The system control microprocessor decodes the signal and executes the command.

Power LED

The power on LED indicator is powered from the 9-V run supply and not a control line from the system control microprocessor. When the set is turned off, the 9-V supply discharges, leaving no power for the LED.

AUX1/AUX2 controls

These lines control the switch positions of the video selection IC1401. These lines do not select the audio source that goes with the video.

Other communications bus operations

The communications bus performs many other operations in these RCA/GE color television projection sets, and these are as follows:

- Television on control
- Picture control
- Channel change/AFT REF (sync kill)
- Television PIX and OSD (on-screen display) position
- OSD position
- Television speakers off/on control
- Rf switch/mono
- Volume control
- Vertical kill/degauss
- Fault detect
- On-screen display (OSD)

DPIP and operation guide

In the following picture-in-a-picture (PIP) information we will give you a brief overview of how some of the most frequently used DPIP and SPIP features operate. Refer to Fig. 10-7 for the required keypresses to operate these PIP features.

PIP, swap, freeze, and move

The buttons shown in Fig. 10-7*a* are all the functions performed by the DPIP and SPIP modules. These functions are also performed by the FF-SPIP module in addition to the features shown in Fig. 10-7*b–e*.

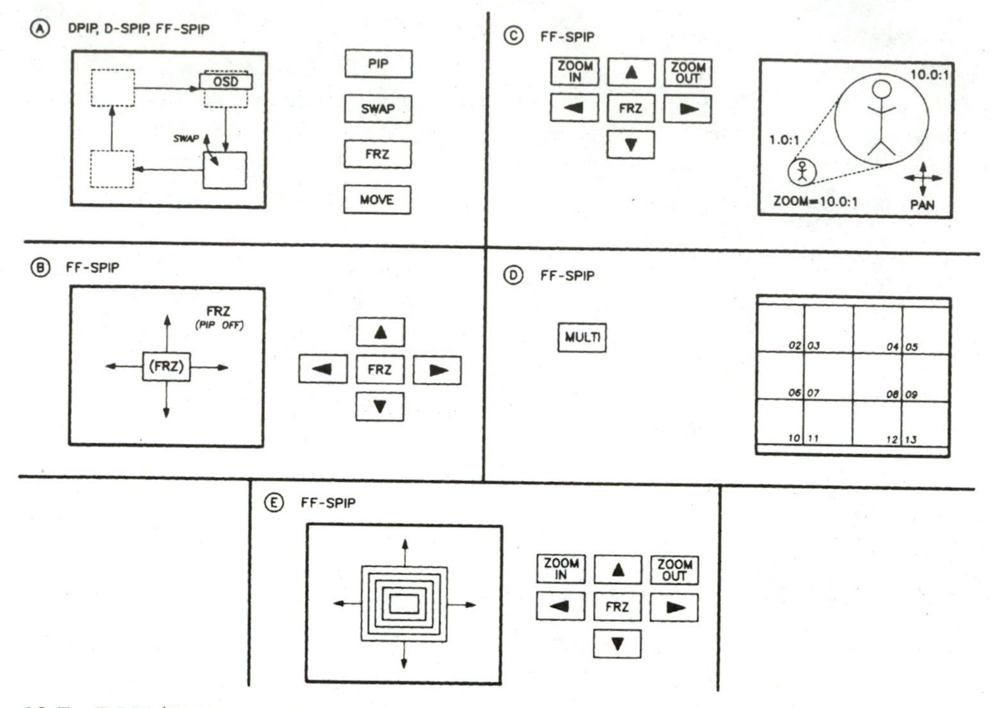

10-7 DPIP/SPIP operation guide. (*Courtesy of Thomson Electronics Corporation.*)

PIP mode.

Pressing the PIP button activates the pix-in-pix feature. The small pix will appear on the screen and contain the same video information as the big pix. The audio you hear will always belong to the big pix.

Pressing the television button will display the word *PIP,* followed by the channel number of the insert video. After a few seconds, the display will change to the big pix channel number. Press the PIP button again to turn off the PIP feature.

Swap mode.

Pressing the swap button exchanges the bit and small pictures. The video previously in the big pix will enter the small pix, while the video previously in the small pix becomes the video in the big pix. The audio also swaps to keep the big pix paired with its corresponding audio source. Pressing the swap a second time returns the big and small pix to their original source.

Freeze mode.

Pressing the FRZ button with PIP turned on freezes (still picture) the small pix video. Pressing the FRZ button again unfreezes the small pix.

Move mode.

Each press of the move button steps the small pix in a clockwise direction to one of the four positions shown in the diagram.

Variable move and freeze (FF-SPIP only)

With the variable move, the small pix can be moved to any position on the screen by using the four arrow buttons shown in Fig. 10-7*b*. Pressing and holding one of the four arrow buttons scrolls the small pix in the direction of the arrow. The small pix continues to move until the arrow key is released or the edge of the screen is reached.

Zoom in (PIP on, FF-SPIP only)

With the PIP enabled, pressing the zoom in button increases the size of the small pix to one of five preset sizes. Each press of the zoom in key advances the size to the next-larger increment until the largest size is reached. The arrow keys still move the small pix as described earlier. The FRZ key freezes the small pix regardless of its size.

Zoom out mode

With PIP enabled, pressing the zoom out button decreases the size of the small pix to one of five preset sizes. Each press of the zoom out key decreases the size to the next-smaller increment until the smallest size is reached.

Zoom in/out and pan (PIP off, FF-SPIP only)

Zoom in.

With PIP turned off, pressing the zoom in button increases the size or zooms in on the content of the big pix. The zoom ratio begins at 1:1 and increases to 10:1. The zoomed picture can be frozen by pressing the FRZ button just as during nonzoom operations.

Zoom out.

Pressing the zoom out button decreases the zoom effect and eventually brings the picture back to the 1:1 (normal viewing) ratio.

Panning mode.

While in the zoom mode, you may search or pan through the picture content by pressing one of the four arrow keys. Then the screen image will move across different areas of the picture.

Multichannel mode (FF-SPIP only)

Pressing the multi button activates the multichannel mode, which cycles through the scan list and displays a frozen pix from each channel on the screen. One active and 11 frozen channels can be displayed on the screen at once. The feature will continue to cycle through the scan list until the multi button is pressed again. The multichannel display can be frozen and zoomed.

SPIP interconnections

Understanding SPIP operation is a little easier if you know how the module is interfaced to the rest of the chassis. We will look at how the signals go to and from the

D-SPIP and FF-SPIP modules. The interconnects for the full and defeatured SPIP modules are the same. Please refer to Fig. 10-8.

External video inputs and outputs are as follows:

- Composite video inputs, J8201
- Composite video output, J8201-1
- Y/C inputs, J8202-3,1
- Selected video out (with PIP), J8201-7
- PIP luma/chroma output, P8202-6,8

SPIP module block diagram

Figure 10-9 shows the basic circuits contained within the SPIP module.

Input selection and Y/C separation.

This stage selects the desired video source to fill the big and small pix viewed on the screen. The television video and switched video inputs are composite and must be separated into component luminance and chrominance signals. The inserted Y/C signals are applied to the decoder stage and the PIP processor.

Y/C to R-Y/B-Y decoder.

The inserted luma and chroma signals enter the decoder stage consisting of U8301 for conversion into R-Y and B-Y color difference signals for application to the PIP processor U8501. Composite sync is separated from the luminance signal and is applied to the horizontal and vertical processing stages.

3.58-MHz oscillator, U8401.

The 3.58-MHz oscillator provides a continuous 3.58-MHz signal to the encoder stage and to the PIP processor. The oscillator is locked to the color burst of the big pix chroma signal.

The PIP processor uses the 3.58-MHz signal during multichannel mode to phase lock its 20-MHz voltage-controlled oscillator (VCO). Encoder U8402 uses the signal to produce the 3.58-MHz chroma signal for the output stage.

Horizontal and vertical sync signals.

The horizontal and vertical sync outputs of the horizontal and vertical processor stages are locked to the composite sync output of the decoder stage. The horizontal and vertical processor free runs in the absence of composite sync to maintain horizontal and vertical pulses to the PIP processor.

Horizontal and vertical sync pulses from the television deflection circuits are also applied to the PIP processor. The PIP processor contains a voltage-controlled 20-MHz oscillator to synchronize its internal timing with external horizontal sync signals.

The horizontal blanking and vertical reset signals from deflection are also used by the PIP processor as timing signals to read small pix information from the VRAM.

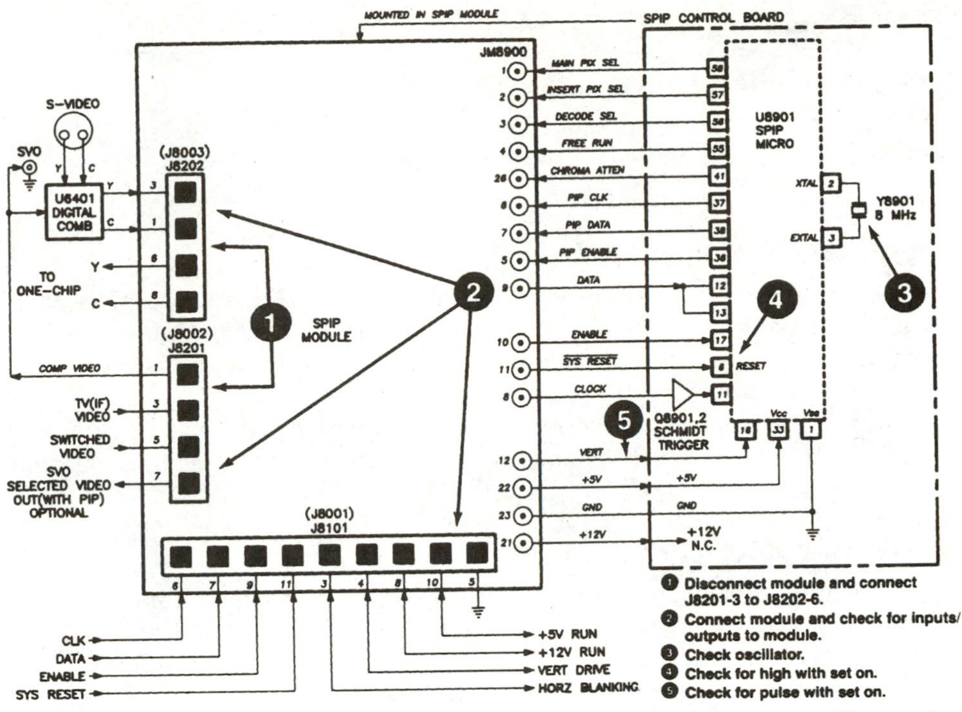

10-8 SPIP interconnect and control board. (*Courtesy of Thomson Electronics Corporation.*)

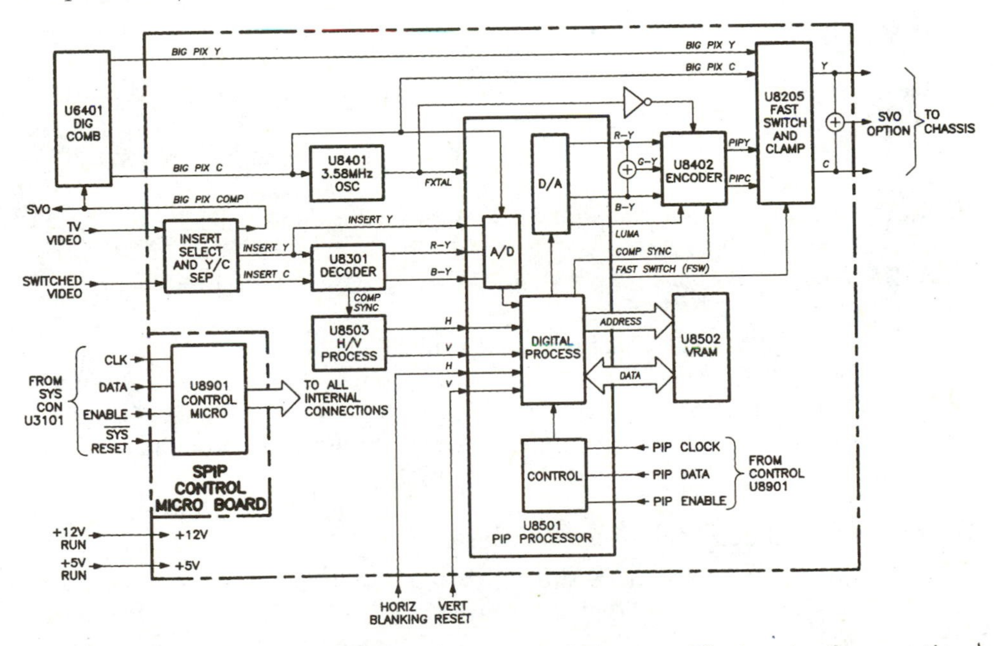

10-9 SPIP module block diagram. (*Courtesy of Thomson Electronics Corporation.*)

PIP processor (8501)

Analog-to-digital conversion

The analog-to-digital converter within PIP processor U8501 converts the analog PIP luma, B-Y, and R-Y signals into digital information. The big pix chroma signal is used as a reference to establish the chroma level for the small pix.

Digital processing and control

The digital processing stage receives the digital video information from the analog-to-digital converter and stores two fields of video into VRAM U8502. Once in random access memory (RAM), the digital data can be recalled and manipulated by the digital processing stage.

The control stage receives commands from control microprocessor U8901 through the three-wire serial communications bus consisting of the PIP clock, data, and enabler.

Digital-to-analog conversion

The manipulated digital data from the digital processing stage enters the digital-to-analog converter to produce the analog luma, R-Y, and B-Y signals that are fed to the output stages of the SPIP module. These video signals contain the information of the small pix. During full field effects such as freeze or zoom (FF-SPIP only), these signals are used to produce the picture on the entire screen.

SPIP input selection circuit

Figure 10-10 shows the input selection circuits used within the SPIP module. The functions of this stage are as follows:

1. Selects the composite video source for the main (big) and the insert (small) pix.
2. Separates the composite video sources for the big and small pix into separate Y and C signals.
3. Selects the desired small pix Y and C signals to be used to produce the small pix.

The two composite video inputs to the module at J8201 are the television and switched video signals at pins 3 and 5. The television video input is always from the television tuner in the chassis. The switched video source is from U1401 of the video input selection circuit and can contain video from either the television tuner, input 1, or input 2.

Big pix signal path (television video)

The main (big) pix signal is selected by switch U8201. It selects either the video from the television tuner or one of the sources from switched video U1401 on the chassis. Switch U8201 is controlled by pin 2 by the inversion of the main pix select line from SPIP control microprocessor U8901. A high at pin 2 selects the switched

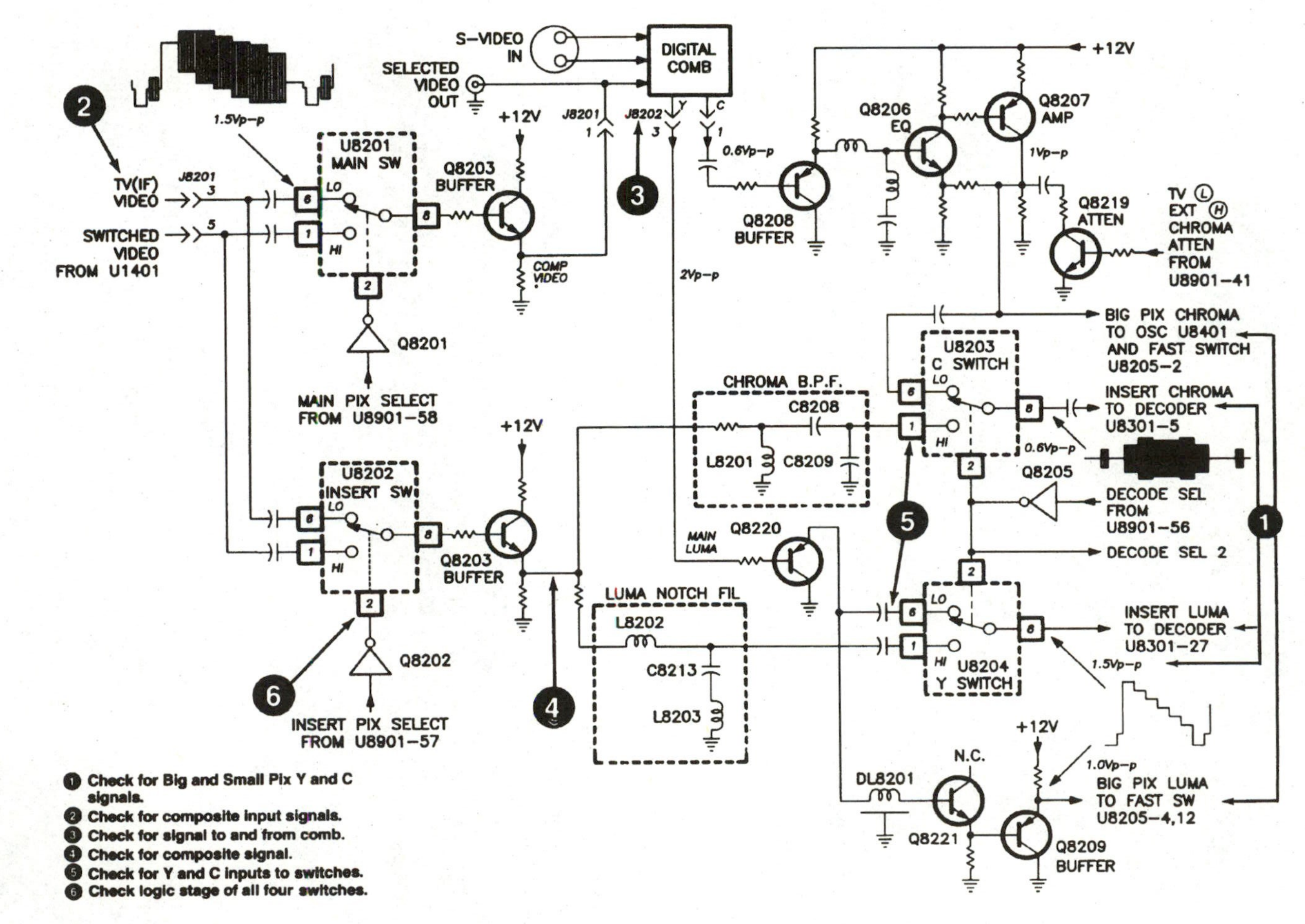

10-10 Input selection circuit (SPIP) diagram. (*Courtesy of Thomson Electronics Corporation.*)

video source, while a low selects the television tuner.

The following table lists the logic level of the main pix select line at U8901-58 under various modes of PIP operation:

Main pix	Insert pix	Main pix select
Tuner	Off	Low
Tuner	Tuner	High
Tuner	Ext	High
Ext	Off	Low
Ext	Tuner	Low

Big pix luma/chroma separation

The output of switch U8201 is buffered by Q8203 and applied to the input of the comb filter (either digital or analog) on the chassis. The signal is also applied to the selected video output jack on the back of the set. Notice that no PIP information is present in the selected video output when taken from this point.

Projection television picture tube arcing and intermittent shutdown

Symptom

Television set turns itself off momentarily and then back on (sometimes accompanied by a snapping or popping sound)

Corrective action

This symptom is normally caused by a picture tube arc. A picture tube arc is an electrostatic discharge occurring inside the picture tube and is commonly experienced during operation of a new television receiver. Arcs are typically caused by minute particles within the tube that exist in normal production. These particles dislodge during shipment and usually cause an arc within the first few hours of operation. The occurrence of these arcs diminishes quickly as the set is operated. Many sets will never experience an arc, while others of the same model may experience several.

Picture tube arcs are common with new tubes from all television manufacturers. Most television chassis are designed to tolerate these electrostatic discharges. However, a major concern is microcomputer lockup. If this occurs, control of operation is lost because the microcomputer becomes "confused" and/or memory contents are altered. To correct lockup, the set must remain unplugged for several minutes. To minimize the probability of a microcomputer lockup and the inconvenience of a customer of having to unplug the set, the chassis is designed to perform a restart function if an arc occurs. The reset circuit momentarily powers down the set immediately after an arc occurs.

When trying to troubleshoot a set for an intermittent shutdown condition, it is necessary to determine how long the set has been in use and how often the shutdown occurs. Shutdowns caused by picture tube arcing are infrequent and decrease in frequency as the television set is used or "burned in." It is unusual to have a picture tube arc after the set has been in use more than a couple of weeks. If the set is new, advise the customer that arcs are not uncommon and will diminish and finally stop as the set is operated. If there is another cause for the momentary shutdowns, the problem normally will get worse as the set is used. If this is the case, then the technician must look for the faulty circuit components.

Zenith PV4543 projection receiver

High-definition, liquid-cooled projection tubes are used to provide a bright, high-resolution, self-converged picture display. Optical coupling is used between the projection tubes and the projection optics for display contrast enhancement. A screen with high-gain contrast and an extended viewer acceptance angle is utilized. Fault mode sensing and electronic shutdown circuits are provided to protect the receiver in the event of a fault mode or CRT arc.

Projection system details

For optics, the television receiver uses three U.S. Precision Lens (USPL) compact delta 7 lenses. This is a new lens design by USPL that incorporates a lightpath fold or bend within the lens assembly. This is done with a front surface mirror with a lightpath bend angle of 72 degrees. Because of this lightpath bend, the outward appearance of the lens resembles, somewhat, that of the upper section of a periscope. The lens elements and the mirror are mounted in a plastic housing. Optical focusing is accomplished by rotating a focus handle with wing nut lock provisions. Rotation of the focus handle changes the longitudinal position of the lens' B element.

Lightpath profile

A side view of the receiver lightpath profile is shown in Fig. 10-11. Note the tight tuck of the light path provided by the Delta 7 compact optics. For comparison purposes, the lightpath profile of the earlier SN4545 set is shown in Fig. 10-12.

Liquid-cooled projection tubes

These receivers use three projection tubes (R, G, and B) arranged in a horizontal in-line configuration. The configuration uses two (R, B) slant-face tubes and one (G) straight-face tube. All tubes are fitted with a metal jacket housing with a clear glass window. The space between the clear glass window and the tubes faceplate is filled with an optical clear liquid. The liquid, which is heat-sinked to the outside world, prevents faceplate temperature rise and thermal gradient differentials from forming across the faceplate when under high-power drive signals. With liquid-cooled tubes, the actual safe power driving level can be essentially doubled over that of the non-liquid-cooled tubes. This is highly desirable in terms of the system's picture brightness. The receivers will be set up for an 18-W drive level instead of the previous 8.5-W drive level.

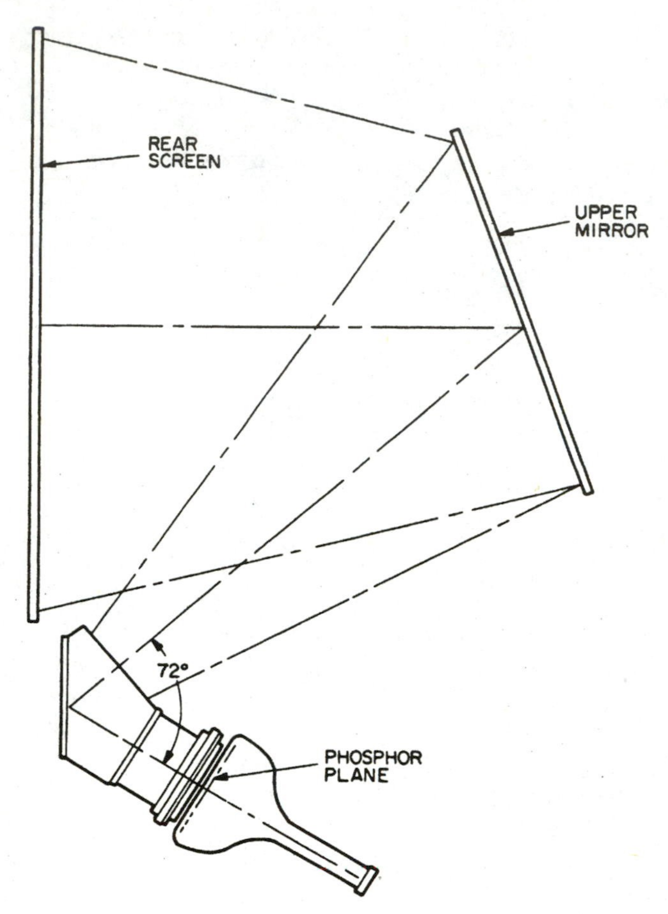

10-11 Side view of projection set light path. (*Courtesy of Zenith Corporation.*)

A side view of the jacket/tube assembly is shown in Fig. 10-13. The metal jacket shell extends back, well over the panel to the funnel seal and thereby functions as an effective x-ray shield. The metal jacket also serves as the mechanical mounting and support for the picture tube assembly. Figure 10-14 shows a front view of the metal jacket. Note that the front of the jacket is elongated and that mounting holes are placed in the elongated sections. This is purposely done to permit the tightest possible tube-to-tube spacing for in-line tube configurations. The projection tubes used in the liquid-cooled assembly are the same 6-in round projection tubes as used in previous projection systems.

Optical coupling

A pliable optical silicone separator is mounted between the glass window on the liquid-cooled jacket assembly and the rear element of the Delta 7 lens. When under mounting pressure, the silicone separator makes intimate contact with these two lightpath interconnecting surfaces.

Self-convergence design

In the receiver, tilted faceplace red and blue tubes, in combination with shifted R and B pointing angles, are image offset and are used to provide for three-image convergence. This combination is required because of the shorter TLC in the Delta 7 lens design and its incompatibility with existing faceplate tilt angles. Since the receiver is a self-converged system, registration of only the three images will be required. This is accomplished with the 9-180 raster registration PC board.

Deflection yoke

The deflection yoke used in the receiver will be the 95-3464-01. This is the same yoke as in the 94-3464 but is supplied with Molex connectors.

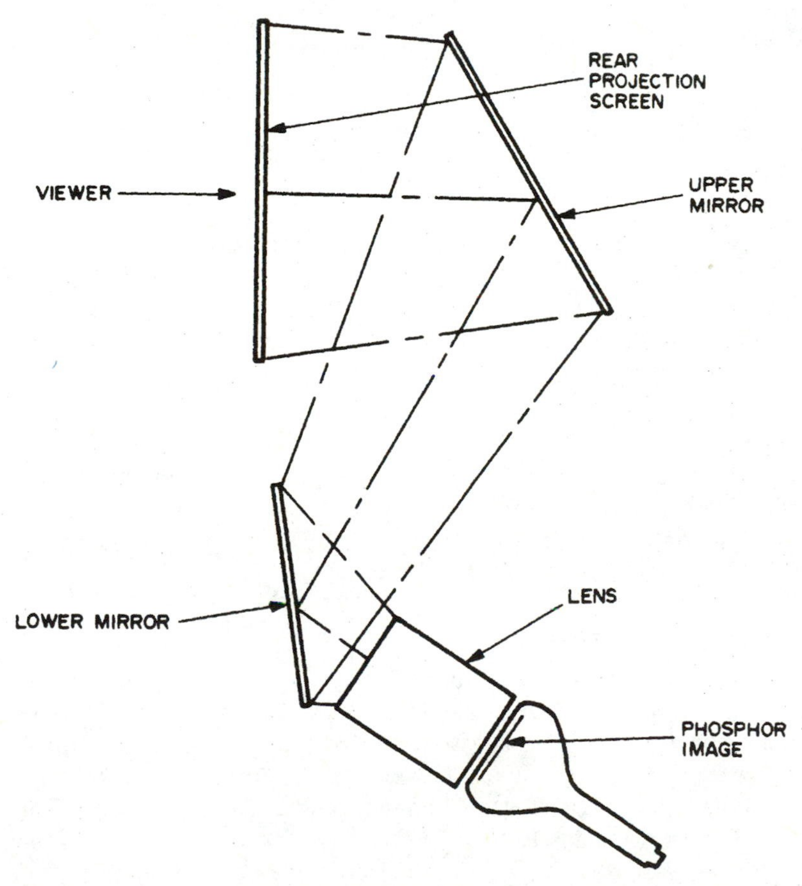

10-12 Side view of Zenith SN4545 light path. (*Courtesy of Zenith Corporation.*)

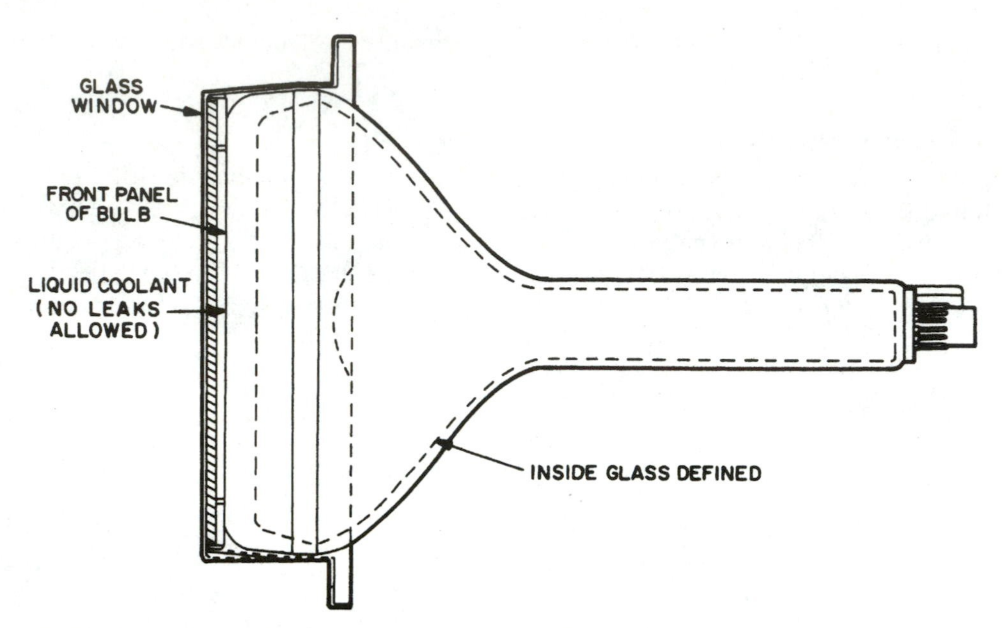

10-13 Liquid-cooled CRT and jacket assembly. (*Courtesy of Zenith Corporation.*)

Projection screen

The projection screen for the receiver will be a two-piece assembly. The front (viewer side) piece will be a vertical lenticular black-striped section. The rear piece is a vertical off-centered Fresnel section. The black striping not only improves initial contrast but also enhances picture brightness and quality for greater viewer pleasure under typical room ambient lighting conditions.

Picture brightness

The receivers demonstrate increased picture brightness over previous Zenith projection models. This is realized by the use of liquid-cooled projection tubes and their ability to accommodate higher-power drive signals. The improvement will be substantial but probably not as great as a 2:1 ratio.

9-323 Switch-mode power supply

The new Zenith projection television incorporating the 9-323 switch-mode power supply and the 9-153-08 sweep module differs from the projection set using the VRT. In the old system, the VRT, along with the 9-154 low-voltage power supply, supplies 132 V dc for the sweep, 62 V dc for the vertical, 25 V dc, 24.5 dc, and 12 V dc for the horizontal drive, 6.05 V rms for the filaments, and an 8-V pilot supply. The space command transformer supplies 12 V dc for the space command tuner, and the audio 60-Hz transformer supplies power to the audio amplifier. The other supply voltages are developed from the sweep system.

These receivers do not use the VRT space command transformer and audio transformer, replacing them with the 9-323 switch-mode power supply, as shown in

Fig. 10-15. The switch-mode power supply is always running as long as the set is plugged into an ac source.

This new system is not a Wessel system, however. There is no longer a regulation transformer as with the old system. The main B+ for the sweep is now regulated by the switch-mode power supply.

The manner in which the switch-mode power supply operates is as follows: When the set is plugged in, the switch mode turns on and remains in standby mode. The 12-V standby is supplied to the space command module. The 132 V is initially unregulated and will be approximately 150 V. This is so because the sweep circuitry is not active and no load is present to load down the supply voltage.

An off/on pulse from the tuner goes to the base of Q3402. Initially it is in the off state, and the base of Q3402 is at 0 V. Transistor Q3402 is also in the off state. This causes Q3403 and Q3404 to be in the off state as well, thus supplying on voltage to the 35-V and the switching 12-V supplies. When the on/off buttons is activated, the pulse goes to a high state, supplying current to the base of Q3402 and thus turning

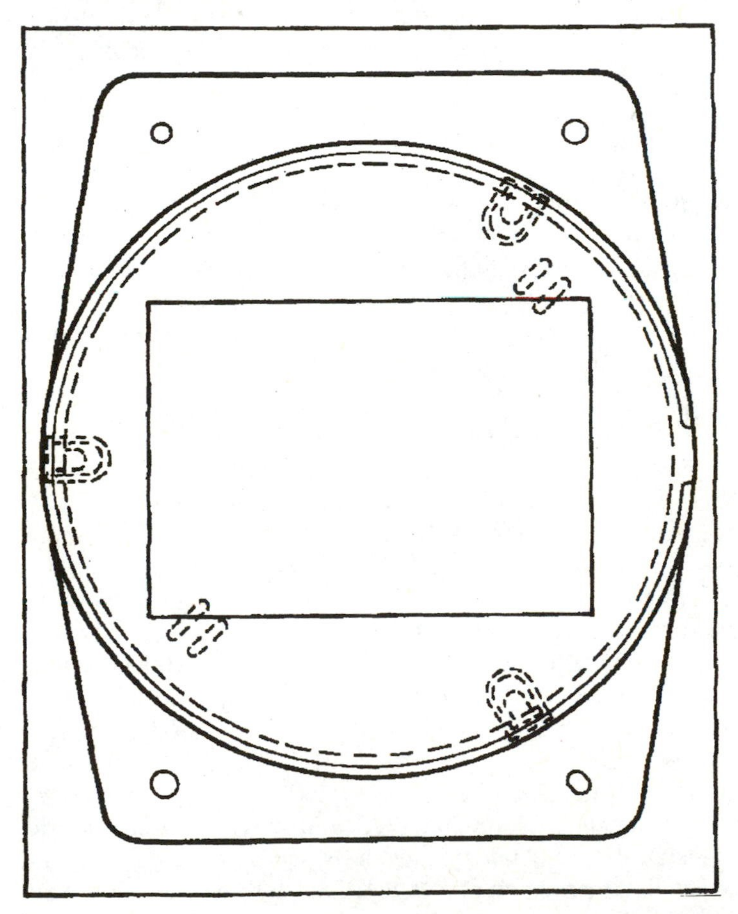

10-14 Front view of CRT mounting jacket. (*Courtesy of Zenith Corporation.*)

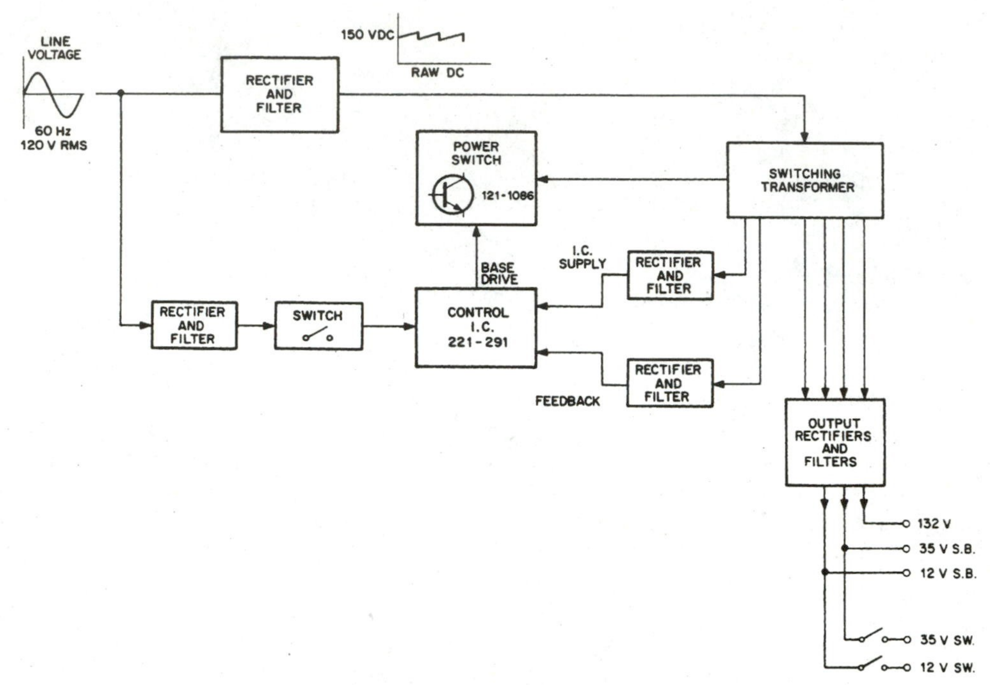

10-15 9-323 switch-mode power supply. (*Courtesy of Zenith Corporation.*)

on Q3402, Q3403, and Q3404. The 12-V switched and the 35-V supplies will come on. When this happens, the 12-V switched supply will begin to supply power to the horizontal drive. The horizontal drive will begin to drive the sweep, and the high voltage will come up, thus powering up the projection television receiver.

Coupling pad installation on CRTs

For replacement of 192-727-04 optical coupling pad (for green CRT) and 192-727-05 optical coupling pad (for red and green CRTs). Field replacement of liquid-cooled projection CRTs require replacement of the optical coupling pad, since the original pad will stick to the defective CRT and cannot be salvaged. Optical coupling pads are supplied with a clear Mylar protective sheet covering both sides. Identification and positioning instructions are printed on these Mylar sheets.

The pads are made of silicone with one tacky side that is always installed against the CRT glass. See Fig. 10-16 for construction of the optical coupling device.

The 192-727-04 pad is for the green CRT and is of equal thickness. The center of the Mylar cover on the "firm" side will be printed "Green CRT side." The reverse or "gummy" side will be printed "Green lens side."

The 192-727-05 pads are for the red and blue CRTs and are thicker on one side. The thick end is always installed on the inboard side of the CRT. Refer to Fig. 10-17. The cover on the "firm" side has printed on one edge "Blue CRT top" and on the opposite edge "Red CRT top." The reverse or "gummy" side is printed "Blue lens top" and on the opposite edge "Red lens top." Refer to Fig. 10-18.

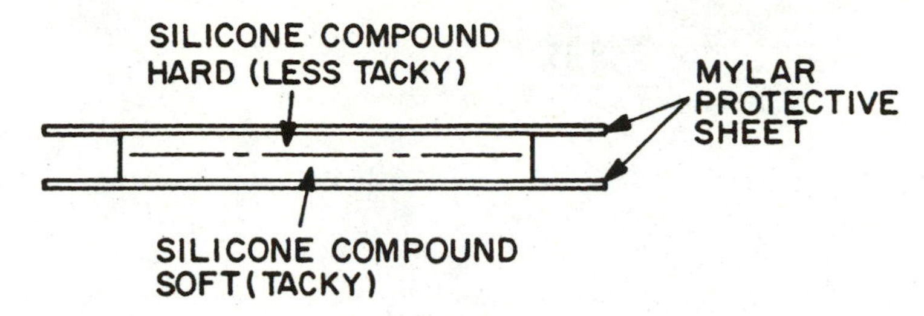

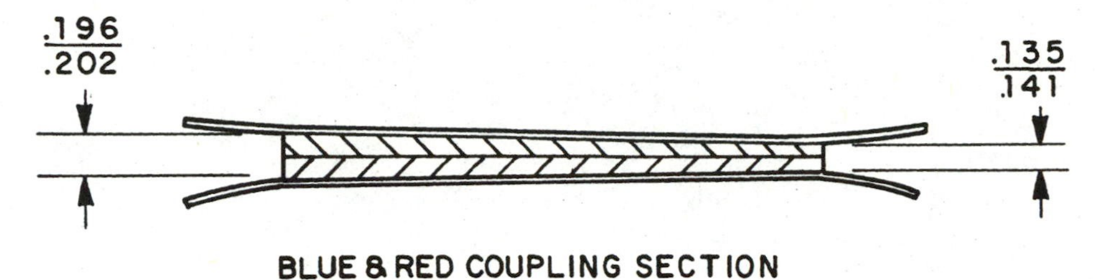

10-16 Blue and red coupling section. (*Courtesy of Zenith Corporation.*)

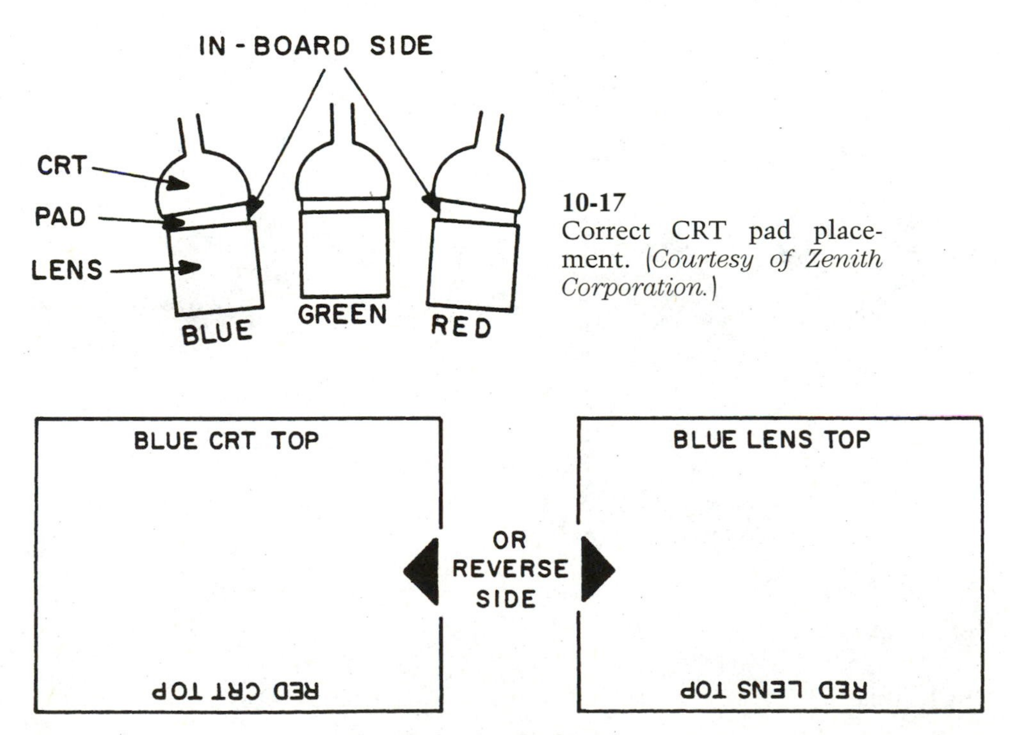

10-17 Correct CRT pad placement. (*Courtesy of Zenith Corporation.*)

10-18 Correct blue and red pad placement. (*Courtesy of Zenith Corporation.*)

Coupling pad replacement tips

1. CRTs requiring new optical coupling pads should be installed on a CRT mounting plate without a coupling pad.
2. Remove front access panel from the set, and remove projection lens from the CRT mounting plate.
3. Remove the Mylar sheet from the "firm" side of pad (if a red or blue CRT is being replaced, position the thick side of pad to the inboard side of tube), center the pad along one edge of the metal aperture on the CRT faceplate, and with a rolling motion, roll the pad onto the CRT faceplate. Smooth out any wrinkles or bubbles, and then remove the lens side Mylar sheet. (A properly installed pad should be flat to the glass inside the metal aperture opening on the CRT faceplate.)
4. Replace the projection lens by carefully aligning the positioning notch on the lens with the notch on the CRT mounting plate. Once the lens is touching the optical pad, it cannot be turned or twisted for alignment, since it would tear or wrinkle the pad.
5. Tighten the lens mounting screws in diagonal sequence to obtain even compression of the pad between the lens and the CRT. The complete lens and CRT assembly is shown in Fig. 10-19.

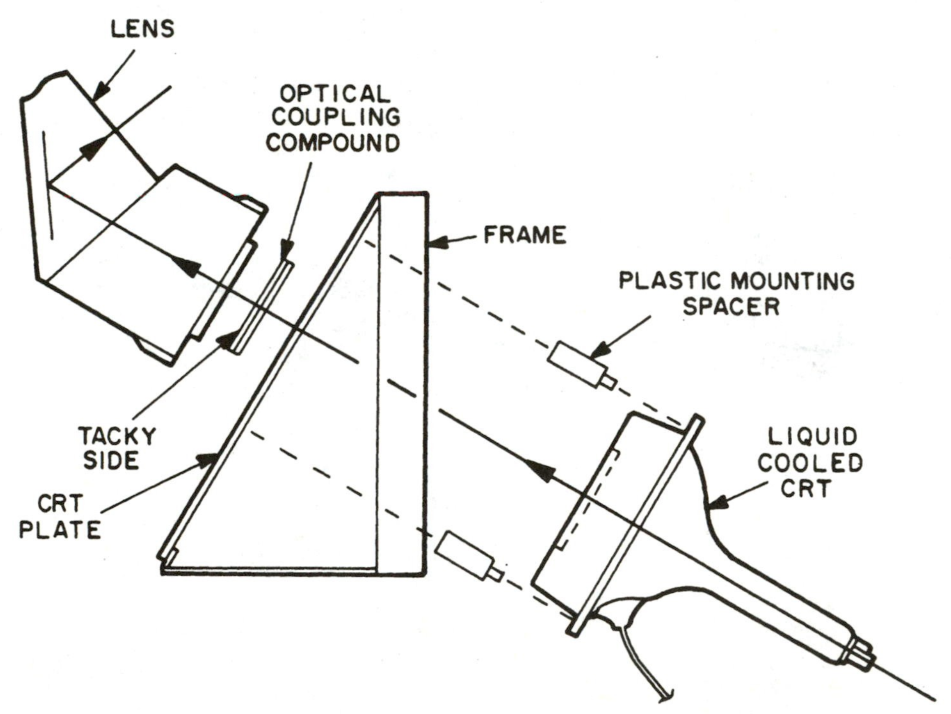

10-19 Lens, CRT, and optical compound assay. (*Courtesy of Zenith Corporation.*)

11 CHAPTER

Audio circuits and digital audio compression standards

This chapter begins with an explanation of audio circuit operations of late-model color television receivers. The chapter continues with some “sound” service tips, how to perform fault detector troubleshooting, and other technical tips. The chapter concludes with a review of how the new digital audio compression system operates and the standards required.

Digital audio IC overview

The following digital audio circuits are found in late-model RCA and GE color television receivers. All audio functions except power amplification in the RCA CTC168 chassis are performed by the digital audio IC U1600. It performs audio source selection, MTS/SAP decoding and selection, expanded stereo operation, volume, tone, and balance. All audio enters the IC in analog form, is converted to digital for processing, and is reconverted to analog prior to power amplification. Looking at Fig. 11-1, you will notice that there is only one alignment, wide-band audio level. The IC is controlled by a three-wire serial communications bus from the system control IC.

Input selection

U1600 selects one of three audio sources: the wide-band audio from the internal tuner and two external stereo audio inputs from the backpanel. The input selection is controlled by the serial clock, data, and enable lines from the system control microprocessor. There are no individual control lines for audio selection from the microprocessor as in the video selection stages of this chassis.

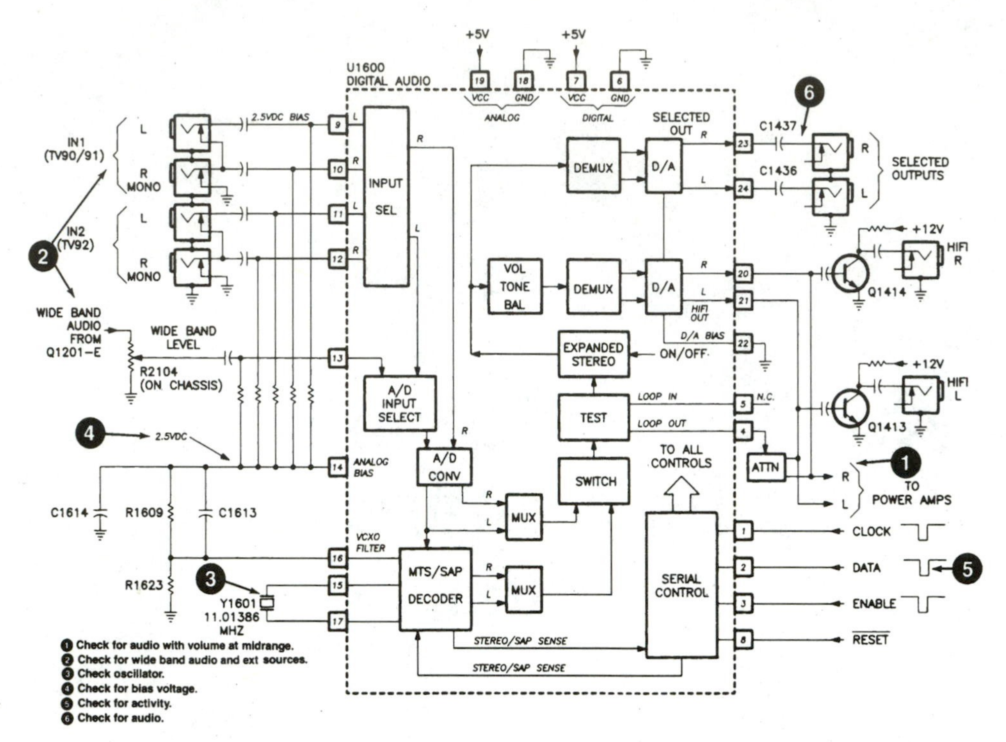

11-1 Block diagram of a digital audio IC. (*Courtesy of Thomson Electronics Corporation.*)

Analog-to-digital conversion

The audio signals are all biased to 2.5 V dc before entering the IC at pins 9 through 13. The bias voltage is provided by pin 14 of U1600. The input 1 audio source is heard by selecting television channel 91 or 90 (S-video). Input 2 audio is heard by selecting television channel 92. The right output of the input select stage is applied directly to an analog-to-digital (A/D) converter. The left channel goes to the A/D input select stage, which chooses between the left external channel or the wide-band audio signal from the tuner. The output of the A/D input select stage is applied to the A/D converter.

MTS/SAP decoder

When the television tuner is selected, the digitized wide-band audio signal is decoded by the MTS/SAP decoder. The 11.01386-MHz oscillator runs at 700 times the stereo pilot signal and is phase locked to it during stereo reception. The VCXO filter at pin 16 affects the phase-locking operation. The VCXO pin 16 is biased to 2.5 V dc from the analog bias pin 14. C1614 is the filter cap for the analog bias voltage. If this cap is defective, you will hear loud "motorboating" from the speakers. C1613 is a filter cap for the VCXO pin 16. If it is defective, the VCXO will not lock the pilot signal and will give you random stereo reception accompanied by static in the audio.

The free-run frequency of the oscillator at pin 15 should be 11,014,000±500 Hz with a mono station tuned and mono mode selected in the audio menu. The VCXO cannot lock to the pilot signal if the free-run frequency is too far off.

The system control microprocessor checks for the availability of stereo and SAP about four times per second through the serial control bus. If stereo or SAP is present and is selected in the audio menu, U1600 is commanded by the system control microprocessor to select stereo or SAP.

The right and left outputs of the MTS/SAP decoder are multiplexed together and applied to a switch that selects between the tuner audio or the external audio sources. The output of the switch passes through a test stage. The loop of the test stage will be used in some chassis to attenuate the audio signal at low volume levels to reduce the noise content of the audio signal.

Volume, tone, and balance and high-fidelity outputs

The output of the expanded stereo stage follows two paths. One path is through the volume, tone, and balance stage. All three settings are controlled inputs. Muting is also performed at this stage during channel change and when the viewer presses the mute button on the remote control.

After the volume, tone, and balance control, the audio signal is demultiplexed to produce individual right and left channel digital information. The digital-to-analog converter stage then produces right and left analog signals at pins 20 and 21. These signals are buffered by Q1413 and Q1414 and passed to the high-fidelity output jacks on the back cover.

Selected audio outputs

The other path for the output of the expanded stereo stage is to a demultiplexer stage and a digital-to-analog (D/A) converter. The right and left outputs of this D/A converter are at pins 23 and 24 and are capacitively coupled to the selected output jacks on the back of the set. The audio sources for these jacks are the same as for the high-fidelity output jacks except they are not affected by the volume, tone, and balance controls.

Clock, data, enable, and reset

As mentioned previously, all commands are delivered to U1600 through the serial communications bus. You will find more information on this subject under serial bus operation and troubleshooting circuit operation.

Digital audio IC troubleshooting

Symptom: No audio

1. Make sure the volume is not at minimum and the speakers are turned on in the audio menu. Connect an audio source to one of the external audio input jacks. If audio can be heard, the problem must lie in the wide-band audio from the tuner or in the MTS/SAP decoder section. If no audio can be heard from either internal or external, then go on to step 2.

2. Check for audio at pins 20 and 21 of U1600. If present, the problem may just be in the power amplifier stage. If audio is not present at pins 20 and 21 but is present at pins 23 and 24, the problem must be in the volume control section of U1600 or in the serial communications bus.

Symptom: OSD stereo indicator never shows stereo on a known stereo station

1. Select mono mode in the audio menu.
2. Connect a 10× low-capacity probe to pin 15 of U1600, and measure the frequency with a frequency counter. It should indicate 11,014,000±500 Hz.
3. Check for defective C1613 or C1614.
4. Check for 2.5 V dc at U1600-14.
5. Check wide-band audio level and bandwidth.

Symptom: Poor stereo separation

Adjust wide-band audio level according to service data.

Power on/off audio mute circuit

The speaker on/off control prevents pops in the internal speakers of the set during power on/off but has no effect on the selected and high-fidelity audio out jacks that may be connected to an external audio amplifier. Figure 11-2 shows the circuit used to prevent these pops from occurring. The circuit may be mounted on an adapter board or may be on the audio sip board. The circuit operates the same regardless of where it is located.

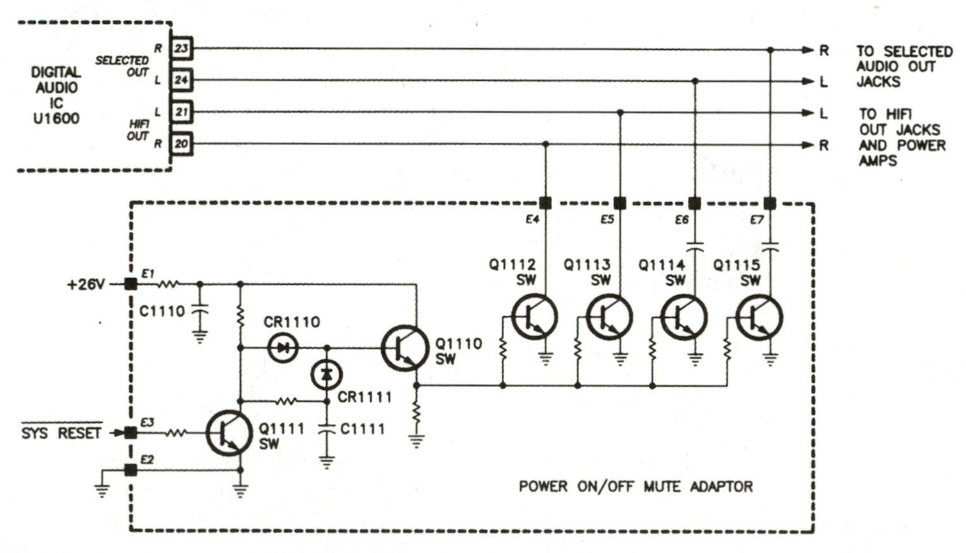

11-2 Set power on/off and audio mute circuit. (*Courtesy of Thomson Electronics Corporation.*)

Set-off mode

When the set is turned off, the 26-V supply is available to charge C1111. The system reset line is low and keeps Q1111 off. The 26-V supply also biases Q1110 on through CR1110. With Q1110 conducting, its emitter voltage turns on Q1112 through Q1115. These transistors mute the right and left audio signals from the selected and high-fidelity outputs of U1600.

Set off to on transitions

When the set is first turned on, the system reset line stays low for about 1 s and then goes high until the set is turned off. After the system reset line goes high, Q1111 turns on to ground the anode of CR1110.

Set on to off transition

When the set enters the off state, the system reset line goes low. This turns off Q1111 and applies base bias from the 26-V supply to Q1110. Q1110 turns on to supply voltage to its emitter to turn on all the mute transistors Q1112 through Q1115.

Troubleshooting tip

If you encounter a "no audio" symptom, check Q1110 in the power on/off mute circuit. There should be no voltage at the emitter at Q1110 after the set is on for at least 5 s. If there is voltage at the emitter, Q1112 through Q1115 will be turned on to mute the audio.

Audio power amplifier: 1 and 5 W

The right and left volume, tone, and balance-controlled audio outputs from the digital audio IC U1600 are capacitively coupled to power amplifier U1900 (Fig. 11-3). The power amplifiers for both the right and left channels are contained within U1900. U1900 can be operated as a 1- or 5-W amplifier. The capacity of the output coupling caps is increased in the 5-W system. The speaker impedance in the 5-W system is lowered to 8 instead of 32 Ω, as used in the 1-W system. Both systems run U1900 at 24 V dc at pin 9.

Low-volume attenuator

During low volume levels, the attenuator circuit tied to the right and left outputs of U1600 decreases the level of the audio signal to decrease noise in the audio output. As mentioned previously, early production sets may not have the software and hardware installed to activate the attenuator.

Speaker on/off control

The speakers, both internal and external, may be turned off from the audio menu. The system control microprocessor U3101 drives the speakers off line at pin 27 high to turn the speakers off. The Hi turns Q1901 on, which pulls the 13-V bias at U1900-3 to 0 V. This in turn disables the power amplifier.

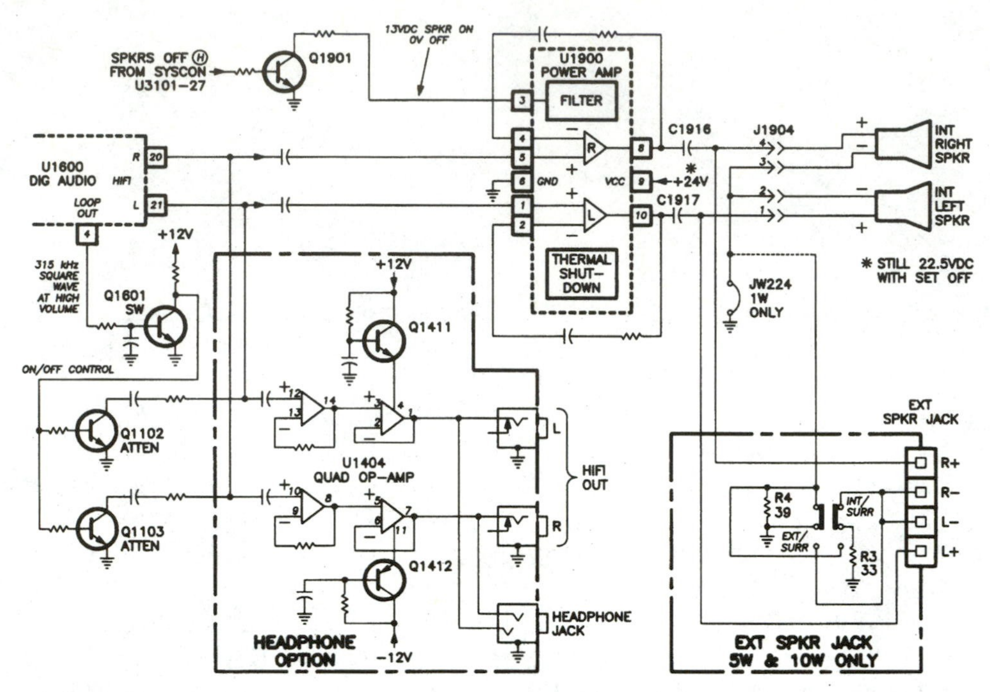

11-3 RCA television receiver audio output circuit. (*Courtesy of Thomson Electronics Corporation.*)

When troubleshooting loss of audio symptoms, be sure to check for the 13 V dc at pin 3 of U1900 before assuming the IC is defective.

External speaker jack and surround switch

The output of the power amplifier is applied to the positive side of the internal speakers and to the positive terminals of the external speaker jacks. The external speaker jack is only available in 5- and 10-W systems.

Headphone option

Some models are equipped with a headphone jack, as shown in Fig. 11-3. The volume, tone, and balance-controlled outputs of U1600 are buffered by the quad OP AMP U1404 to provide the drive for the headphones.

Audio power amplifier troubleshooting

Use the following steps for symptoms of no audio output from the speakers.

1. Check for audio at U1600 pins 20 and 21. If present, go on to step 2. If audio is not present, troubleshoot the U1600 chip.
2. Check for signal at pins 5 and 1 of U1900. If present, check for 24 V dc at pin 9. If the heat sink is extremely hot, turn off the set and let it cool down. Turn the set on to see if audio returns. If not, go on to step 3.

3. Check for about 13 V dc at U1900-3. If 0 V, check for proper operation of speakers off control line at system control U3101-27 or defective Q1901. If all checks out OK, go on to step 4.
4. Check output coupling caps and speaker continuity.

10-W audio power amplifier with fault detect

The circuit in Fig. 11-4 shows the 10 W per channel audio output stage used in all CTC169 projection sets. The power amplifiers consist of separate ICs, U1901 and U1902. Each IC is powered by a +17- and −17-V dc supply as opposed to +24 V for the lower-wattage amplifiers. Also note that the outputs of the power amplifiers are direct coupled (no capacitors) to the speakers.

Fault detect circuit

Since the 10-W audio system is direct coupled to the speakers, a dc fault detector is required to prevent high dc voltages at the amplifier outputs from destroying the power amplifiers and speakers. The fault detector is designed to turn the television off when a high positive or negative dc voltage is detected on the audio lines.

Fault detector troubleshooting

If you suspect the set is shutting down due to the fault detector, place the INT/SURR switch to the EXT/SURR position and disconnect any external speakers. Ground the base of Q1907, and turn the set back on. If it still shuts down, the shutdown must be caused by loss of the 9-V run supply. If the shutdown stops, immediately turn the set off and check the audio output stage for shorts or defective components. Before returning the set to the customer, be sure to place the INT/SURR switch to the internal position. Check the voltage on the collector of Q1907. If it goes low when the set is on, the fault detector is causing the shutdown. If it never goes low, your problem is elsewhere.

If the fault detector is causing the shutdown, check the right and left power amplifier outputs for high dc offsets that could fool the fault detector into being tripped. If the outputs appear OK, the problem may be in the fault detector itself.

No audio

1. Make sure there is a good signal from the tuner (clear picture). If not, troubleshoot the tuner and i.f. circuits.
2. Look to see if the stereo indicator responds to stereo broadcasts. If it does, this indicates that the wide-band audio is making it from the i.f. to the decoder stages. If it does not, check for the wide-band audio signal at pin 13 of U1600 in the digital audio circuit and pin 2 of U1600 on the analog audio circuit.
3. Apply line level audio to the AUX 1 or AUX 2 and select these sources.

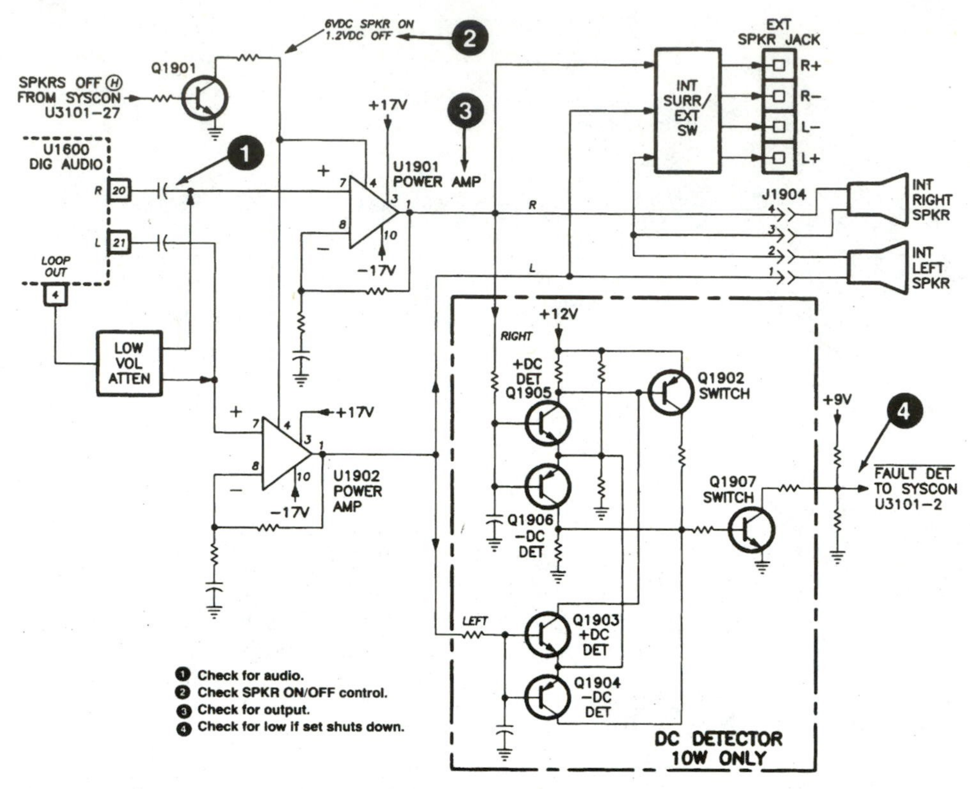

11-4 Audio output circuit with fault detector. (*Courtesy of Thomson Electronics Corporation.*)

Analog audio circuit

4. If the audio returns, the audio circuit from U1410 to the audio output ICs has been confirmed. If it does not, check the switching logic to U1410 pins 6, 9, and 10.

Digital audio circuit

If the audio returns, U1600 is operating for the most part. A peripheral component, such as a capacitor, is causing a decoding problem. Refer to Fig. 11.1.

5. If the audio does not return, use an oscilloscope to trace the audio through the decoder circuitry and isolate the point where the audio stops.
6. Perform dc voltage checks on all the pins of the ICs to isolate the defective component. The block diagram for the analog audio circuit is shown in Fig. 11-5.

Distorted audio tips

One of the more difficult audio problems to troubleshoot and isolate is distorted audio. In order to effectively diagnose such a problem, an oscilloscope, an audio generator, and an rf/i.f. generator are required.

1. Confirm a good signal from the tuner. Make sure the picture is clear with no beats in it.
2. Apply an audio signal to the input of AUX 1 or AUX 2. If the distortion continues, the problem is after the decoder stages. If there is no distortion, the problem is most likely in the i.f. stages.
3. If the problem is after the decoder stage, apply a sine-wave signal to one of the AUX inputs and trace the signal with an oscilloscope. Locate the point where the distortion first appears, and check the associated components.

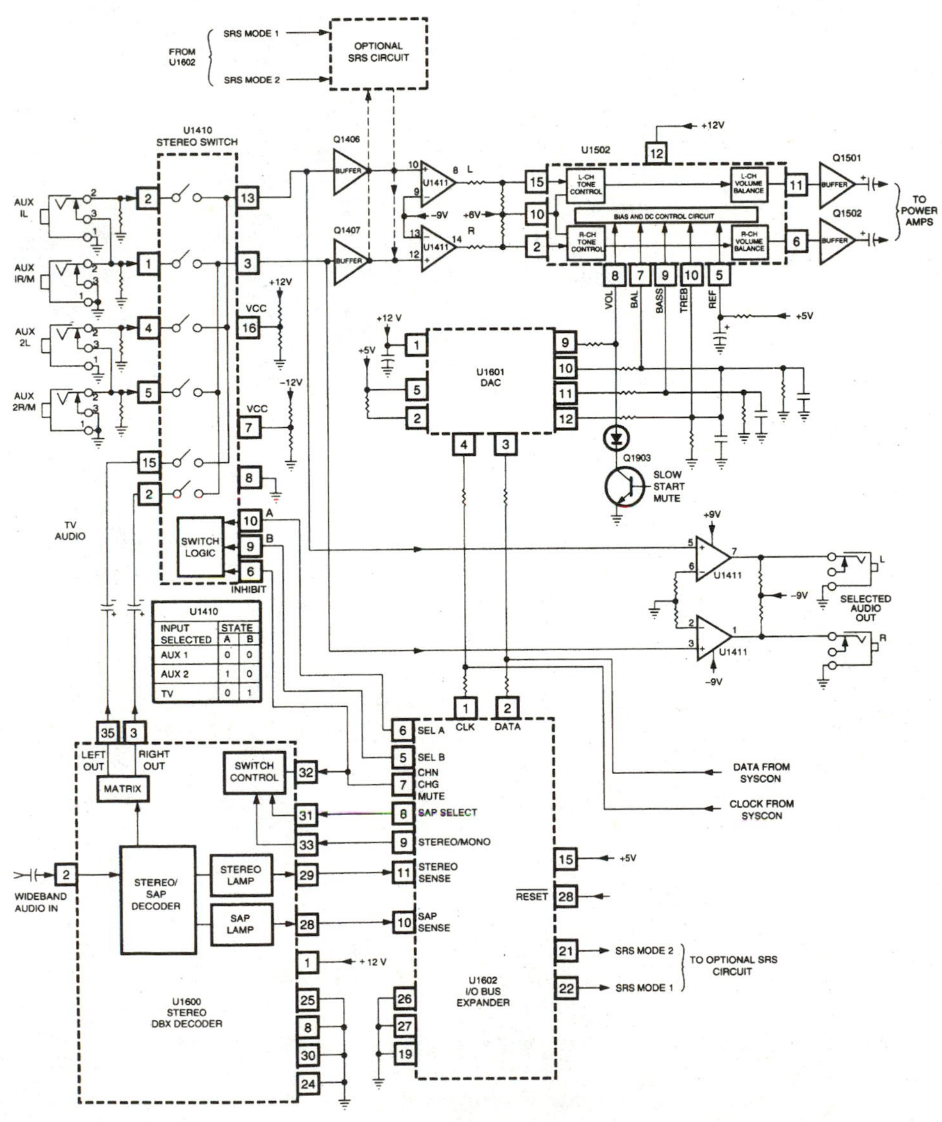

11-5 Analog audio circuit block diagram. (*Courtesy of Thomson Electronics Corporation.*)

4. If the distortion is originating in the audio i.f. stages, apply an appropriate rf or i.f. signal with a sine-wave–modulated signal. Trace the signal to the point where the distortion begins, and check out the associated components. Typical i.f. problems include bad ceramic filters, detuned coils, and leaky capacitors.

Technical tips

The symptoms are the mono audio sound is low, but stereo is normal. When in mono, the audio level is half the level of stereo. Also, the audio is distorted. In finding the solution, a scope shows that the wide-band audio from pin 52 of U1001 is normal. The fault was traced to C1203 (connected to the wiper of R1204, WBA level adjustment).

The symptom is a dead television set. Removing P1903 allows the set to come on and operate normally. The solution is to replace the U1902 chip, which returns the set to normal operation.

Another symptom is audio distortion when the volume is set at a low level. The solution is found with a scope when checking the wide-band audio from the CTV processor U1001 pin 52, which reveals some distortion. Adjustment of the audio detector coil offers no improvement. Replacing the U1001 chip restores the set to normal audio.

Audio problems and solutions

The audio problems are audio hiss, audio "pop," and nonlinear volume control action.

Symptom: Some of the early production RCA CTC168/CTC169 chassis may have potentially objectional audio performance. Audio hiss (background noise) becomes noticeable as the volume level exceeds the midpoint on the volume level display. An audio "pop" may be heard at power turn-on. The volume control taper is nonlinear, and this causes nominal volume to occur at a high control setting.

Corrective action: The audio board and the system control microprocessor must be replaced to correct the symptom. It will be necessary to replace both the audio circuit board and the U3101 system control IC in all chassis listed below.

Never perform the modification unless both parts are changed at the same time. The volume control circuit has been changed, and the new audio board will not operate with the old microcomputer. It is necessary to check the chassis version to determine the proper parts needed. The chassis version is printed on the label located on the heat sink for the horizontal output transistor. Table 11-1 gives the stock numbers for the various combinations of audio boards and ICs. This modification is required only on models equipped with the specific chassis versions listed in Table 11-1.

Introduction to digital audio compression

The reasons for using *digital audio compression* are to more efficiently broadcast or record audio signals and to reduce the amount of information required to represent audio signals. In the case of digital audio signals, the amount of digital

Table 11-1. Models and chassis that require modification

Model(s)	Chassis	Stock no.	Description
20XT9006	CTC168C	205999	Audio circuit board
FX209002		206395	U3101 microcomputer
F20705	CTC168E	205991	Audio circuit board
		206395	U3101 microcomputer
PS20110	CTC168F	205994	Audio circuit board
		206395	U3101 microcomputer
F27187	CTC169AB	205991	Audio circuit board
F27188		206395	U3101 microcomputer
F27191	CTC169AC	205991	Audio circuit board
F27193		205995	U3101 microcomputer
G27390			
G27391			

SOURCE: Thomson Electronics Corporation.

information required to reproduce the original pulse-code-modulated (PCM) samples accurately may be reduced by applying a digital compression algorithm, resulting in a digitally compressed representation of the original signal. The goal of the digital compression algorithm is to produce a digital representation of an audio signal which, when decoded and reproduced, sounds the same as the original signal while using a minimum of digital information (bit rate) for the compressed (or encoded) representation. The AC-3 digital compression algorithm can encode from 1 to 5.1 channels of source audio from a PCM signal into a serial bit stream at data rates ranging from 32 to 640 kb/s. The 0.1 channel refers to a fractional bandwidth channel intended to convey only low-frequency (subwoofer) signals.

A typical application of the algorithm is shown in Fig. 11-6. In this example, a 5.1-channel audio program is converted from a PCM representation requiring more than 55 Mb/s into a 384-kb/s serial bit stream by the AC-3 encoder. Satellite transmission equipment converts this bit stream to an rf transmission that is directed to a satellite transponder. The amount of bandwidth and power required by the transmission have been reduced by more than a factor of 13 by the AC-3 digital compression. The signal received from the satellite is demodulated back into the 384-kb/s serial bit stream and decoded by the AC-3 decoder. The result is the original 5.1-channel audio program.

Digital compression of audio is useful wherever there is an economic benefit to be obtained by reducing the amount of digital information required to represent the audio. Typical applications are in satellite or terrestrial audio broadcasting, delivery of audio over metallic or optical cables, or storage of audio on magnetic, optical, semiconductor, or other storage media.

Encoding process

The AC-3 encoder accepts PCM audio and produces an encoded bit stream consistent with this standard. The specifics of the audio encoding process are not nor-

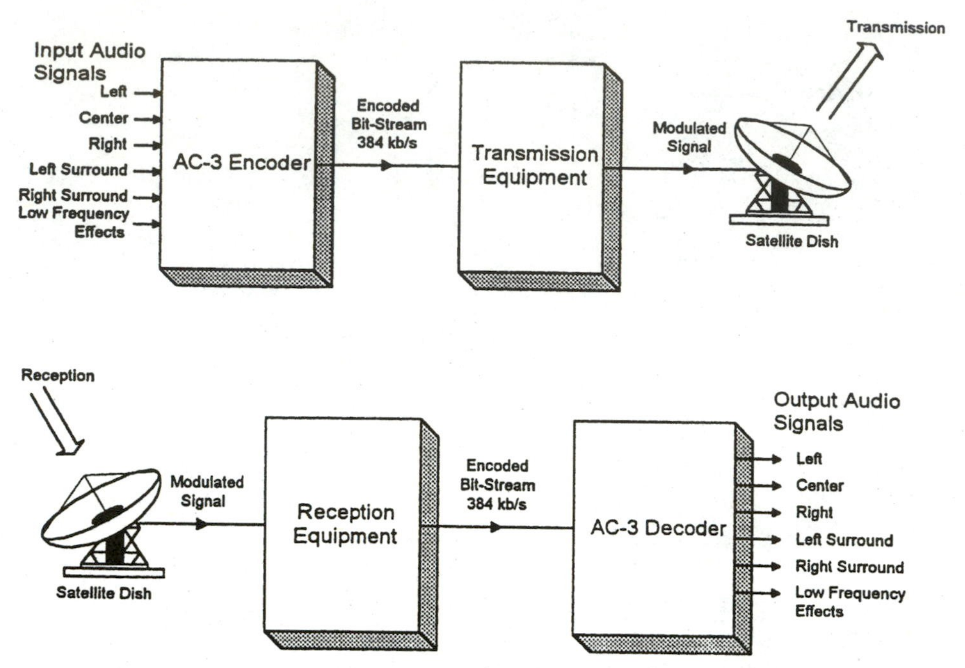

11-6 AC-3 to satellite audio transmission. (*Courtesy of Thomson Electronics Corporation.*)

mal requirements of this standard. However, the encoder must produce a bit stream matching the prescribed syntax, which, when decoded, must produce audio of sufficient quality for the intended application. Let us now take a brief look at the encoding process.

The AC-3 algorithm achieves high coding gain (the ratio of the input bit rate to the output bit rate) by coarsely quantizing a frequency-domain representation of the audio signal. A block diagram of this process is shown in Fig. 11-7. The first step in the encoding process is to transform the representation of audio from a sequence of PCM time samples into a sequence of blocks of frequency coefficients. This is done in the analysis filter bank. Overlapping blocks of 512 time samples are multiplied by a time window and transformed into the frequency domain. Because of the overlapping blocks, each PCM input sample is represented in two sequential transformed blocks. The frequency-domain representation may then be decimated by a factor of 2 so that each block contains 256 frequency coefficients. The individual frequency coefficients are represented in binary exponential notation as a binary exponent and a mantissa. The set of exponents is encoded into a coarse representation of the signal spectrum that is referred to as the *spectral envelope.* This spectral envelope is used by the core bit allocation routine, which determines how many bits to use to encode each individual mantissa. The spectral envelope and the coarsely quantized mantissa for 6 audio blocks (1536 audio samples) are formatted into an AC-3 frame. The AC-3 bit stream is a sequence of AC-3 frames.

The actual AC-3 encoder is more complex than indicated in Fig. 11-7. The following functions not shown above are also included:

1. A frame header is attached that contains information (bit rate, sample rate, number of encoded channels, etc.) required to synchronize to and decode the encoded bit stream.
2. Error-detection codes are inserted to allow the decoder to verify that a received frame of data is error-free.
3. The analysis filter bank spectral resolution may be altered dynamically so as to better match the time/frequency characteristic of each audio block.
4. The spectral envelope may be encoded with variable time/frequency resolution.
5. A more complex bit allocation may be performed, and parameters of the core bit allocation routine may be modified so as to produce a more optimal bit allocation.
6. The channels may be coupled together at high frequencies to achieve higher coding gain for operation at lower bit rates.
7. In the 2-channel mode, a rematrixing process may be performed selectively to provide additional coding gain and to allow improved results to be obtained in the event that the 2-channel signal is decoded with a matrix surround decoder.

Decoding process

The decoding process is basically the inverse of the encoding process. The decoder, shown in Fig. 11-8, must synchronize to the encoded bit stream, check for errors, and deformat the various types of data such as the encoded spectral envelope

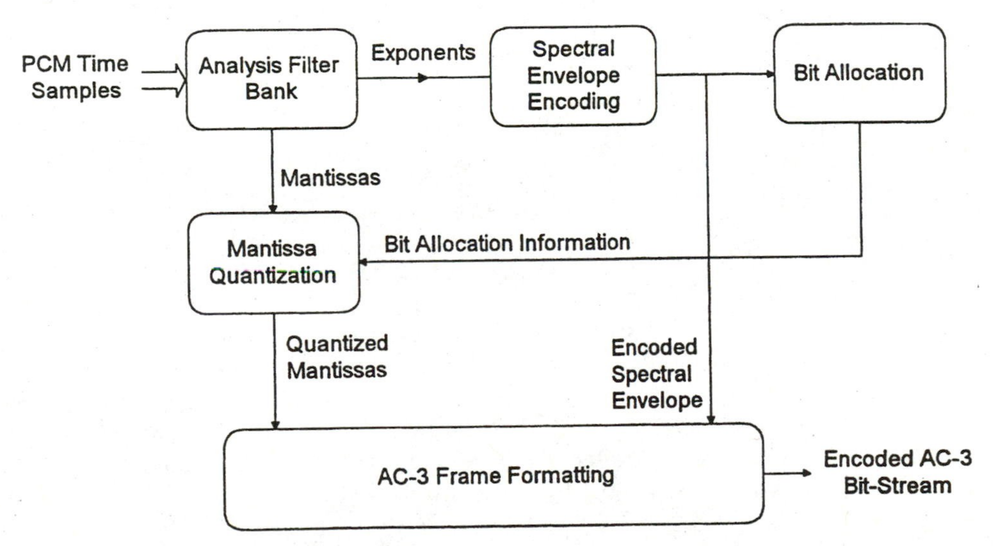

11-7 Blocked-out example of the AC-3 encoder. (*Courtesy of Thomson Electronics Corporation.*)

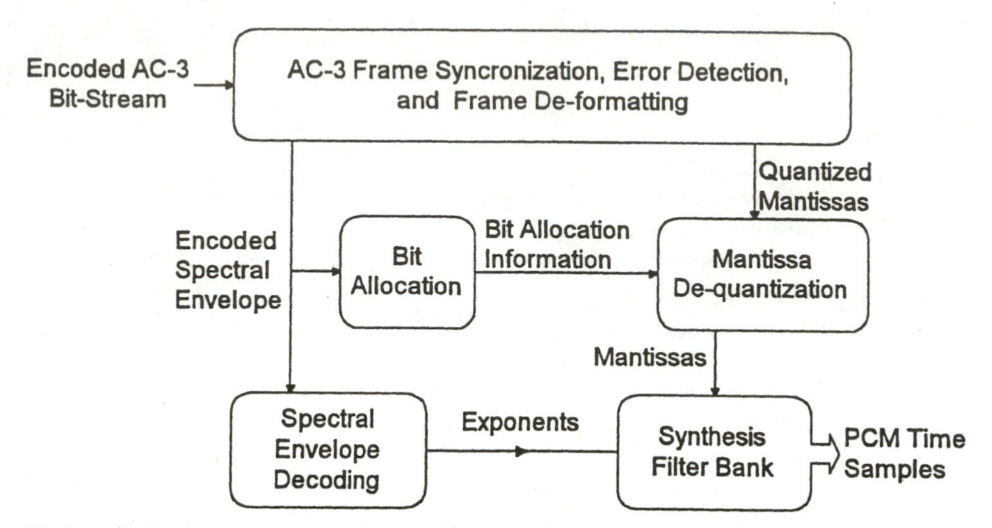

11-8 Blocked-out example of the AC-3 decoder. (*Courtesy of Thomson Electronics Corporation.*)

and the quantized mantissas. The bit allocation routine is run, and the results are used to unpack and dequantize the mantissas. The spectral envelope is decoded to produce the exponents. The exponents and mantissas are transformed back into the time domain to produce the decoded PCM time samples.

The actual AC-3 decoder is more complex than indicated in Fig. 11-8. The following functions not shown above are included:

1. Error concealment or muting may be applied in case a data error is detected.
2. Channels that have had their high-frequency content coupled together must be decoupled.
3. Dematrixing must be applied (in the 2-channel mode) whenever the channels have been rematrixed.
4. The synthesis filter bank resolution must be altered dynamically in the same manner as the encoder analysis filter bank had been during the encoding process.

Digital coding notations

The normal portions of this standard specify a coded representation of audio information and also specify the decoding process. Informative information on the encoding process is included. The coded representation specified herein is suitable for use in digital audio transmissions and storage applications. The coded representation may convey from 1 to 5 full bandwidth audio channels, along with a low-frequency enhancement channel. A wide range of encoded bit rates is supported by this specification. A short-form designation of this audio coding algorithm is *AC-3.*

Synchronization and error detection

The AC-3 bit stream format allows rapid synchronization. The 16-bit sync word has a low probability of false detection. With no input stream alignment, the probability of false detection of the sync word is 0.0015 percent per input stream bit position. For a bit rate of 384 kb/s, the probability of false sync word detection is 19 percent per frame. Byte alignment of the input stream drops this probability to 2.5 percent, and word alignment drops it to 1.2 percent.

When a sync pattern is detected, the decoder may be estimated to be in sync, and one of the CRC words (crc1 or crc2) may be checked. Since crc1 comes first and covers the first ⅝ of the frame, the result of a crc1 check may be available after only ⅝ of the frame has been received. Alternatively, the entire frame size can be received and crc2 checked. If either CRC word checks, the decoder may be presumed to be in sync, and decoding and reproduction of audio may proceed. The chance of false sync in this case would be the concatenation of the probabilities of a false sync word detection and a CRC word misdetection of error. The CRC check is reliable to 0.0015 percent. This probability, concatenated with the probability of a false sync detection in a byte-aligned input bit stream, yields a probability of a false synchronization of 0.000035 percent (or about 1 in 3 million synchronization attempts).

If this small probability of false sync is too large for an application, there are several methods to reduce it. The decoder may only presume correct sync in the case that both CRC words check properly. The decoder may require multiple sync words to be received with the proper alignment. If the data transmission or storage system is aware that data are in error, this information may be made known to the decoder.

Unpack BSI, side information

Inherent to the decoding process is the unpacking (demultiplexing) of the various types of information included in the bit stream. Some of these items may be copied from the input buffer to dedicated registers, some may be copied to specific working memory locations, and some simply may be located in the input buffer with pointers to them saved to another location for use when the information is required.

Decoder exponents.

The exponents are delivered in the bit stream in an encoded form. In order to unpack and decode the exponents, two types of side information are required. First, the number of exponents must be known. For *fbw* channels this may be determined from either *chbwcod {ch}* (for uncoupled channels) or *cplbegf* (for coupled channels). For the coupling channel, the number of exponents may be determined from *cplbegf* and *cplendf*. For the *lfe* channel (when on), there are always 7 exponents. Second, the exponent strategy in use (D15, etc.) by each channel must be known.

Bit allocation.

The bit-allocation computation reveals how many bits are used for each mantissa. The inputs to the bit-allocation computation are the decoded exponents and the bit-allocation side information. The outputs of the bit-allocation computations are a set of bit-allocation pointers (*baps*), one *bap* for each coded mantissa. The *bap* indicates the quantizer used for the mantissa and how many bits in the bit stream were used for each mantissa.

Process mantissas

The coarsely quantized mantissas make up the bulk of the AC-3 data stream. Each mantissa is quantized to a level of precision indicated by the corresponding *bap*. In order to pack the mantissa data more efficiently, some mantissas are grouped together into a single transmitted value. For instance, two 11-level quantized values are conveyed in a single 7-bit code (3.5 bits per value) in the bit stream.

The mantissa data are unpacked by peeling off groups of bits, as indicated by the *baps*. Grouped mantissas must be ungrouped. The individual coded mantissa values are converted into a dequantized value. Mantissas that are indicated as having zero bits may be reproduced as either zero or random dither value (under control of the dither flag).

Decoupling

When coupling is in use, the channels that are coupled must be decoupled. Decoupling involves reconstructing the high-frequency section (exponents and mantissas) of each coupled channel from the common coupling channel and the coupling coordinates for the individual channels. Within each coupling band, the coupling channel coefficients (exponent and mantissa) are multiplied by the individual channel coupling coordinates.

Rematrixing

In the 2/0 audio coding mode, rematrixing may be employed as indicated by the rematrix flags. Where the flag indicates a band is rematrixed, the coefficients encoded in the bit stream are sum and difference values instead of left and right values.

Dynamic range compression

For each block of audio, a dynamic range control value (*dynrng*) may be included in the bit stream. The decoder, by default, uses this value to alter the magnitude of the coefficient. The flow diagram blocks for the audio decoding process are shown in Fig. 11-9.

Inverse transform

The decoding steps described above will result in a set of frequency coefficients for each encoded channel. The inverse transform converts the blocks of frequency coefficients into blocks of time samples.

Window, overlap/add

The individual blocks of time samples must be windowed, and adjacent blocks must be overlapped and added together in order to reconstruct the final continuous time output PCM audio signal.

Downmixing

If the number of channels required at the decoder output is smaller than the number of channels encoded in the bit stream, then downmixing is required. Downmixing

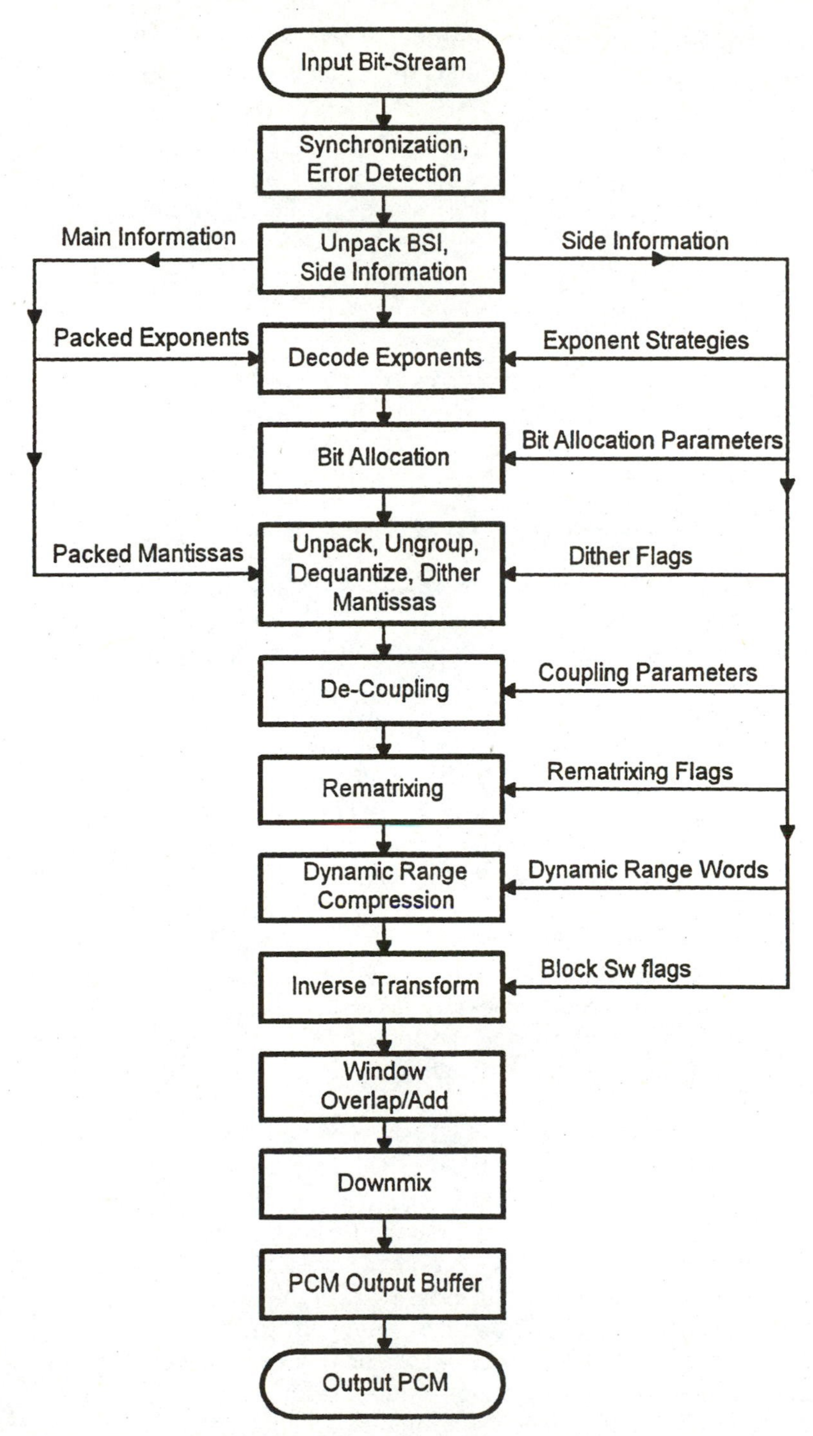

11-9 Flow diagram of the decoder process. (*Courtesy of Thomson Electronics Corporation.*)

in the time domain is shown in the example decoder. Since the inverse transform is a linear operation, it is also possible to downmix in the frequency domain prior to transformation.

PCM output buffer

Typical decoders will provide PCM output samples at the PCM sampling rate. Since blocks of samples result from the decoding process, an output buffer is typically required.

Output PCM

The output PCM samples may be delivered in a form suitable for interconnection to a digital-to-analog (D/A) converter or in any form. This standard does not specify the output PCM format.

Exponent coding overview

The actual audio information conveyed by the AC-3 bit stream consists of the quantized frequency coefficients. The coefficients are delivered in floating-point form, with each coefficient consisting of an exponent and a mantissa. Let us now see how the exponents are encoded and packed into the bit stream.

Exponents are 5-bit values that indicate the number of leading zeros in the binary representation of a frequency coefficient. The exponent acts as a scale factor for each mantissa, equal to 2−exp. Exponent values are allowed to range from 0 (for the largest coefficients with no leading zeros) to 24. Exponents for coefficients that have more than 24 leading zeros are fixed at 24, and the corresponding mantissas are allowed to have leading zeros. Exponents require 5 bits in order to represent all allowed values.

AC-3 bit streams contain coded exponents for all independent channels, all coupled channels, and the coupling and low-frequency effects channels (when they are enabled). Since audio information is not shared across frames, block 0 of every frame will include new exponents for every channel. Exponent information may be shared across blocks within a frame, so blocks 1 through 5 may reuse exponents from previous blocks.

AC-3 exponent transmission uses differential coding, in which the exponents for a channel are differentially coded across frequency. The first exponent of an *fbw* or *lfe* channel is always sent as a 4-bit absolute value, ranging from 0 to 15. The value indicates the number of leading zeros of the first (dc term) transform coefficient. Successive (going higher in frequency) exponents are sent as differential values that must be added to the prior exponent value in order to form the next absolute value.

The differential exponents are combined into groups in the audio block. The grouping is done by one of three methods, D15, D25, or D45, which are referred to as *exponent strategies*. The number of grouped differential exponents placed in the audio block for a particular channel depends on the exponent strategy and on the frequency bandwidth information for that channel. The number of exponents in each group depends only on the exponent strategy.

An AC-3 audio block contains two types of fields with exponent information. The first type defines the exponent coding strategy for each channel, and the second type contains the actual coded exponents for channels requiring new exponents. For independent channels, frequency bandwidth information is included along with the exponent strategy fields. For coupled channels and the coupling channel, the frequency information is found in the coupling strategy fields.

Exponent strategy

Exponent strategy information for every channel is included in every AC-3 audio block. Information is never shared across frames, so block 0 will always contain a strategy indication (D15, D25, or D45) for each channel. Blocks 1 through 5 may indicate reuse of the prior (within the same frame) exponents and their frequency resolution. The D15 mode provides the finest frequency resolution, and the D45 mode requires the least amount of data. In all three modes, a number of differential exponents are combined into 7-bit words when coded into an audio block. The main difference between the modes is how many differential exponents are combined together.

The absolute exponents found in the bit stream at the beginning of the differentially coded exponent sets are sent as 4-bit values that have been limited in either range or resolution in order to save one bit. For *fbw* and *lfe* channels, the initial 4-bit absolute exponent represents a value from 0 to 15. Exponent values larger than 15 are limited to a value of 15. For the coupled channel, the 5-bit absolute exponent is limited to even values, and the *lsb* is not transmitted. The resolution has been limited to valid values of 0, 2, and 4 through 24. Each differential exponent can take on one of five values: −2, −1, 0, +1, and +2. This allows deltas of up to ±2 (±12 dB) between exponents. These five values are mapped into the values 0, 1, 2, 3, and 4 before being grouped, as shown in Table 11-2.

In the D15 mode, Table 11-2 mapping is applied to each individual differential exponent for coding into the bit stream. In the D25 mode, each *pair* of differential exponents is represented by a single mapped value in the bit stream. In this mode the second differential exponent of each pair is implied as a delta of 0 from the first element of the pair, as indicated in Table 11-3. The D45 mode is similar to the D25

Table 11-2. Mapping of differential exponent values, D15 mode

Differential exponent	Mapped value
+2	4
+1	3
0	2
−1	1
−2	0

Note: Mapped value=differential exponent+2; differential exponent=mapped value−2.
SOURCE: Thomson Electronics Corporation.

Table 11-3. Mapping of differential exponent values, D25 mode

Differential exponent n	Differential exponent n+1	Mapped value
+2	0	4
+1	0	3
0	0	2
−1	0	1
−2	0	0

SOURCE: Thomson Electronics Corporation.

Table 11-4. Mapping of differential exponent values, D45 mode

Differential exponent n	Differential exponent n+1	Differential exponent n+2	Differential exponent n+3	Mapped value
+2	0	0	0	4
+1	0	0	0	3
0	0	0	0	2
−1	0	0	0	1
−2	0	0	0	0

SOURCE: Thomson Electronics Corporation.

mode except that *quads* of differential exponents are represented by a single mapped value, as indicated in Table 11-4.

Since a single exponent is effectively shared by two or four different mantissas, encoders must ensure that the exponent chosen for the pair or quad is the minimum absolute value (corresponding to the largest exponent) needed to represent all the mantissas. For all modes, sets of three adjacent (in frequency) mapped values (M_1, M_2, and M_3) are grouped together and coded as a 7-bit value according to the following formula:

$$\text{Coded 7-bit grouped value} = (25 \times M_1) + (5 \times M_2) + M_3$$

The exponent field for a given channel in an AC-3 audio block consists of a single absolute exponent followed by a number of these grouped values.

Exponent decoding

The exponent strategy for each coupled and independent channel is included in a set of 2-bit fields designated *chexpstr [ch]*. When the coupling channel is present, a *cplexpstr* strategy code is also included. Table 11-5 shows the mapping from exponent strategy code into exponent strategy.

When the low-frequency effects channel is enabled, the *lfeexpstr* field is present. It is decoded as shown in Table 11-6.

Table 11-5. Exponent strategy coding

chexpstr[ch], cplexpstr	**Exponent strategy**	**Exponents per group**
'00'	Reuse prior exponents	0
'01'	D15	3
'10'	D25	6
'11'	D45	12

SOURCE: Thomson Electronics Corporation.

Table 11-6. *lfe* **Channel exponent strategy coding**

lfeexpstr	**Exponent strategy**	**Exponents per group**
'0'	Reuse prior exponents	0
'1'	D15	3

SOURCE: Thomson Electronics Corporation.

12
CHAPTER

Remote control systems

This chapter discusses basic television remote control operations. This will include functions such as channel search, channel skip, and channel selection. In addition, system operation and microprocessor control circuits are discussed. This chapter continues with system control troubleshooting and system control troubleshooting charts. Next, we will look at learning unified remote control and Advanced System-3 on-screen menu and screen setups. This will include macro control, setup, and audio and features menus. The chapter concludes with information on the world system teletext, Zenith CRT computer brain, and some intermittent remote control troubleshooting tips.

Basic remote control operations

The following description is based on using the remote control to operate the television receiver. Similar operation may be done using the front controls on the television set, except for those operations which use the timer and the clock. Refer to the remote control hand unit drawing in Fig. 12-1.

Television on mode

Press the power button (1) to turn the television set on. The picture appears in a few seconds. The channel number is shown for a few seconds. If the time clock is set, then the time will appear with the channel number. The television has last channel memory, so when the set is turned back on, that channel will be shown.

TV/CATV band

The tuning band determines the channel selection capability of the tuner. Press the TV/CATV button (2) to toggle through the TV-STD-HRC-IRC selections. TV is used for standard "over the air" broadcasts. STD is the standard CATV mode. HRC is for cable systems that use the harmonically related carrier. IRC is used for cable

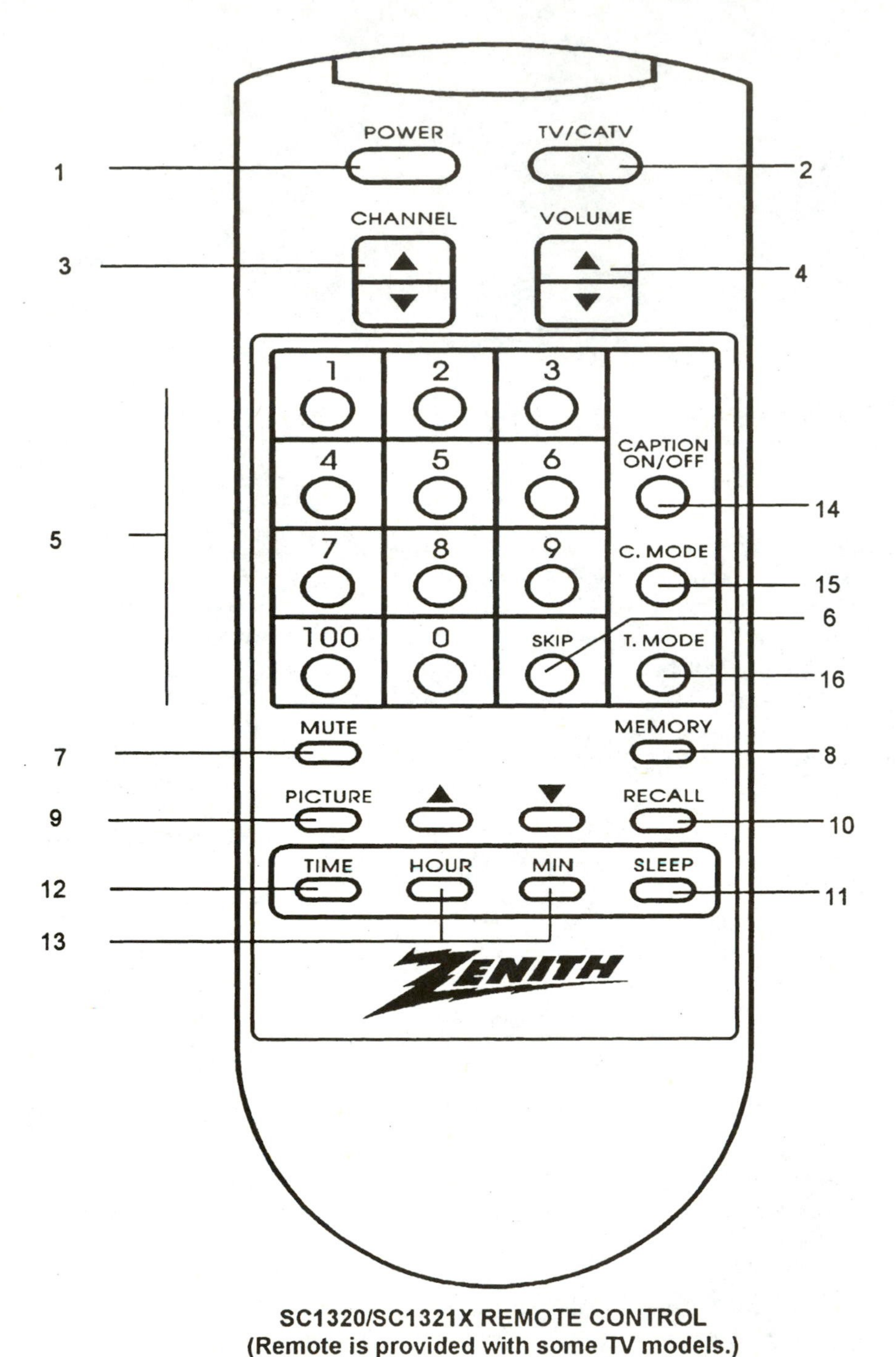

SC1320/SC1321X REMOTE CONTROL
(Remote is provided with some TV models.)

12-1 Zenith SC1320/SC1321X remote control hand unit. (*Courtesy of Zenith Corporation.*)

systems that use the incrementally related carrier. Sometimes it is called ICC, for incrementally coherent carrier.

After the band has been selected, press the memory button (8) to save the selection. The color of the read-out will change from red to green when it is saved.

Note: Band selection may be lost if power is interrupted to the television.

Channel search

Some television sets do not have an auto search system, and channels are placed in memory manually, using the channel search and memory features. Select the lowest channel in the tuning band. Press the skip button (6) until the search option is selected. Press the channel up button (3) to activate the search operation. The channel numbers will flash on the screen until an active channel is found. When this happens, the search stops, and the picture is displayed. Now press the memory button (8) to save this channel in the memory. The color of the read-out will change from red to green, indicating that the channel has been saved. Press the channel up button (3) again to find the next active channel. Repeat this procedure until all the active channels have been found and are saved in the memory.

Note: Channel search may need to be repeated if power is interrupted to the television set.

Channel skip

Channels saved in memory can be scanned rapidly using the channel skip feature. When the skip feature is selected, the television will sequence only the saved channels when the channel scanning feature is used. Unused channels will not be previewed. After channel search is completed, press the skip button (6) to bring up the skip feature. Press the memory button (8) to save the skip feature. The television remembers it is in the skip mode.

Note: Skip may be lost if the power is interrupted to the television set.

Channel selection

You can select any channel in the previously selected tuning band by direct number entry (5). Simply enter a channel number, and the television selects that channel automatically.

Single-digit channels.

Press 0 followed by the channel number for immediate selection. If only the channel number is pressed, the channel selection will occur in about 2 seconds.

Two-digit channels.
Press the two numbers for the desired channel. Both numbers must be pressed within 2 seconds of each other or a single-digit channel will be selected.

Up/down scan.
Press the channel up or down button (3) to scan channels saved in memory.

Three-digit channels.
Press the 100 button (5) followed by the one- or two-digit channel number to access a cable channel over 100.

Control adjustment

All picture adjustments are made by using the picture (9) and up/down buttons (5). Press the picture button to select the desired adjustment. Each press of the button will toggle through brightness, color, tint, and contrast. Use the up/down buttons to make the adjustments. The red block on the on-screen display will show the current setting.

Volume adjust

Press the volume up/down button (4) to increase or decrease the volume. The adjustment will be shown on the on-screen display with a colored display. The red block will indicate the current level adjustment.

Sound mute

Press the mute button (7) to turn off the sound while the picture continues. The red block on the on-screen display will be at the left side of the display indicating that the sound is muted. Press the mute button again to restore the sound.

Recall mode

Press the recall button (10) to show the current channel selection and the present time of day. Press the recall button again to remove the channel/time display. The display produced when recall is pressed continues to be shown until recall is pressed again. This display is not removed automatically after a few seconds, as with other displays.

Clock adjust

To set the clock, press the time button (12) repeatedly until the clock options are shown. Press the hour and min (minutes) buttons (13) separately to set the current time of day. Press the memory button (8) to save the current time setting in memory. The color of the display changes each time memory is pressed. The time is set and the clock starts when the clock display is shown in green.

Note: It may be necessary to reset the clock if ac power failure occurs.

Sleep timer

Some television sets have a sleep timer that may be used to turn off the television automatically after 10 minutes or up to 90 minutes in 10-minute intervals. Press the sleep button (11) repeatedly to select the desired turn-off time. The time can be extended at any time before turn-off by selecting a longer time using the sleep button. To clear the sleep timer, press the sleep button until 0 time appears.

On/off timer

Some television sets have a built-in timer that may be used to turn the television on and off automatically. Press the time button (12) repeatedly until the desired on time is shown. Press the hour and min (minutes) buttons (13) separately to set the desired on time. Do the same with the off time. Press the memory button (8) to save the timer setting in memory. The color of the time display changes each time memory is pressed. The time is saved and the time starts when the time display is shown in green.

Closed captions and text

The receiver will decode and display the closed caption and informational text that is broadcast with some television shows. Captions can be subtitles for the hearing impaired or translations into another language. Information text can be the daily program schedule for the television station or any special announcements.

The C caption or text mode may be selected for viewing. There are four different text modes. Not all these modes are being used at this time. You may have to experiment with different channels to determine which captions and text modes are available.

Press the C.mode (15) or T.mode (16) buttons repeatedly to select the desired mode. Wait 2 or 3 seconds to see if captions or text is available before making the next mode selections.

IC1502 microprocessor control circuit

The IC1502 chip is a 42-pin CMOS processor in a DIP package. Refer to the circuit diagram of this chip in Fig. 12-2. The keyboard commands are tied to pins 10 through 18. Infrared (ir) commands come in on pin 35. Control of contrast, brightness, color, and tint come in at pins 2, 3, 4, and 5. These are PWM signals that are coupled to pins 39, 31, 36, and 41 of the video processor IC1501.

U3101 IC system control

In most cases, the system control circuit either works or does not work. In rare instances, an isolated failure such as "no remote control function" will be encountered; however, the exclusiveness of the failure itself isolates the problem down to a few possibilities. In the majority of cases, a system control circuit failure results in a dead set. It is therefore important to determine if U3101 is defective or whether it is

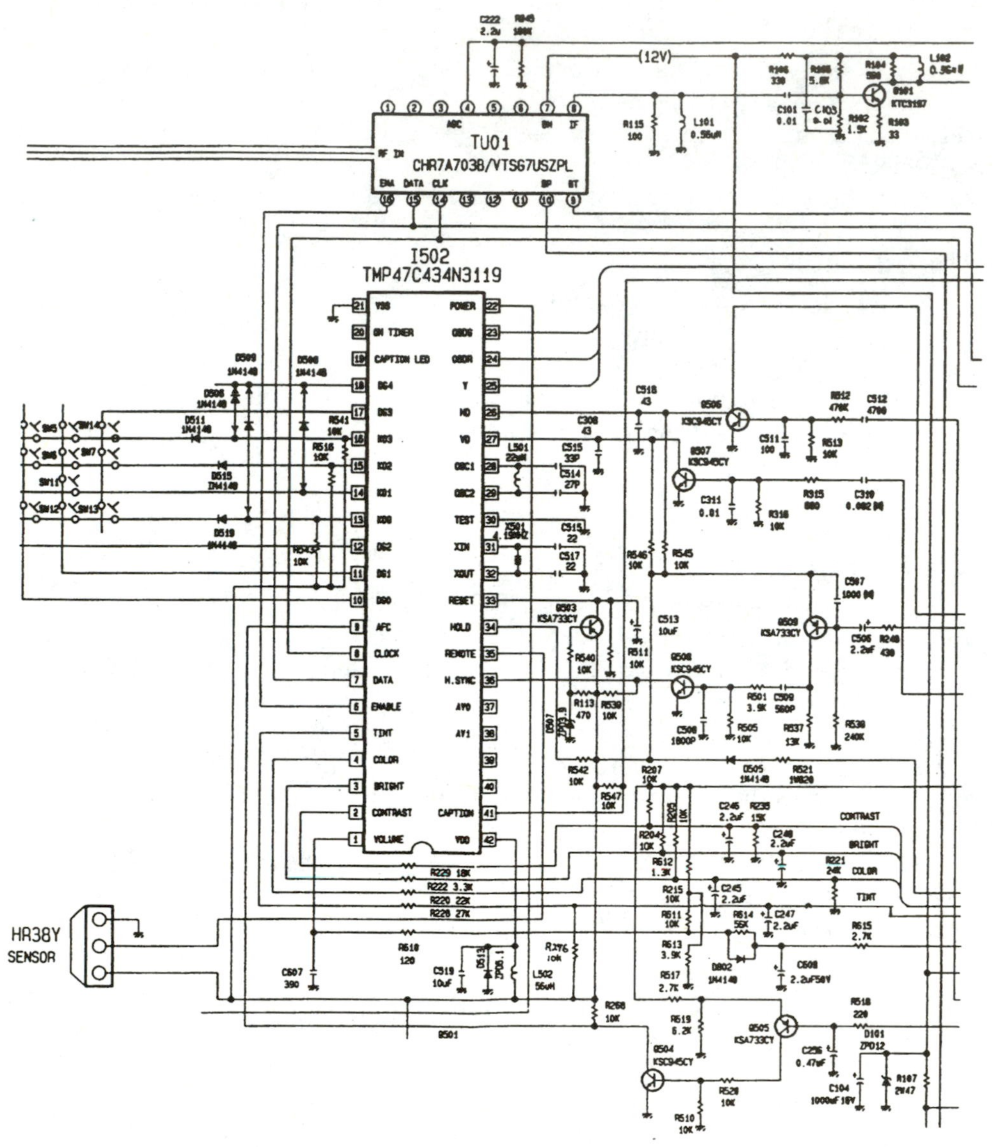

12-2 Zenith IC1502 microprocessor systems chip. (*Courtesy of Zenith Corporation.*)

malfunctioning due to a peripheral component or circuit failure. All too often microprocessors are replaced only to discover that the new IC reacts in the same way as the original. While it is not possible to completely test and confirm the microprocessor at the service shop level, there are procedures that will confirm whether or not the IC is functioning in the majority of cases. For these checks and procedures, refer to circuit in Fig. 12-3.

1. Check for +5-V VDD on pin 19. This voltage *must* be within a few tenths of a volt of 5 V or U3101 will not operate properly. An analog meter is not

accurate enough for this measurement. Make sure the dc is clean with an oscilloscope. Noise riding on the dc +5-V line will cause problems. If you see noise with the scope, check power supply regulation and filters.

2. Check for the +5-V reset on pin 1. This voltage should go high immediately after pin 19 goes high.
3. Make sure pin 20 is connected to ground.
4. Check for the presence of the 4-MHz oscillator at pins 39 and 40. The signal should be 5 V peak-to-peak when measured with a ×10 scope probe. A ×1 probe will load the oscillator down. If the +5-V VDD and reset are present on pins 19 and 1 and the oscillator will not run, check the components off of pins 39 and 40. A defective crystal or a leaky capacitor will kill the oscillator. If the components check good, the IC is defective.
5. If all the previous measurements check out OK but the microprocessor will not operate the set, perform voltage and resistance measurements on all the IC pins and compare them with the service data or the voltage waveform chart for this IC. Again, voltage must be within a few tenths of the specified voltage or the IC will not function correctly. A leaky capacitor pulling a pin low or a bad connection on a pin can lock up U3101.

Only after performing the preceding steps should U3101 be considered to be defective. Refer to Fig. 12-4 for U3101 IC pin-out information for troubleshooting.

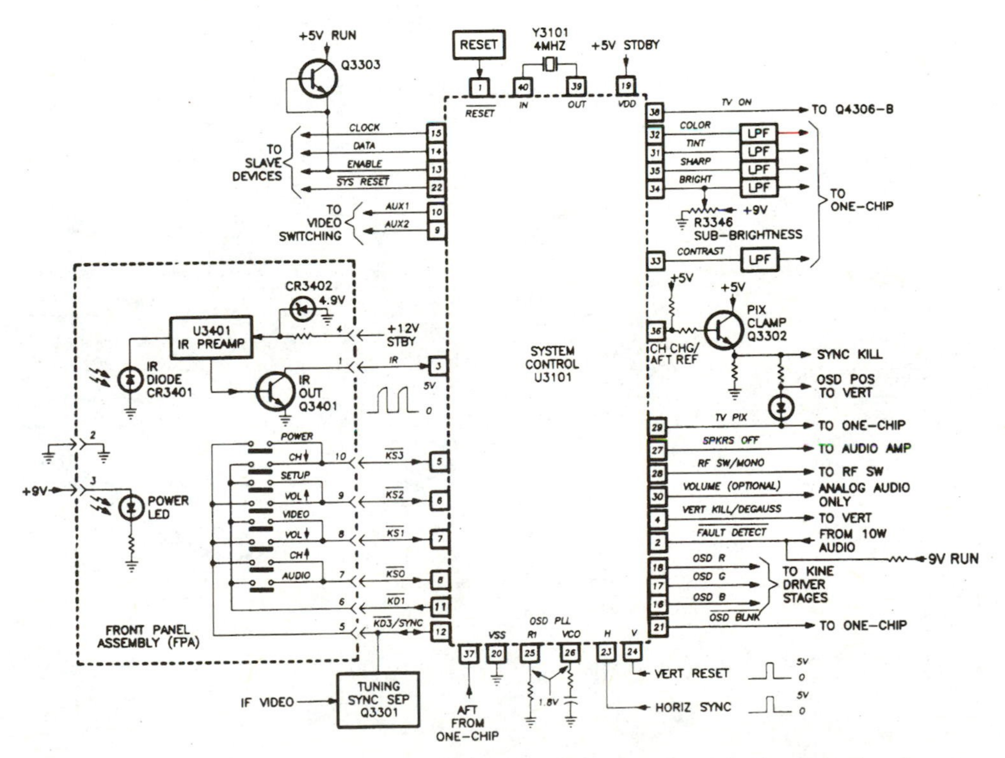

12-3 Zenith U3101 system control chip. (*Courtesy of Zenith Corporation.*)

U3101 SYSTEM CONTROL/OSD MICRO

PIN NO.	I/O	SIGNAL NAME	IN CKT RES.	DESCRIPTION
1	I	PWR-ON RESET	10K	Micro reset - Active Low.
2	I	FAULT DET	15K	When a low is sensed at this input, system control turns the set off for two seconds and turns it back on. If pin 2 goes low three times in one minute, system control keeps the set off.
3	I	IR	38K	Receives 5 Vp-p IR signal from remote receiver.
4	O	VERT KILL/ DEGAUSS	>20M	Goes high to kill vertical deflection during degaussing and the service line in direct view sets.
5	I	KS3	900K	Keyboard scan input.
6	I	KS2	900K	Keyboard scan input.
7	I	KS1	900K	Keyboard scan input.
8	I	KS0	900K	Keyboard scan input.
9	O	AUX 2	>20M	Video select control line.
10	O	AUX 1	>20M	Video select control line.
11	O	KD1	10K	Keyboard scan output.
12	I/O	KD3/ TUNING SYNC	9K	Keyboard scan output/tuning sync input.
13	O	ENABLE	>20M	Serial communications control line which goes high during data transmission and low during address transmission.
14	I/O	DATA	111K	Serial communications data line.
15	O	CLOCK	>20M	Serial communications clock line.
16	I	BLUE	16M	Blue OSD output. Active high.
17	I	GREEN	14M	Green OSD output. Active high.
18	I	RED	11M	Red OSD output. Active high.
19	-	VDD	1.4M	+5 VDC.
20	-	VSS	0	GND.
21	O	BLNK	>20M	OSD black surround out. Low = Black.
22	O	SYS RST	>20M	System reset line connected to bus devices. Goes low when set is off and high when set is on.
23	I	H-SYNC	16K	Horizontal timing input for OSD.
24	I	V-SYNC	11K	Vertical timing input for OSD.
25	-	R1	10K	OSD PLL external control pin.
26	-	VCO	>20M	OSD VCO external control pin.
27	O	SPKRS OFF	110K	Goes high to turn speakers off and low to turn them on.
28	O	RFSW/MONO	>20M	RF switch control line. Low selects Ant. A. Mono function currently not used.
29	O	TV PIX	>20M	Goes high when TV tuner is selected and low when external video is selected.
30	O	VOLUME	>20M	PWM for Volume control (currently not used).
31	O	TINT	11K	PWM for Tint control.
32	O	COLOR	19K	PWM for Color control.
33	O	CONTRAST	33K	PWM for Contrast control.
34	O	BRIGHTNESS	97K	PWM for Brightness control.
35	O	SHARPNESS	12K	PWM for Sharpness control.
36	I/0	CH CHG/AFT REF	105K	At start of channel change, voltage at pin 36 is read by micro for use in AFT A/D converter. During channel change, line goes high until channel change is executed.
37	I	AFT	73K	Automatic fine tuning input. Crossover point detected at 2.5 VDC.
38	O	TV ON	>20M	Power ON/OFF control. High = ON, Low = OFF.
39	O	OSC OUT	5M	4 MHz oscillator output.
40	I	OSC IN	4M	4 MHz oscillator input.

12-4 U3101 system troubleshooting information. (*Courtesy of Zenith Corporation.*)

Refer to Fig. 12-5 for U1001 CVT IC microprocessor pin-out information and circuit troubleshooting information.

Zenith learning unified remote control unit

The LR4800 learning remote, found with all Zenith Advanced System-3 television sets, allows the customer to operate most brands of remote-controlled televisions, VCRs, and cable decoders. In addition, the unit also operates other remote-controlled equipment such as audio components and lighting systems.

Like the MBR3020, the LR4800 shown in Fig. 12-6 is preprogrammed to operate any Zenith infrared remote-controlled television, VCR, or cable decoder. The most effective demonstration of the LR4800 is seen in how easy it is to "teach" and "learn." To do this you will need a remote-controlled product, the remote control dedicated to that product, and the LR4800 learning remote control. For this demonstration, use a remote-controlled VCR that is not a Zenith product, and perform the following steps:

1. If the learning remote is not in the desired mode, change to this mode by pressing the VCR button.
2. Align the learning remote and the VCR's dedicated remote control (teaching remote) head to head and about 1 in apart.
3. Press and hold down the enter and zero buttons until all three mode indicator lights come on.
4. Press the 9 button twice and then enter within 5 seconds. The active mode indicator light will come on and stay on. To change the active mode, touch the desired mode button, TV, VCR, or AUX.
5. Press any button on the "teaching" remote. The indicator of the current mode should flash to confirm that the LR4800 is receiving codes. Realign the remotes until flashing occurs.
6. Press the button of the LR4800 that you wish to reprogram. For this demonstration, reprogram the VCR power on by pressing the off/on button. The three mode indicators will light, confirming that the LR4800 is ready to "learn."
7. Press and hold the power on button on the "teaching" remote until the indicator lights of the LR4800 turn off and the selected mode flashes.
8. Press the macro button on the LR4800 once.

The LR4800 will now turn on or off the non-Zenith VCR. This "learning" process can be used to reprogram any button on the LR4800 to operate the functions of the "teaching" remote.

Resetting the LR4800 to its original codes is equally easy to do:

1. Press and hold down the enter and zero buttons until all three mode indicator lights come on.

U1001 CTV PROCESSOR (ONE-CHIP)

PIN NO.	I/O	SIGNAL NAME	IN CKT RES.	DESCRIPTION
1	O	AUDIO OUT	>20M	Not used.
2	O	RF AGC OUT	69K	RF gain control voltage for tuner.
3	I	RF AGC IN	2.3K	Adjusts the operation point of the RF AGC circuit.
4	-	SIF DET	>20M	Connection for audio detection tank.
5	-	PIF AGC1	>20M	High frequency IF AGC filter.
6	I	EXT AUDIO IN	37K	Not used.
7	-	PIF AGC2	>20M	Low frequency IF AGC filter and IF defeat connection point.
8	-	P/S GND	0	PIF/SIF ground.
9	I	IF IN	20K	IF input from SAW filter.
10	I	IF IN	20K	IF input from SAW filter.
11	I	FC ADJUSTMENT	41K	Chroma oscillator and filter adjustment.
12	-	APC FILTER	>20M	Controls phase (tint) of chroma signal.
13	-	X'TAL 3.58	>20M	3.58 MHz chroma oscillator.
14	I	V/C/D VCC	1.8K	9 VDC VCC for video, chroma, and deflection circuits within U1001.
15	O	R-Y	4.3K	R-Y color difference signal output.
16	O	-Y	1.5M	Luminance output.
17	O	G-Y	4.2K	G-Y color difference signal output.
18	O	B-Y	4.2K	B-Y color difference signal output.
19	I	RED OSD	>20M	Red OSD input which is connected to the Blank (bar) line from system control to produce the black border around OSD characters.
20	I	GREEN OSD	>20M	Not used.
21	I	F B PULSE	>20M	Input for chroma burst amp and horizontal centering control.
22	I	X-RAY PROTECT	32K	When voltage at this pin reaches approximately 1.5 VDC, X-Ray protect activates and stops horizontal drive pulses from pin 23 of U1001.
23	O	H OUT	440ohm	Horizontal output pulses which are applied to the horizontal deflection stage.
24	I	H AFC	>20M	Horizontal AFC input.
25	-	32H OSC	>20M	503 kHz oscillator for horizontal countdown stage.
26	I	H VCC	81K	VCC for sync stages.
27	O	V OUT	1K	Not used.
28	-	V NFB	49K	Negative feedback for vertical ramp stage.
29	0	V RAMP	3.2K	Vertical reset output applied to vertical deflection stage.
30	-	COLOR KILLER	2.7M	Disables chroma circuits when chroma burst is not received. Color killer activates at about 6.5 VDC at pin 30.
31	I	TV/EXT SW CHROMA IN	40K	Chroma input in addition to control line for TV/EXT video switching at pins 40 and 43. DC level of chroma signal is used to perform the switching. With TV video selected, DC offset is about 4.5 VDC and 1.2 VDC with external video selected.
32	I	S-VHS SW	0	Not used.
33	-	OSD BRIGHT	1.8K	Voltage at this pin is transferred to the -Y output of U1001 whenever the Red OSD pin 19 is pulled low. This technique is used to produce the black border around the OSD characters.
34	I	SHARP	17K	DC control and high frequency luma input for sharpness control. 5.8 VDC to 7.9 VDC min to max (approx).
35	I	DELAYED VIDEO	210K	Delayed low frequency luma input for sharpness control circuit.

12-5 U1001 CVT processor troubleshooting readings. (*Courtesy of Zenith Corporation.*)

2. Press the zero button and then press enter again.

The LR4800 is now fully reset to use the original Zenith codes.

Macro control

Customers familiar with computer operations will recognize *macro* as a term used when operating a whole sequence of key presses with one or two strokes. How-

12-6
Zenith LR4800 remote control hand unit. (*Courtesy of Zenith Corporation.*)

ever, many of your customers will not understand this term, and it is up to you to explain the feature and its convenience.

The macro feature of the LR4800 learning remote can be programmed to operate a sequence of up to five operations, much like a telephone can be programmed to "speed dial" a seven-digit phone number with one press of the button. The LR4800 learning remote can send up to five different instructions to the television, VCR, or auxiliary operations simply by pressing the macro button and then A, B, or C. Figure 12-7 illustrates the macro button locations.

On-screen menus

The menu button on the LR4800 learning remote will call up the available menus just as the MBR3020 and SC3825 remote units. The LR4800, however, has a second select button that enables the customer to move the cursor up or down to take the shortest path to any particular feature on the menu, as shown in Fig. 12-8. This

12-7
Remote unit control button features. (*Courtesy of Zenith Corporation.*)

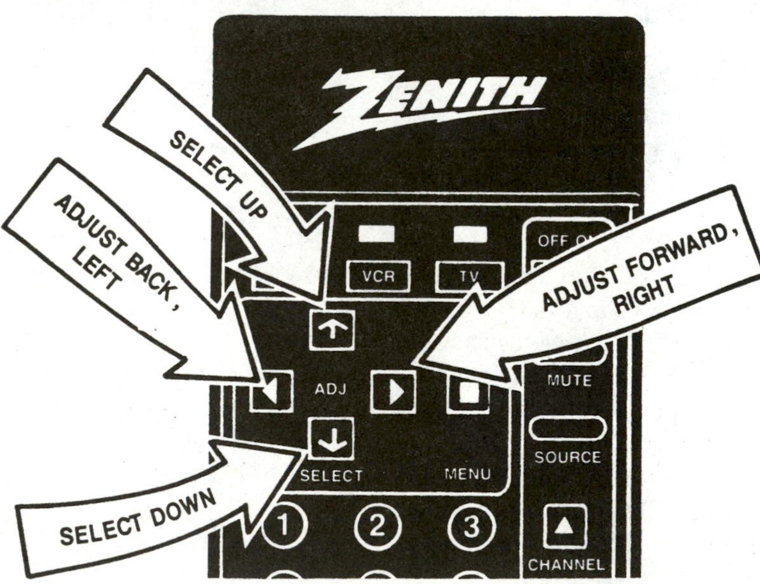

12-8 Remote control unit adjust features. (*Courtesy of Zenith Corporation.*)

makes the feature selection process even easier than Sentry 2 or Sentry 3. The familiar sequence of menu and select and adjust remains the same.

The source menu

The Advanced System-3 family of televisions provides the customer with even more source selection capability. As shown in Fig. 12-9, these sources can all be accessed through the advanced source menu of the on-screen display.

LR5000 learning remote control

The LR5000 learning remote control and the advanced on-screen menus, available in all Digital System-3 models, offer the customer the ultimate in remote-controlled features selections and adjustments. The LR5000 is very similar to the LR4800 learning remote control found in the Advanced System-3 sets with teletext and picture-in-picture feature controls unique to the Digital System-3 family (Fig. 12-10).

On-screen menus

By pressing the menu button, all four menus available in the digital Zenith television sets are displayed like file cards overlapping each other, as shown in Fig. 12-11.

12-9 Television screen showing input selections. (*Courtesy of Zenith Corporation.*)

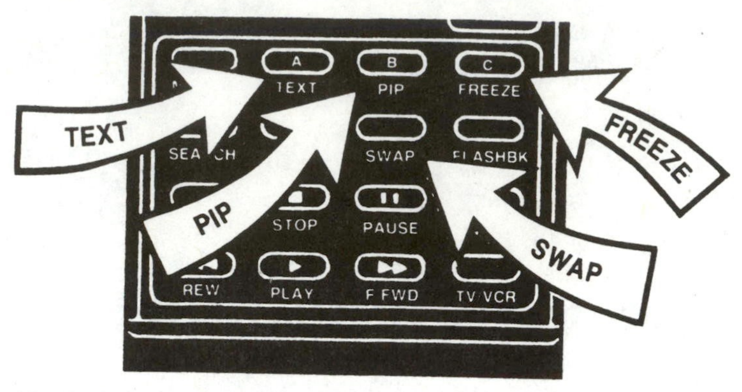

12-10 Remote control on-screen menu selections. (*Courtesy of Zenith Corporation.*)

12-11 Television screen showing system install and other selections. (*Courtesy of Zenith Corporation.*)

Successive presses of the menu button cycle the front menu card to the back and move each of the other menu pages one step nearer the front. Once again, the features contained on each menu are accessed using the same three-step process of menu, select, and adjust.

To familiarize your customer with the Digital System-3 menus and their features, use the auto demonstration mode built into the television set. Here's how: Press and hold the menu button for about 5 seconds or until the auto demonstration display appears on screen. The set will cycle through each menu and feature automatically. You can cancel the demonstration at any time by pressing any of the television control buttons.

Setup menu

The setup menu of the digital Zenith steps up to a special auto installation feature. This should definitely be a part of your demonstration to the television owner. The multiple source capability of the digital Zenith televisions may lead your customers to believe that these models are more difficult to install. The system install feature takes the customer through the installation step by step and even displays sample hookups with additional components on the screen (Fig. 12-12).

To demonstrate this feature, press the menu button of the LR5000 remote control repeatedly until the setup menu appears on screen. Press the select button up or down to highlight the system install feature, and then press adjust. Follow the directions on the screen, and take your customer through each display to show just how easy the installation can be done.

Video menu

The video menu of the Digital System-3 sets, like those found in the other Zenith television systems, contains features used to adjust the picture (Fig. 12-13). In addition to the standard picture controls, Digital System-3 models include a feature in the video menu called *pix enhance*. This picture adjustment feature works in much the same way as a video filter turned off and on. A third setting, called *wide band*, has been added for use with a source offering high-resolution capability.

Audio menu

The audio only feature, unique to the Digital System-3 audio menu, allows the customer to enjoy the premium audio performance of the set without the picture

12-12 Screen showing system in the install mode. (*Courtesy of Zenith Corporation.*)

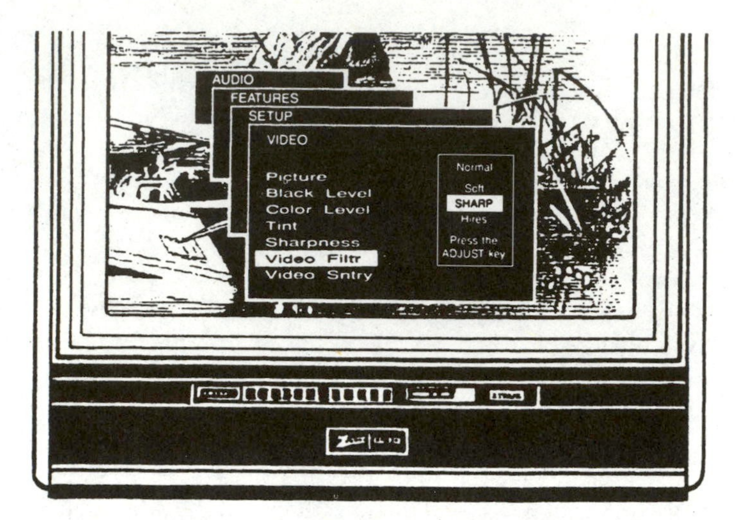

12-13 Video menu for the Digital System-3 television receivers. (*Courtesy of Zenith Corporation.*)

being turned on. The current technology used to record high-fidelity music on videotape or laser disc, combined with the premium audio system available in many digital Zenith models, enables the customer to use their digital Zenith as an audio component system. Audio only turns off the picture tube so the television actually becomes just like an additional audio component to your customer's home entertainment system (Fig. 12-14).

Features menu

The Zenith digital models step up to an advanced features menu with parental control, channel label, and teletext options (Fig. 12-15). Parental control enables the customer to use a personal code to lock out an unwanted channel. The channel can only be viewed after the code has been entered to unlock it. Channel label allows the customer to display the station identification call letters such as ABC, CBS, or NBC along with the channel number and time. Figure 12-16 shows this screen display information.

World System Teletext is also accessed through the features menu. Teletext decodes various electronic magazines available in select markets and over most cable systems. Customers can view continuously updated news, sports, weather, or business information at the touch of a button. On some channels closed captioning is being transcoded and simulcast as teletext so that the captions can be viewed by activating the teletext decoder built into every Digital System-3 set. See this teletext display in Fig. 12-17.

Channel display

The channels are displayed on the same LED as the tuning bands in either two (68/157ch) or three (178ch) digits. Tuners with 157 channels use 1 to 99 plus channel 0 and 00 to accomplish the 101 cable channel capability.

Favorite channel scan

Most Zenith tuners feature favorite channel scan. This feature makes it possible for customers to program the tuner to scan only the channels they usually view. To program favorite channels into the scan, simply call up the channel number by pressing the channel number button on the set or the remote control or by using the all channel scan up/down buttons. Once the channel is displayed, press the add or enter button on the set. That channel is now programmed into scan. To remove a channel from the scan, call up the channel you want deleted and press the skip button. The letters *PO* will appear on the LED display to confirm that the channel has been programmed out of the scan.

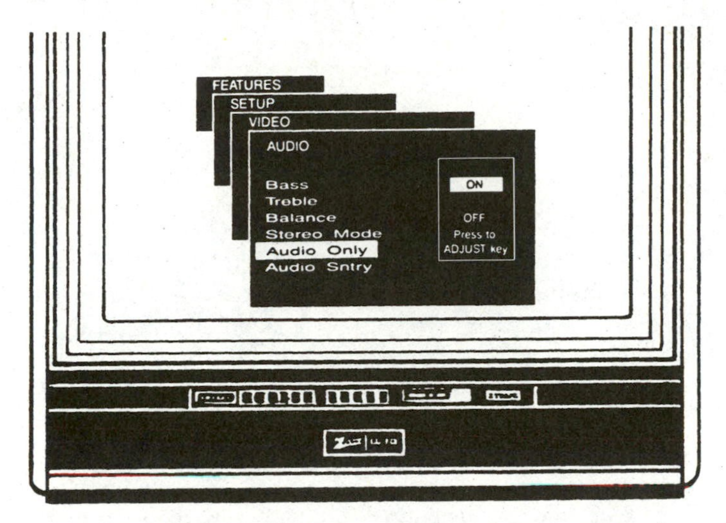

12-14 Audio menu. (*Courtesy of Zenith Corporation.*)

12-15 Zenith digital television screen showing features menu. (*Courtesy of Zenith Corporation.*)

12-16 Portrayal of screen "channel label." (*Courtesy of Zenith Corporation.*)

Direct access tuning

Direct access tuning is available on all customer series models through either a keyboard on the set or the remote control unit. By using direct access, the customer can tune channels without going through all the channels in the scan. Customers can access channels directly the "quick way" by pressing the channel numbers and pressing enter or the "easy way" by pressing the channel numbers and letting the tuner handle the rest. In this method, the selected channel will appear after a 2-second delay. This delay gives the customer enough time to enter a second or a third digit of a higher-numbered channel. Directly accessing a single-digit channel does not require pressing zero first, as applies to some other brands. Whether the customer selects the "quick," "easy," or "scan" method of channel tuning, Zenith owners can be assured of precise tuning every time.

Remote controls

Selected custom series models offer the convenience of remote control operation. The nine-function SC3300 remote control combines large, easy-to-read graphics with large, rubber, rounded-edge buttons for easy, one-hand operation. The SC3350 remote control adds a sleep timer function to its list of various operations.

World System Teletext

A feature unique in the United States to Zenith and Digital System-3 is a built-in decoder for accessing World System Teletext. World System Teletext is a relatively new technology used to transmit a variety of electronic magazines and display them on the television screen. One such magazine, called "Electra," is available on most cable systems via superstation WTBS, in Atlanta, Georgia. "Electra" provides Zenith digital viewers with the latest news, sports, weather, and business reports updated continually from 6 A.M. seven days a week. Note the "Electra" screen shown in Fig. 12-18.

The teletext features can be accessed easily through the LR4900 or the LR5000 learning remote controls. Simply press the text button on the remote to access the teletext menu. Next, press the select up/down buttons to choose the desired option on the menu. Once the option has been selected, press the adjust right button to increase the page number, page by page. Press adjust left to decrease the page number, page by page.

Computer brain

The electron gun in the picture tube of a Zenith color television set makes 525 scans across the tinted phosphors 30 times a second to create the color picture on the screen. The output of these guns can become unbalanced with age and can create pictures with an unnatural blue, red, or green tint. Normally, this problem would require an expensive, time-consuming adjustment or repair.

All System-3 sets feature a computer brain that automatically monitors the electron gun illuminating the phosphors. Should the gun's output become unbalanced, the computer's brain makes any necessary correction. This built-in service adjustment assures the customer of lifelike color reception day after day, year after year.

"Odd ball" intermittent remote control operation

Should you have an odd VCR or television remote control intermittent operation problem and you have some of these new "power saver" fluorescent tubes that screw

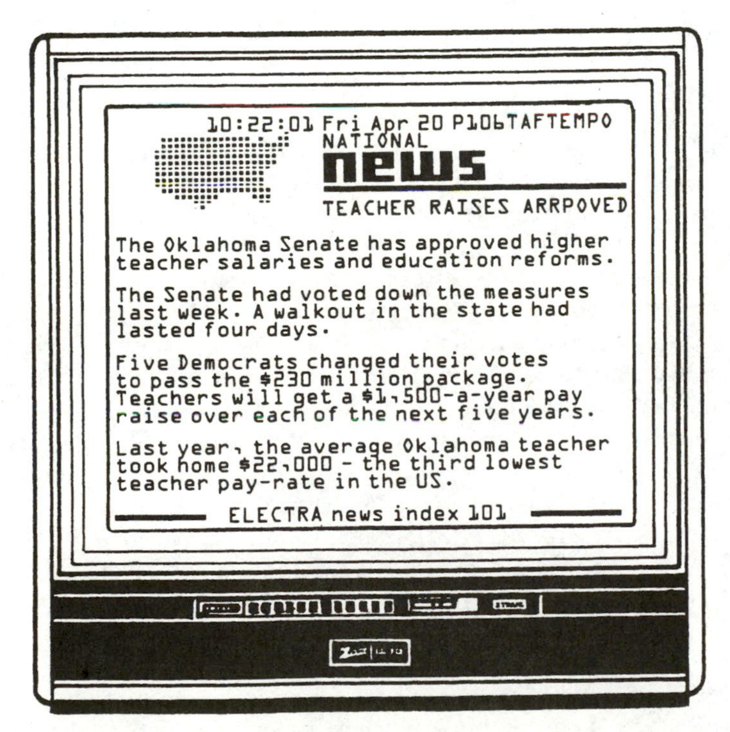

12-17 World System Teletext screen. (*Courtesy of Zenith Corporation.*)

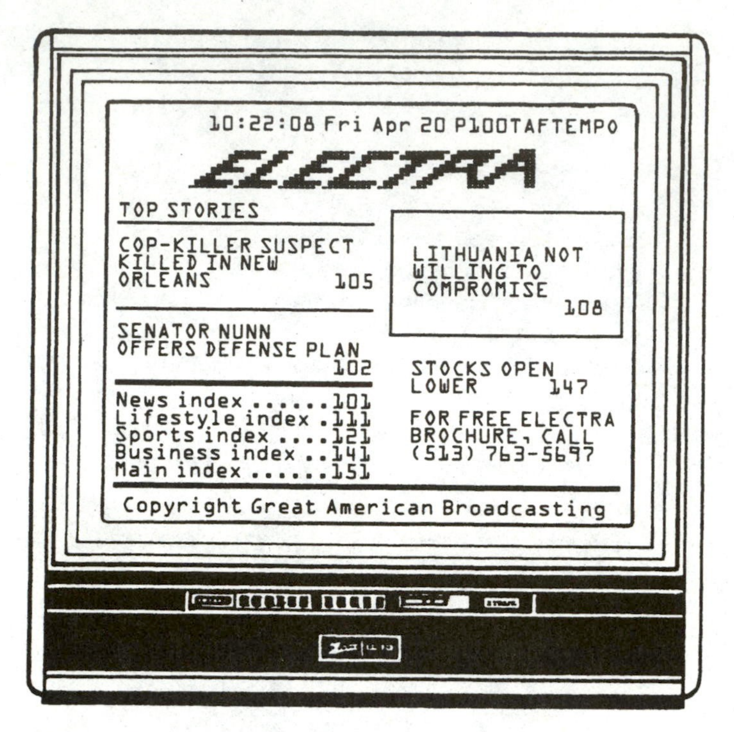

12-18 "Electra" screen layout mode. (*Courtesy of Zenith Corporation.*)

into a standard light socket, this might be your problem. These lights have been known to interfere with the infrared pulses emitted by the remote unit.

Most remote control units transmit pulses of infrared energy. These coded pulses can vary, but the remote control circuits are much the same. They usually transmit pulses of infrared energy at a frequency of 56 kHz. This happens each time a button is pushed, thus producing a group of coded pulses.

A code usually contains 24 bits of binary data. The first 4 bits determine which one of 16 types of equipment is to be controlled. Generally, the 0 is for a television set, 1 is for number 1 VCR, 2 is for the number 2 VCR, and 3 for a compact disc player. The next 8 bits in this sequence are used to identify the pressed key. Now these 8 bits permit a total of 256 key control codes. The last 12 bits are checksum bits that are compared with the first 12 bits to detect any errors.

When the television/VCR remote receiver detects an infrared pulse train, it compares it with a clock pulse operating at the same frequency. If the signal is high, a binary 1 is produced. If not, a binary 0 is produced. After the code pulses are processed, the VCR/television equipment responds for the proper action.

Compact fluorescent tubes operate at high frequencies. These high frequencies modulate the tube's infrared output and cause interference for remote controls based on the infrared emission. The big and heavy ballast used with standard fluorescent tubes limits the flow of 60-Hz current through the tube. This ballast produces a flicker that can be annoying when working under these older fluorescent lights. This flicker rate is 120 Hz, which is twice the power line frequency rate.

It is this nonvisible flicker of a switched infrared pulse signal that will interfere with the infrared remote control units. The remote control units for controlling VCRs, television sets, stereos, dish receivers, and cable boxes also switch in the 50- to 100-kHz range to transmit these binary data to the equipment's receiver remote circuits. Therefore, if you are having remote control problems and have these types of fluorescent lights, this might be the cause of the intermittent remote control problem. If you turn off these lights and your remote now works OK, that is the problem.

Glossary

4:2:2 Term commonly used for component digital video as specified by CCIR-601. The numbers refer to the ratio of luminance sampling to chrominance sampling. There are four luminance samples to every two chrominance samples.

4fsc Term commonly used for composite digital video. The name is used because the sampling is done at four times the subcarrier frequency, normally 13.3 MHz for NTSC and 17.7 MHz for PAL.

ACATS Advisory Committee on Advanced Television Service.

ACC (automatic color control) Used to maintain constant color signal levels.

Accelerating electrode Otherwise known as the second anode of a cathode-ray tube (CRT), this electrode serves to increase the velocity of the electron beam so that light is emitted when it strikes the screen.

Access unit A coded representation of a presentation unit. In the case of audio, an access unit is the coded representation of an audio frame. In the case of video, an access unit includes all the coded data for a picture and any stuffing that follows it, up to but not including the start of the next access unit. If a picture is not preceded by a group_start_code or a sequence_header_code, the access unit begins with the picture start code. If a picture is preceded by a group_start_code and/or a sequence_header_code, the access unit begins with the first byte of the first of these start codes. If it is the last picture preceding a sequence_end_code in the bit stream, all bytes between the last byte of the coded picture and the sequence_end_code (including the sequence_end_code) belong to the access unit.

ACK (automatic color killer) Turns off the color control if the 3.58-MHz color burst is missing.

Active elements Those components in a circuit which have a gain or through which dc flows, such as diodes and transistors.

Active lines Those lines, of the 525 possible lines in a television image, which appear on the screen. The inactive lines are blanked out during the time that the beam is returning from the bottom of the picture to start the next frame.

Active state The state in which the logic circuit performs its particular function. For example, the active state of the power set (initializer) circuit provides a reset function when initial power is applied; therefore, the logic levels should be shown for this active state.

Activity The presence of active signals. These signals are detected in the activity checker, which generates an activity signal.

A/D convertor (analog-to-digital-convertor) The device that converts an analog waveform to a digital representation of the signal by sampling.

Adder, logical Switching circuits that generate an output (sum and carry bits) representing the arithmetic sum of two inputs.

Adder, resistive A method for converting multibit digital information to an analog level by sourcing/sinking current from a fixed-voltage supply through a resistor network.

Address A code designating the particular unit (vehicle, decoder, teleprinter, radiotelephone, etc.) that must respond to an incoming message. When used with computers, it is a code that designates the location of information or instructions in the main storage, peripherals, etc.

Adjacent sound carrier frequency The sound carrier in the television channel on the low-frequency side of the channel.

AES/EBU audio Name commonly given to a digital audio standard that has been defined jointly by the Audio Engineering Society (AES) and the European Broadcasting Union (EBU).

AIU The copy is to come.

Aliasing Undesirable effect of spurious waveforms caused by sampling at a frequency below the Nyquist limit. Also used to describe the jagged diagonal lines produced by computer images.

Align To adjust two or more tuned circuits in ratio so that they respond to the same frequency.

Alternating current (ac) An electric current that reverses its direction of flow at regular intervals.

Amplitude Modulation (AM) A method of conveying information by changing the amplitude of a radiofrequency carrier.

Amplitude separation The process of separating the synchronizing signal from the video information in the composite television signal by using the difference in their amplitude levels.

Analog Term describing a system or device that operates on a continuously varying scale rather than on increasing or decreasing fixed steps.

Analog computer A continuous-variable computer or nondigital computer. A differential analyzer. Measures the effect of changes in one variable on all other variables in a system. Its operation is analogous to a slide rule.

Analog switch A device that allows the transfer of signals in both directions whenever the switch is enabled. When the switch is not enabled, it offers a high impedance.

Anchor frame A video frame that is used for prediction. I-frames and P-frames are generally used as anchor frames, but B-frames are never anchor frames.

Ancillary data Term used to describe data, not related to the picture content, embedded in the data stream of a digital video signal. Ancillary data commonly are used to transport digital audio, time-code, or other picture-related auxiliary data.

AND gate See *Gate, AND.*

ANSI American National Standards Institute.

Antenna (ant.) The portion, usually wires or rods, of a radio or television station or receiving set for radiating waves into space or receiving them from space. Also called *aerial.*

Antenna array An arrangement of two or more antennas (or reflectors) coupled to improve transmission or reception in a given direction.

Aspect ratio The ratio of picture width to picture height, 4:3 in the current television system.

Associated sound carrier frequency The sound carrier in the television channel under consideration.

Astable multivibrator A free-running electronic circuit that generates pulses that can be used as turning signals or other similar signals.

Astigmatism A condition of electron beam focus in which the spot is not perfectly round, resulting in different trace widths when the beam is deflected from the center of the screen.

Asynchronous 1. Operation of a switching network of free-running signals that triggers successive instructions; the completion of one instruction triggers the next. **2.** Lacking synchronization. In data transfer, the term relates to a signal distributed without an accompanying reference clock. Synchronizing information is embedded in the data stream.

Asynchronous transfer mode (ATM) A digital signal protocol for efficient transport of both constant-rate and bursty information in broadband digital networks. The ATM digital stream consists of fixed-length packets called *cells. Each contains 53 eight-bit bytes—a 5-byte header and a 48-byte information payload.*

ATEL Advanced Television Evaluation Laboratory.

ATTC Advanced Television Test Center.

ATV The U.S. advanced television system.

Audio The Latin word for "hear." Used synonymously with the word *sound.*

Audio carrier The frequency-modulated rf signal that carries the sound information.

Audio frequency (af) A frequency corresponding to a normally audible sound wave (about 20 to 14,000 cycles per second).

Automatic brightness control A circuit that automatically controls the average brightness in the received image so that it corresponds with that being transmitted.

Automatic frequency control (AFC) A circuit that keeps a radio receiver from drifting from the frequency to which it is tuned.

Automatic volume control A circuit that varies the amplification of a receiver so that its output remains constant despite changes in signal-strength input.

Average brightness The average illumination in the television picture.

Bandpass filter A filter that passes a group of continuous frequencies and rejects all others.

Bandwidth 1. The range of frequencies over which a system or circuit can function with minimal loss, usually less than 3 dB. The bandwidth of an NTSC television system is 4.2 MHz; in a PAL television system, it is 5.5 MHz.
2. The number of continuous frequencies required to convey the information being transmitted, either visual or aural. The bandwidth of a television channel is 6 mc.

Beam The stream of electrons that travels from the electron gun toward the screen in a cathode-ray tube.

Beat frequency A frequency resulting from a combination of two frequencies.

Bidirectional pictures or B-pictures or B-frames Pictures that use both future and past pictures as a reference. This technique is termed *bidirectional* *prediction.* B-pictures provide the most compression. B-pictures do not propagate coding errors because they are never used as a reference.

Binary A system of numerical representation that uses only two symbols: 0 and 1.

Binary-coded decimal (BCD) Four bits of binary information that are used as a unit to encode one decimal digit. When a decimal digit is encoded in this way, it is called a *binary-coded decimal (BCD).*

Bistable See *Flip-flop.*

Bit **1.** Abbreviation for binary digit.
2. The smallest unit of information in a binary notation system. An abbreviation for binary digit. A bit is a single 0 or 1. A group of bits used simultaneously by a digital system (usually 8 bits) is called a *byte*. In digital video, a group of 8 or 10 bits often is called a *word.*

Bit parallel A transmission standard in digital video where whole words are sent simultaneously, each bit on its own pair of wires. This standard is defined by SMPTE 125M.

Bit rate The rate at which the compressed bit stream is delivered from the channel to the input of a decoder.

Bit serial A transmission standard in digital video for sending words 1 bit at a time, one after another, down one pair of wires or, more typically, a coaxial cable. This standard is defined by SMPTE 259M.

Bit slip A condition in bit serial digital video transmission where word framing is lost. When this occurs, order of the bits is no longer correct, and the data are corrupted. This can occur in systems with excessive jitter on the bit stream, which prevents correct decoding of the data.

Bit stream A series of bits continuously flowing in an unbroken stream.

Black box A description used for an electronic circuit concerned only with the input and output, ignoring the interior elements, discrete or integrated.

Black level The level that represents the darkest part of a television image. In a properly adjusted television system, no picture information is at a level lower than this level.

Blacker-than-black level The portion of the television signal devoted to the synchronizing pulses. These synchronizing signals are transmitted at higher amplitudes than those representing the blackest part of the picture.

Blanking **1.** The process of cutting off the picture tube beam during the time it is not forming the picture. This occurs when the spot returns from the far right to begin the next line and from the bottom to the top of the next picture.
2. The process of turning off the electron beam in a CRT after it reaches the right edge of the screen and is returning to the left edge to begin the next scan line. That is done by setting the level to black or, in some systems, slightly below black. If the beam was not blanked, it would leave a trace as it returned to its starting position and degrade the image.

Blanking pulse The pulse used to blank out the electron beam in both the camera and the picture tube during the blanking interval.

Block A block is an 8×8 array of pel values or DCT coefficients representing luminance or chrominance information.

Blocking oscillator A type of oscillator used for triggering horizontal and vertical sweep generators.

Blooming The defocusing of the white regions of a television picture when an excess of electrons increases the spot size.

Brightness control A receiver control used to regulate the overall brightness of the picture.

Buffer A noninverting stage, such as an emitter follower, that provides isolation and sometimes is used to handle a large fan-out or to convert input and output levels.

Burst (more correctly, color burst) The portion of a composite video waveform that is placed between the horizontal sync pulse and the beginning of active video information. It is made up of several cycles of wave size at the frequency of the color subcarrier. It is used by the receiver as a reference to ensure correct decoding of the chrominance information in the composite signal.

Burst gate A circuit that is keyed to conduct during the time of the color burst. It is timed by a flyback pulse.

Bus transceiver A three-state device that provides two-way, asynchronous data transfer. Data transfer capability is in either direction but not both simultaneously. Direction of data transfer depends on the logic level of the direction input.

Byte A group of 8 bits considered as an entity. Instruction sets for 8-bit microprocessors are defined in multiples of bytes.

Byte-aligned A bit in a coded bit stream is byte-aligned if its position is a multiple of 8 bits from the first bit in the stream.

Cable equalization The process whereby the frequency response of an amplifier is altered to compensate for high-frequency losses in a cable.

Camera The unit that houses the optical system and light-sensitive pickup tube that converts the visual image into electrical impulses.

Camera tube A cathode-ray tube used to transform an image into electrical impulses.

Carrier frequency The frequency of an unmodulated radio signal produced by a transmitter.

Cathode-ray tube (CRT) An electron tube in which a stream of electrons from a cathode is formed into a narrow beam and deflected by means of electrostatic or magnetic fields over a target, usually a mosaic or fluorescent screen that glows wherever the beam strikes. The iconoscope, kinescope, picture tube, etc., fall into this category.

CCIR (The Comité Consulatif International de Radio, the International Radio Consultative Committee) The international body that codified the standards for digital video. The CCIR has been replaced by the ITU-R (International Telecommunication Union, Radio Sector).

CCIR-601 A recommendation developed by the CCIR for digitizing the component color video signal. It defines the sample rate, horizontal resolution, and filters to limit the bandwidth of the signal.

CCIR-656 A recommendation developed by the CCIR that defines the physical and electrical interface for the exchange of digital video according to CCIR-601. It describes both a parallel and serial interface, but only the parallel interface has been accepted widely. The parallel interface is described in SMPTE 125M. An interface described in SMPTE 259M has largely replaced the serial interface.

Center frequency As applied to frequency modulation, it is the frequency of the unmodulated carrier. With modulation, the instantaneous frequency swings above and below the center frequency.

Centering controls The controls used to move an image in vertical and horizontal directions to properly center it on a television screen.

Channel 1. A band of frequencies assigned to a station for the transmission of television and sound signals.
2. A digital medium that stores or transports a digital television stream.

Chip A single substrate on which all the active and passive elements of an electronic circuit have been fabricated using the semiconductor technologies of diffusion, passivation, masking, photo resist, and epitaxial growth. A chip is not ready for use until it is packaged and provided with terminals for connection to the outside world.

Chroma A shortened form of chrominance, which is the color part of a video signal. It expresses the hue and saturation of a color but not the luminance or brightness. In a component system, the R-Y, B-Y, or Cr and Cb signals carry the chroma information. In a composite system, such as PAL, SECAM, or NTSC, the chroma information is encoded.

Chrominance See *Chroma.*

Clear To restore a memory or storage device, counter, etc., to a standard state, usually the 0 state. Also called *reset.*

Clipper A vacuum-tube circuit that removes a portion of a signal above or below a fixed-amplitude level. In television, it refers to the stage that separates the video and sync signals.

Clipping level The amplitude level of a signal above or below the part of a signal that a clipper removes.

Clock A pulse generator that controls the timing of switching circuits and memory states and determines the speed at which the major portion of the computer or timing circuits operate. (The pulse generated is referred to as the *clock pulse,* sometimes shortened to *clock.)*

Clocked RS flip-flop The clocked RS flip-flop has two conditioning inputs that control the state to which the flip-flop will go at the arrival of the clock pulse. If the S (set) input is enabled, the flip-flop goes to the 1 condition when clocked. If the R (reset) input is enabled, the flip-flop goes to the 0 condition when checked. The clock pulse is required to change the state of the flip-flop.

Clock jitter A variation of digital signal transitions from their ideal position in time. It is the instantaneous difference in phase of a signal to a stable and jitter-free primary clock.

Coaxial cable A type of conductor that efficiently transmits a wide range of frequencies. Such a cable, in its simplest form, consists of a hollow metallic conductor with a single wire accurately centered along the axis of the hollow conductor and held in position by a suitable insulating material.

Coded representation A data element as represented in its encoded form.

Color bars A test pattern used to properly calibrate a video system. Brightness, hue, and saturation of a signal can be verified with this pattern and proper test equipment.

Color burst An 8- to 10-cycle reference of the 3.58-MHz color subcarrier that occurs right after the horizontal sync pulse. It is used as the reference for phase comparison to keep the color oscillator locked to the broadcast signal. Also, it references the ACC and ACK circuits.

Color subcarrier A 3.579545-MHz signal that is modulated to produce the color sidebands that carry the color information. It is suppressed during transmission.

Comb filter An electronic circuit with a pass response that resembles a comb. It separates the video luminance from the color based on their phase relationship.

Comparator A device used to determine if 2 bits of information are in the same state (both 0 or both 1).

Complement The complement of a variable or function is the binary opposite of that variable or function. If a variable or function is 1, the complement will be 0. If a variable or function is 0, the complement will be 1. The complement of 011010 is 100101.

Component analog video (CAV) A video format in which separate video signals represent the luminance and chrominance information. The signals are made up of varying analog voltages that represent the picture content. The signals can represent red, green, and blue (RGB) output from an image-producing device, or they may be transformed into a luminance signal and chrominance signal or signals, as in Y, R-Y, B-Y.

Component digital video A digital video signal consisting of signal components that are sampled and transmitted in separate channels. It is generally the digitized version of the analog signal of a Y, R-Y, or B-Y component but also can be a digitized version of RGB.

Composite analog video An analog video format containing a mixture of luminance, chrominance, sync, and color burst that have been combined by one of the encoding-process standards: NTSC, PAL, or SECAM.

Composite digital video A digital video signal that results from digitizing a composite analog video signal. The composite analog signal usually is sampled at four times the frequency of the color subcarrier for the standard being used. Standards have been defined for both PAL and NTSC composite digital videos.

Composite sync Signal that contains vertical and horizontal sync information. Also called *V&H sync.*

Composite sync signal The portion of a television signal that consists of the horizontal, vertical, and equalizing pulses.

Composite television signal The television signal composed of the video information and the synchronizing and blanking pulses.

Composite video Signal consisting of luminance, chroma, color burst, composite sync pulses, and blanking pulses. The luminance and chroma signals are interleaved.

Compression Reduction in the number of bits used to represent an item of data.

Constant bit rate Operation where the bit rate is constant from start to finish of the compressed bit stream.

Contrast The difference in brightness between black-and-white portions of a picture. Pictures having high contrast have deep blacks and brilliant whites, while a picture with low contrast has an overall gray appearance.

Contrast control The control that varies the contrast of a television picture by changing the gain of the video stages. It corresponds to the volume control in an aural receiver.

Control electrode The metal structure adjacent to the cathode in a cathode-ray tube to which a voltage is applied to regulate the electron flow. Sometimes called the *grid*, this electrode controls the light intensity of the image on the screen.

Conventional definition television (CDTV) This term is used to signify the *analog* NTSC television system. Also see *Standard definition television*.

Counter A device capable of changing states in a specified sequence on receiving appropriate input signals. Or a circuit that provides an output pulse or other indication after receiving a specified number of input pulses (see specific counters).

Counter, binary A series of flip-flops having a single input. Each time a pulse appears at the input, the flip-flop changes state; sometimes called a *T flip-flop*.

Counter, nonsynchronous (ripple) A number of series-connected flip-flops that divide (counts down) the input into a series output occurring at a lower rate. The actual division depends on the number of flip-flops and their specific connections.

Counter, programmable An integrated circuit containing series-connected flip-flops into which a binary number (within the counter range) can be programmed. The counter then counts the input pulses starting at the programmed number until the counter is full and resets. The cycle then repeats.

Counter, ring A loop or circuit of interconnected flip-flops so arranged that only one is on at any given time. As input signals are received, the position of the on state moves in sequence from one flip-flop to another around the loop.

Counter, shift A number of clocked series-connected flip-flops in which the flip-flops do not change with each clock. Instead, each flip-flop only changes once for each cycle of the counter. That is, in a three-stage shift counter, the flip-flops change only once every three clocks, once every four clocks in a four-stage shift counter, etc.

Counter, synchronous A number of series-connected flip-flops in which the next state of each depends on the current state of the previous and in which all state changes occur simultaneously with a clock pulse. The output (or equivalent count of input pulses) can be taken from the counter in parallel form.

CRC The cyclic redundancy check to verify the correctness of data.

Crossover point The point, between the grid and preaccelerator, of a cathode-ray tube where the electrons are emitted by the cathode coverage.

D1 A recording standard, conforming to the CCIR 601 and 656 standards, for a digital videotape component. It refers to the use of a cassette containing 19-mm-wide magnetic tape. The term *D1* sometimes is used incorrectly to indicate the signal or interface format defined by CCIR 601 and 656.

D2 A recording standard, conforming to SMPTE 244M standards, for a composite digital-video videotape. It refers to the use of a cassette containing 19-mm-wide magnetic tape. Data are recorded with 8-bit resolution. The term *D2* sometimes is used incorrectly to indicate composite digital video.

D3 A recording standard, conforming to SMPTE 244M, for a composite digital-video videotape. It refers to the use of a cassette containing ½-in-wide magnetic tape.

D5 A recording standard, conforming to the CCIR 601 and 656 standards, for a component digital-video videotape. It refers to the use of a cassette containing ½-in-wide magnetic tape.

D/A converter (digital-to-analog converter) The device that converts a digital data stream to an analog signal.

D-frame Frame coded according to an MPEG-1 mode that uses dc coefficients only.

D-type flip-flop A D-type flip-flop propagates whatever information is at the D (data) conditioning input, prior to the clock pulse, to the Q output on the leading edge of a clock pulse. If the input is 1, the Q output becomes 1 on the leading edge of the next clock pulse. If the input is 0, the Q output becomes 0 on the next clock.

Damper diode A fast-switching high-current diode in parallel with the horizontal output transistor. It is forward-biased when the horizontal output transistor is reverse-biased and completes the resonant current path for the flyback primary and deflection yoke.

Damping tube A vacuum tube used in horizontal sweep circuits to prevent transient oscillations.

Data element An item of data as represented before encoding and after decoding.

Dc restorer A circuit used to reinsert the dc component of a video signal that was lost during amplification. The dc component determines the average brightness of the received image.

Dc transmission A system of transmission that retains the dc component of a signal.

DCT A recording standard, conforming to CCIR 601 and 656 standards, for a component digital-video videotape. It is manufactured by AMPEX.

DCTL (direct-coupled transistor logic) Logic that is performed by transistors.

Decimal A system of numerical representation that uses the symbols 0, 1, 3,…9.

Decoded stream The decoded reconstruction of a compressed bit stream.

Decoder A device used to convert information from a coded form into a more usable form, i.e., binary-to-decimal, binary-to-BCD, BCD-to-decimal, etc.

Decoding (process) The process defined in the *Digital Television Standard* that reads an input coded bit stream and outputs decoded pictures or audio samples.

Decoding time stamp (DTS) A field that may be present in a PES packet header that indicates the time that an access unit is decoded in the system target samples.

Definition The ability of a system to reproduce small details in an image.

Deflecting plates Two pairs of metal plates used in an electrostatic cathode-ray tube.

Deflection The moving of the cathode-ray beam by electrostatic or magnetic fields.

Delay Undesirable delay effects are caused by rise and fall times that reduce circuit speed, but intentional delay units may be used to prevent inputs from changing while clock pulses are present. The delay time normally is less than the clock-pulse interval.

Demodulation The process of removing the modulating signal from a modulated radiofrequency carrier.

Demultiplexer A device or circuit that separates two or more signals that previously were combined by a multiplexer and sent over a single channel.

Deserializer A device or circuit that converts serial digital data to parallel form.

Diathermy interference Interference that results from signals generated by diathermy machines operated by doctors or hospitals near a television receiver.

Differentiating circuit A circuit used to separate the high-frequency horizontal sync pulses from the low-frequency vertical sync pulses.

Digit A digit is one character in a number. There are 10 digits in the decimal number system. There are 2 digits in the binary number system.

Digital circuit A circuit that operates like a switch; it is either on or off.

Digital storage media (DSM) A digital storage or transmission device or system.

Digital word A group of bits that compose a unit for the purpose of signal treatment in a system. Each value of a sample in a digital video system is expressed by 8 or 10 bits, which make up a digital word for that sample.

DIP Digital inline pin.

Diplexer A coupling unit that allows two transmitters to operate simultaneously or separately from the same antenna.

Dipole A simple antenna, the total length of which is equal to one-half the wavelength of the frequency for which it is tuned.

Direct-view receiver A television receiver in which the image is viewed on the face of the picture tube.

Directional antenna An antenna designed to receive radio signals better from some directions than from others.

Director A rod slightly shorter than a dipole that is placed in front of the dipole to provide greater direction.

Discharge tube A tube used in sawtooth-generating circuits to discharge a capacitor.

Discrete Electronic circuits built of separate finished components, such as resistors, capacitors, transistors, etc.

Discrete cosine transform (DCT) A mathematical transform that can be perfectly undone and which is useful in image compression.

Discriminator A circuit used in FM receivers to convert the frequency-modulated signal into an audiofrequency signal.

Dissector tube A pickup tube containing a continuous photosensitive cathode on which an electron image is formed.

Dissolve A camera technique whereby two images from different cameras are momentarily overlapped and then one is gradually faded.

Divider, programmable See *Counter, programmable.*

Double-sideband transmission A system of transmission in which the sum and difference frequencies of the modulating and carrier signals are transmitted.

DSM-CC Digital storage media command and control.

DS1 A telephone format for digital transmission. It has the capacity to transmit 24 voice circuits simultaneously. It operates at a speed of 1.544 Mb/s.

DS3 A telephone format for digital transmission. It has the capacity to transmit 672 voice circuits simultaneously. It operates at a speed of 44.735 Mb/s. DS3 circuits currently are used to send compressed digital video over telephone lines.

DTL (diode transistor logic) Logic performed by diodes. The transistor acts as an amplifier, and the output is inverted.

DVCR Digital video cassette recorder.

EAV (end of active video) The timing reference signal (see *TRS*) that indicates that the active (picture) area of a video line has ended and the time to begin the retrace process has begun.

EBU (European Broadcast Union) An organization of European broadcasters that standardizes technical documents relating to European television systems. These documents usually concern 625-line, 50-field-per-second systems and their associated encoded formats, PAL and SECAM.

EDH (error detection and handling) A check system for detecting and indicating bit errors in a digital video system. It is standardized in SMPTE RP-165.

Editing A process by which one or more compressed bit streams are manipulated to produce a new compressed bit stream. Conforming edited bit streams are understood to meet the requirements defined in the *Digital Television Standard.*

E-E mode (electronic-to-electronic mode) An equipment mode where the incoming signal passes through the equipment for some processing but returns to the output mostly unmodified. In a VTR, the input signal is not recorded but routed directly to the video output of the recorder. Similarly, in some pattern generators, an incoming signal may be routed to the output of the generator in place of a standard pattern.

EEPROM Electrically erasable programmable read-only memory.

Electromagnetic deflection coil A current-carrying coil placed over the neck of a cathode-ray tube. The resulting magnetic field deflects the electron beam. Two sets of coils, the vertical and horizontal, are combined into one case, called a *yoke.*

Electron beam The stream of electrons in a cathode-ray tube. The stream is focused to a sharp point on the fluorescent screen of the tube.

Electron emission The releasing of electrons by the surface of an electrode, usually due to heat.

Electron gun That part of a cathode-ray tube in which the electrons are emitted and focused into a beam.

Electron ions The electromagnetic or electrostatic fields in a cathode-ray tube that cause the electrons to converge into a narrow beam.

Electronic scanning The deflection of an electron beam by means of electromagnetic or electrostatic fields.

Electrostatic focusing The process by which electrons are confined into a thin stream by an electrostatic field.

Elementary stream (ES) A generic term for one of coded video, coded audio, or other coded bit streams. One elementary stream is carried in a sequence of PES packets with one and only one stream.

Elementary stream clock reference (ESCR) A time stamp in the PES stream from which decoders of PES streams may derive timing.

EMM See *Entitlement management message.*

Enable A gate is enabled if the input conditions result in a specific output. The specific output varies for different gating functions. For instance, an AND gate is enabled when its output is the same level as its inputs, whereas a NAND gate is enabled when its output is the complement of its inputs. In some cases, function-enabling inputs allow operation to be executed on a clock pulse after the inputs are enabled with the correct logic level.

Encoder In video, a device that combines the luminance, chrominance, sync, and color burst signals into one composite video signal.

Encoding (process) A process that reads a stream of input pictures or audio samples and produces a valid coded bit stream as defined in the *Digital Television Standard.*

Entitlement control message (ECM) Entitlement control messages are private conditional access information that specifies control words and possibly other stream-specific, scrambling, and/or control parameters.

Entitlement management message (EMM) Entitlement management messages are private conditional access information that specifies the authorization level or the services of specific decoders. Such messages may be addressed to single decoders or groups of decoders.

Entropy coding Variable-length lossless coding of the digital representation of a signal to reduce redundancy.

Entry point Refers to a point in a coded bit stream after which a decoder can become properly initialized and commence syntactically correct decoding. The first transmitted picture after an entry point is either an I-picture or a P-picture. If the first transmitted picture is not an I-picture, the decoder may produce one or more pictures during acquisition.

Equalizing pulses 1. A series of six pulses occurring at twice the horizontal frequency. The equalizing pulses precede and follow the vertical sync pulse and are used to maintain proper interlace.
2. Two groups of pulses, one occurring before the serrated vertical sync pulse and the other occurring after. These pulses occur at twice the normal horizontal line rate. They were implemented to ensure correct interlace operation in early televisions.

Erasable programmable read-only memory (EPROM) Similar to PROM, except that the previous information can be erased and new information can be written in.

Error correction A system for adding data to a digital signal to allow transmission errors to be detected and corrected.

ES See *Elementary stream.*

ESCR See *Elementary stream clock reference.*

Event A collection of elementary streams with a common time base, an associated start time, and an associated end time.

Exclusive OR The output is true only when the two inputs are opposites (complementary) and is false if both inputs are the same.

Exponent of a number The number of times the base number is to be used as a factor.

Fall time A measure of the time required for a circuit to change its output from a high level to a low level.

Fan-in The number of inputs available on a gate.

Fan-out The number of gates that a given gate can drive. The term is applicable only within a given logic family.

Fidelity The ability of a circuit to reproduce faithfully those signals impressed on it.

Field 1. One-half of a television picture or frame. It is composed of one complete scan of the image. It is made up of $262\frac{1}{2}$ lines in the 525-line system. The lines of field 1 are interlaced with field 2 to make up a frame or complete picture.
2. For an interlaced video signal, a *field* is the assembly of alternate lines of a frame. Therefore, an interlaced frame is composed of two fields, a top field and a bottom field.

Field frequency The repetition rate of the field, which, in current systems, is 60 per second, or twice the frame frequency.

Field pickup The televising of remote events by mobile camera and transmitting equipment.

Film pickup The televising of motion picture films.

Fine-tuning control A control on the receiver that varies the frequency of the local oscillator over a small range to compensate for drift and permit fine adjustment to the carrier frequency of a station.

Flicker Objectionable low-frequency variation in intensity of illumination of a television picture.

Flip-flop An electronic circuit having two stable states and the capability to change from one state to the other on the application of a signal in a specified manner. Specific types follow.

Flip-flop, D A flip-flop with output from the input that appeared one pulse earlier. If a 1 appears at its input, the output one pulse later will be 1. Sometimes it is used to produce a one-clock delay. (D stands for data.)

Flip-flop, JK A flip-flop having two inputs designated J and K. At the application of a clock pulse, a 1 on the J input will set the flip-flop to the 1 or on state and 1s simultaneously on both inputs will cause it to change state regardless of what state it has been in. If 0s appear simultaneously on both inputs, the flip-flop state remains unchanged.

Flip-flop, RS A flip-flop having two inputs designated R and S. The application of a 1 on the S input will set the flip-flop to the 1 or on state and 1 on the R input will reset it to the 0 or off state. It is assumed that 1s will never appear simultaneously at both inputs. (In actual practice, the circuit can be designed so that a 0 is required at the S and R inputs.)

Flip-flop, synchronized RS A synchronized RS flip-flop having three inputs, R, S, and clock (stroke, enable, etc.). The R and S inputs produce states as described for the RS flip-flop. The clock causes the flip-flop to change states.

Flip-flop, T A flip-flop having only one input. A pulse appearing on the input will cause the flip-flop to change states. A series of these flip-flops makes up a binary ripple counter.

Fluorescent screen The chemical coating on the inside face of a cathode-ray tube that emits light when struck by electrons.

Flyback An abbreviated reference to either flyback transformer or flyback time.

Flyback time 1. Time when the horizontal output transistor is turned off. The magnetic field in the output stage suddenly collapses to produce a large pulse at the collector of the horizontal output transistor and return the electron beam to the left side of the screen.

2. The period during which the electron beam is returning from the end of a scanning line to begin the next line.

Flyback transformer A specially designed transformer that operates at the horizontal frequency. The primary is driven by the pulses that occur at the horizontal output transistor during flyback. It has several secondary windings that provide scan-derived voltages.

Flywheel synchronization Another term for *automatic frequency control* of a scanning circuit. In such a system, the sweep oscillator responds to the average timing of the sync pulses and not to each individual pulse.

Focus In a cathode-ray tube, this refers to the size of the spot of light on the fluorescent screen. The tube is said to be focused when the spot is smallest. This term also refers to the optical focusing of camera lenses.

Focusing control The potentiometer control on the receiver that varies the first anode voltage of an electrostatic tube (or the focus-coil current of a magnetic tube) and focuses the electron beam.

Focusing electrode A metal cylinder in the electron gun, sometimes called the *first anode*. The electrostatic field produced by this electrode, in combination with the control electrode and the accelerating electrode, acts to focus the electron beam in a small spot on the screen.

Forbidden This term, when used in clauses defining coded bit stream, indicates that the value shall never be used. This is usually to avoid emulation of start codes.

FPLL Frequency and phase-locked loop.

Frame A complete television picture, consisting of two fields or all 525 lines in the system of the United States. In the NTSC system, 29.97 frames are scanned each second. PAL and SECAM systems scan 25 frames per second.

Frame A frame contains lines of spatial information of a video signal. For progressive video, these lines contain samples starting from one time instant and continuing through successive lines to the bottom of the frame. For interlaced video, a frame consists of two fields, a top field and a bottom field. One of these fields will commence one field later than the other.

Frame frequency The number of times per second the picture area is completely scanned. This frequency is 30 times per second in the current television system.

Frequency modulation A system in which the frequency of a radio signal is varied in proportion to the modulating signal in order to transmit intelligence.

Gate A circuit having two or more inputs and one output, the output depending on the combination of logic signals at the inputs. There are five gates, called AND, exclusive OR, OR, NAND, and NOR. The following gate definitions assume that positive logic is used.

Gate, AND All inputs must have 1-state signals to produce a 1-state output. (In actual practice, some gates require 0-state inputs to produce a 0-state output.)

Gate, exclusive OR The output is true only when the two inputs are opposites (complementary) and is false if both inputs are the same.

Gate, NAND All inputs must have 1-state signals to produce a 0-state output. (In actual practice, some gates require 0-state inputs to produce a 1-state output.)

Gate, NOR Any one or more inputs having a 1-state signal will yield a 0-state output. (In actual practice, some gates require 0-state inputs to produce a 1-state output.)

Gate, OR Any one or more inputs having a 1-state signal will yield a 1-state output. (In actual practice, some gates require 0-state inputs to produce a 0-state output.)

Genlock A process whereby a signal is phase locked to a master sync source. Most professional video-generation sources, VTRs, pattern generators, CGs, etc., can be genlocked to the house sync source.

Ghost A secondary picture formed on a television receiver by a signal from the transmitter that has reached the antenna by a longer path. Ghosts usually are caused by reflected signals.

Group of pictures (GOP) A group of pictures consists of one or more pictures in sequence.

Halation The ring of illumination that surrounds the point at which the electron beam strikes the fluorescent screen.

Half-Add The half-add is performed first in doing a two-step binary addition. It adds corresponding bits in two binary numbers, ignoring any carry information.

HDTV (high-definition television) A system of television having significantly more scan lines per image than current systems for increased resolution. In the United States, such a proposed system would have 1080 active vertical lines with 1920 horizontal samples per line and would transmit a component digital-video signal. High-definition television has a resolution of approximately twice that of conventional television in both the horizontal (H) and vertical (V) dimensions and a picture aspect ratio (H×V) of 16:9. Other recommendations defines HDTV quality as the delivery of a television picture that is subjectively identical to the interlaced HDTV studio standard.

Height The vertical dimension of a television image.

Height control The control that varies the vertical size of a television picture.

Heterodyne frequency A frequency that is produced by combining two other frequencies and is the numerical sum or difference of these frequencies.

High level A range of allowed picture parameters defined by the MPEG-2 video coding specifications that corresponds to high-definition television.

High-level AND gate An AND gate that is activated when all inputs have 1-state output.

High-level NAND gate A NAND gate that is activated when all inputs have 1-state signals applied to produce a 0-state output.

High voltage The accelerating potential used to increase the velocity of the electrons in the beam of a cathode-ray tube.

High-voltage multiplier A component that receives high-amplitude pulses from the flyback transformer and rectifies and multiplies them to produce the CRT high voltage and focus voltages. Common multipliers include doublers, triplers, and quadruplers.

Hold control A potentiometer, in either a vertical or horizontal sweep oscillator circuit, that varies the natural frequency of the oscillator and enables it to synchronize with applied sync pulses.

Horizontal The direction of sweep of the electron beam from left to right.

Horizontal blanking The blanking pulse that occurs at the end of each horizontal line and cuts off the electron beam while it is returning to the left side of the screen.

Horizontal centering control The potentiometer used to move the picture in a horizontal direction.

Horizontal frequency The rate at which the electron beam makes one scanning cycle across the screen from left to right and back.

Horizontal hold control A control used to vary the natural frequency of the horizontal sweep oscillator so that it locks with the applied sync pulses.

Horizontal interval The time allotted for the electron beam in the scanning device to return to the left edge of the image after a complete video line has been scanned. It is initiated by the leading edge of the horizontal sync pulse and ends when active video information begins after color burst.

Horizontal output transistor (HOT) A power transistor that switches current from the B+ supply into the horizontal output stage through the flyback transformer primary.

Horizontal resolution The ability of a television system to reproduce small objects in the horizontal plane.

Horizontal retrace 1. The time during which the electron beam returns from the right side of the CRT to the left side.
2. A line on the screen that is formed by the electron beam during the time the spot is returning from the right to the left side of the screen.

Huffman coding A type of source coding that uses codes of different lengths to represent symbols that have unequal likelihood of occurrence.

Iconoscope A camera tube in which a high-velocity electron beam scans a photosensitive mosaic that stores an electrical image.

IEC International Electrotechnical Commission.

IHVT (integrated high-voltage transformer) A component that functions as both a flyback transformer and a high-voltage multiplier. Some IHVTs also provide screen voltage.

Image dissector A television camera tube in which the photoelectrons are moved past a pickup aperture by deflection circuits. See also *Dissector tube*.

Image orthicon A highly sensitive camera tube that combines the principles of the image dissector, orthicon, and image multiplier.

Integrated circuit (IC) 1. The Electronics Industries Association defines semiconductor integrated circuit as "the physical realization of a number of electrical elements inseparably associated on or within a continuous-body semiconductor material to perform the functions of a circuit."
2. A circuit that combines the vertical pulses into a single composite pulse.

Intensifier electrode Otherwise known as the *third anode*. It imparts additional kinetic energy to the electron beam after it has been deflected.

Interface A video scanning system in which the odd- and even-numbered lines of an image are transmitted separately as two interlaced fields. In other words, lines 1, 3, 5, etc., are scanned until the bottom of the image is reached; then 2, 4, 6, etc., which fill in the skipped lines, are scanned and sent as the second field.

Interference Spurious signals that enter a receiver and mar the picture or sound.

Interlace NTSC video scanning method with the raster having two fields of horizontal scan lines. The odd lines are scanned in the first field, and the even lines are scanned in the second field.

Interlaced scanning A system of scanning whereby the odd- and even-numbered lines of a picture are transmitted consecutively as two separate fields that are superimposed to create one frame or complete picture at the receiver. The effect is to double the apparent number of pictures, thus reducing the amount of flicker.

Interleave The process that allows the luminance and chroma signals to occupy the same frequency spectrum in the NTSC video signal. This is a result of the luminance and chroma information being phase locked to different multiples of the horizontal scan frequency.

Intermediate frequency (i.f.) The frequency resulting from the combination of two frequencies in one circuit.

Intracoded pictures or I-frames or I-pictures Pictures that are coded using information present only in the picture itself and not depending on information from other pictures. I-pictures provide a mechanism for random access into the compressed video data. I-pictures employ transform coding of the pel blocks and provide only moderate compression.

Inverter The output is always in the opposite logic state of the input. Also called a *NOT circuit.*

Ion A particle carrying an electric charge. Ions may be positive or negative.

Ion spot A discoloration at the center of the screen of a picture tube due to bombardment of negative ions on the fluorescent material.

Ion trap An electron gun structure and magnetic field that permit electrons to flow toward the screen but divert negative ions, thereby avoiding the formation of an ion spot.

IRE (Institute of Radio Engineers) A unit of measurement for video levels derived by dividing the range from the bottom of the sync pulse to the peak white (1 V peak to peak) into 140 equal units. The actual video information has an amplitude of 100 IRE; the sync pulse, 40 IRE.

ISO International Organization for Standardization.

ITU-R (International Telecommunication Union, Radio Sector) The international standards body who has replaced the CCIR. See *CCIR.*

Jaggies Common name for a stairstep effect on diagonal lines in an image. This effect sometimes is called *aliasing.*

JEC Joint Engineering Committee of EIA and NCTA.

Jitter The tendency of several lines or the entire picture to vibrate because of poor synchronization.

Jitter, digital A variation of digital signal transitions from their ideal position in time. It is the instantaneous difference in phase of a signal from a stable and jitter-free primary clock.

Keystone effect Distortion of a television image that results in a keystone-shaped pattern.

Kickback supply A high-voltage power supply that derives its energy from the pulses occurring in the primary of the horizontal sweep output transformer when the magnetic field collapses during the retrace period.

Layer One of the levels in the data hierarchy of the video and system specifications.

Leading edge The leading edge of a pulse is defined as that edge or transition that occurs first (i.e., the leading edge of a high pulse is the low-to-high transition).

Lens turret A part of a television camera on which several lenses are mounted for rapid selection.

Level A range of allowed picture parameters and combinations of picture parameters.

Limiter The last i.f. stage in FM audio circuits. This state is so biased that it removes amplitude variations above a given level.

Line Short for horizontal scan line. This is one complete unit of video information produced by scanning one horizontal line. It includes the active video information and horizontal sync information. In the United States, there are 15,734 lines per second.

Line-scanning frequency The number of lines scanned each second. In any system, it is equal to the number of scanning lines in each frame multiplied by the frame frequency. Under current standards, this is 525 lines times 30 frames per second, or 15,570 lines per second.

Linearity The relative spacing of picture elements in the television image.

Linearity control A potentiometer in a vertical or horizontal sweep circuit that is used to adjust the spacing and distribution of the picture elements.

Local oscillator The heterodyne oscillator in a superheterodyne receiver.

Lock-in A term describing the condition that exists when a sweep oscillator is synchronized with the applied sync pulses.

Logic A mathematical approach to the solution of complex situations using symbols to define basic concepts. The three basic logic symbols are AND, OR, and NOT. When used in boolean algebra, these symbols are somewhat analogous to addition and multiplication.

Low-level AND gate An AND gate that is activated when all inputs have 0-state signals applied to produce a 0-state output.

Low-level NAND gate A NAND gate that is activated when all inputs have 0-state signals applied to produce a 1-state output.

Low-level NOR gate A NOR gate that is activated when one or more inputs have a 0-state signal applied to produce a 1-state output.

Luminance The black-and-white only portion of the video signal, also called the Y signal. It produces the bright and dark portions of the picture.

Macroblock In the advanced television system, a macroblock consists of four blocks of luminance and one each of Cr and Cb blocks.

Magnetic focus The focusing of the electron beam by means of a magnetic field from a coil placed over the neck of the cathode-ray tube.

Main level A range of allowed picture parameters defined by the MPEG-2 video coding specifications with maximum resolution equivalents.

Main profile A subset of the syntax of the MPEG-2 video coding specifications that is expected to be supported over a large range of applications.

Megacycle One million cycles.

Microsecond One-millionth of a second.

Mixing amplifier An amplifier that combines several signals of different amplitudes and wave shapes into a composite signal. Such an amplifier is used to mix the blanking and sync pulses in the sync generator.

Modulation The variation of the amplitude, phase, or frequency of a radio carrier frequency by a lower-frequency signal.

Modulation grid An electrode, interposed between the cathode and focusing electrodes in a cathode-ray tube, that controls the amount of emission, thereby the brilliance of the spot. This controlling effect is produced by altering the voltage of the grid with respect to the cathode.

Monitor A video-display device that accepts a composite video and/or Y/C signal. It often does not contain a tuner.

Monitor, television test A cathode-ray tube, and associated circuits, used in a television station to check the transmitted picture.

Monoscope A cathode-ray tube that produces a stationary pattern for the testing and adjusting of television equipment.

Mosaic The photosensitive plate in the iconoscope that emits electrons when struck by light.

Motherboard The circuit board on which the CPU, RAM, ROM, and other chips are connected.

Motion vector A pair of numbers that represents the vertical and horizontal displacement of a region of a reference picture for prediction.

Mouse A small, input device that you roll on your desktop to move the cursor (or pointer) on a computer screen and then activate to send commands to the CPU.

MP@HL Main profile at high level.

MP@ML Main profile at main level.

MPEG Refers to standards developed by the ISO/JTC1/SC29 WG11, Moving Picture Experts Group. MPEG also may refer to the group.

MPEG-1 Refers to ISO/IEC standards.

MPEG-2 Refers to ISO/IEC standards.

Multipath reception The condition in which the radio signal from the transmitter travels by more than one route to a receiver antenna, usually because of reflections from obstacles, resulting in ghosts in the picture.

Multivibrator A type of oscillator, using R-C components, commonly used to generate the sawtooth voltages in television receiver circuits.

Negative ghosts Ghosts that appear on the screen with intensity variations opposite to those of the picture.

Negative logic The reverse of positive logic; the more negative voltage represents the 1-state, and the less negative voltage represents the 0-state.

Negative transmission The modulation of the picture carrier by a picture signal, the polarity of which is such that the sync pulses occur in the blacker-than-black level.

Noise Specious impulses that modulate the picture or sound signals.

Noninterlace A scanning method consisting of two fields of horizontal scan lines. The lines from field two are scanned directly on top of the lines from field one.

Nonlinearity The unequal distribution of picture elements in the vertical and/or horizontal direction.

NOT A boolean logic operator indicating negation. A variable designated NOT will be the opposite of its AND or OR functions. A switching function for only one variable.

NRZI (nonreturn to zero inverted) A digital coding scheme that pseudorandomly scrambles the data to render the data stream insensitive to polarity. A change in logic state (from high to low, for example) represents a digital 1, while no change in state represents 0.

NTSC (National Television Systems Committee) The name of the organization that formulated the standards for the color television system in the United States. NTSC has come to denote the television standard. The system employs a 3.578565-MHz subcarrier, with phase that varies with the instantaneous hue of the televised color and amplitude that varies with the instantaneous saturation of the color. The system operates with 525 lines and 9.94-fields per second.

Nyquist frequency The lowest frequency that can be used to sample a waveform without significant aliasing when converting an analog waveform to digital. This frequency usually is considered to be twice the highest frequency to be sampled.

Octal The octal number system is one that has eight distinct digits: 0, 1, 2, 3, 4, 5, 6, and 7.

Odd-line interlace A type of interlace system, such as is now used, in which there are an odd number of lines in each frame.

Open-wire transmission line A transmission line formed by two parallel-spaced wires. The distance between the two wires and their diameters determine the surge impedance of the transmission line.

Orthicon A camera tube in which a low-velocity electron beam scans a photosensitive mosaic.

Orthogonal sampling A sampling method performed in phase with the horizontal line such that the same sample number is taken at the same position in each line. This produces sample positions that are aligned vertically in the image.

Oscillograph An indicating instrument consisting of a cathode-ray tube and a sweep generator for plotting an alternating voltage against time.

OSD On-line systems driver.

Overcoupled circuit A tuned circuit in which the coupling is greater than critical coupling, resulting in a broadband response characteristic.

Pack A pack consists of a pack header followed by zero or more packets. It is a layer in the system coding syntax.

Packet A packet consists of a header followed by a number of contiguous bytes from an elementary data stream. It is a layer in the system coding syntax.

Packet data Contiguous bytes of data from an elementary data stream present in a packet.

Packet identifier (PID) A unique integer value used to associate elementary streams of a program in a single or multiprogram transport stream.

Padding A method to adjust the average length of an audio frame in time to the duration of the corresponding PVM samples by continuously adding a slot to the audio frame.

Pairing A condition of improper interlacing that exists when the lines in alternate fields are superimposed. The fields may pair intermittently or continuously.

PAL (Phase Alternate Line) The name of the color-television system used in most of Europe. It differs from NTSC in several important ways: It uses 725-lines per frame and 50 frames per second; the frequency of the subcarrier is 4.43361875 MHz; and the color burst is shifted 90 degrees in phase from one line to the next in order to minimize hue errors during transmission.

Panning The movement of the camera head from left to right or up and down.

Parallel operation Pertaining to the manipulation of information within logic circuitry so that digits of a word are transmitted simultaneously on separate lines. Parallel operation is faster than serial operation but requires more circuitry.

Parallel transmission A transmission method that uses a separate physical channel for each bit of the digital word, plus a channel for synchronizing the clock.

Parity Parity is a method by which binary numbers can be checked for accuracy. An extra bit, called a *parity bit*, is added to numbers in systems using parity. If even parity is used, the sum of all 1s in a number and its corresponding parity bit is always even. If odd parity is used, the sum of 1s in a number and its corresponding parity bit is always odd.

Passive elements The components in a circuit that have no gain characteristics: capacitors, resistors, and inductors.

Patch panel A panel containing a quantity of receptacles connected to outputs and inputs of equipment. Signals can be routed by manually connecting short cables (called *patch cords*) between the receptacles.

Payload Payload refers to the bytes that follow the header byte in a packet. For example, the payload of a transport stream packet includes the PES_packet_header and its PES_packet_data_bytes or pointer_field and PSI sections or private data. A PES_packet_payload, however, consists only of PES_packet_data_bytes. The transport stream packet header and adaptation field are not payload.

PCM (pulse-code modulation) A method for representing analog values in digital form. Each digital output code represents a finite range of analog values.

PCR See *Program clock reference*.

Peaking coil A small inductive coil placed in an amplifying circuit in order to increase its response at certain frequencies.

Peaking resistor A resistor placed in series with the charging capacitor of the vertical sawtooth generator in order to add a negative peaking pulse to the sawtooth voltage for creation of the waveform required to produce a linear sawtooth current in the yoke.

Pedestal The portion of the television video signal used to blank out the beam as it flies back from the right to the left side of the screen.

Peripheral interface adapter (PIA) A device that provides menus of interfacing peripheral equipment to a microprocessor unit (MPU). All control signals for the PIA are originated by the MPU, and some are supplied via certain lines (chip select) on the address bus.

PES An abbreviation for packetized elementary stream.

PES packet The data structure used to carry elementary stream data. It consists of a packet header followed by PES packet payload.

PES packet header The leading field in a PES packet up to but not including the PES_packet_data_byte fields where the stream is not a padding stream. In the case of a packet stream, the PES packet header is defined as the leading fields in a PES packet up to but not including the padding_byte fields.

PES stream A PES stream consists of PES packets, all of whose payloads consist of data from a single elementary stream and all of which have the same stream_id.

Phase shift A change in relative timing between two signals of the same frequency. It is expressed in degrees; one entire period equals 360 degrees.

Phosphor A chemical compound that fluoresces when struck by electrons. The screen material of cathode-ray tubes.

Photocell A device containing a photosensitive cathode that emits electrons when exposed to light.

Photoelectric emission The discharge of electrons by a photosensitive material when exposed to light.

Pickup tube A camera tube used to transform a light image into an equivalent electrical signal.

Picture Source-coded or reconstructed image data. A source or reconstructed picture consists of three rectangular matrices representing the luminance and two chrominance signals.

Picture element The smallest portion of an image that can be resolved by the electron beam.

Picture tube The receiving cathode-ray tube.

Pixel (picture element or pel) The smallest visual unit in digital-image systems. There is one pixel for each sample taken in the D/A converter. Thus, in component digital video, there are 720 pixels per line.

Polarization The direction of the electrostatic and electromagnetic fields surrounding an antenna.

Positive logic The more positive voltage represents the 1-state; the less positive voltage represents the 0-state.

Preamplifier An auxiliary amplifier usually located near the camera in order to minimize effects of noise pickup.

Predicted pictures or P-pictures or P-frames Pictures that are coded with respect to the nearest *previous* I- or P-picture. This technique is termed *forward prediction*. P-pictures provide more compression than I-pictures and serve as a reference for future P-pictures or B-pictures. P-pictures can propagate coding errors when P-pictures (or B-pictures) are predicted from prior P-pictures where the prediction id is flawed.

Preemphasis The increasing of the relative amplitude of the higher audio frequencies in order to minimize the effects of noise during transmission.

Presentation time stamp (PTS) A field that may be present in a PES packet header that indicates the time that a presentation unit is presented in the system target decoder.

Presentation unit (PU) A decoder audio access unit or a decoded picture.

Profile A defined subset of the syntax specified in the MPEG-2 video coding specifications.

Program A program is a collection of program elements. Program elements may be elementary streams. Program elements need not have any defined time base; those which do have a common time base are intended for synchronized presentation.

Program clock reference (PCR) A time stamp in the transport stream from which decoder timing is derived.

Program element A generic term for one of the elementary streams or other data streams that may be included in a program.

Program-specific information (PSI) PSI consists of normative data that are necessary for the demultiplexing of transport streams and the successful regeneration of programs.

Programmable read-only memory (PROM) A ROM device into which information can be written by means of special equipment.

Projection receiver A television receiver in which the image is optically enlarged and projected onto a screen.

Projection television A combination of lenses and mirrors that project an enlarged television picture onto a screen.

Propagation delay A measure of the time required for a change in logic level to propagate through a chain of circuit elements.

PSI See *Program-specific information.*

PTS See *Presentation time stamp.*

PU See *Presentation unit.*

Pulse A change in voltage or current of some finite duration and amplitude. The duration is called the *pulse width* or *pulse length*; the magnitude of the change is called the *pulse amplitude* or *pulse height.*

Quantization The process of expressing different analog levels with digital levels. Although the analog values vary continuously, the digital values vary in discrete steps, called the *quantizing levels.*

Quantizer A processing step that intentionally reduces the precision of DCT coefficients.

Radix The number of symbols used in the number system. (For example, the decimal system has a radix of 10.)

Random access The process of beginning to read and decode the coded bit stream at an arbitrary point.

Random access memory (RAM) A static or dynamic memory device into or out of which data can be written from a specific location. The specific RAM location is selected by the address applied via the address bus and control lines. Data are stored in such a manner that each bit of information can be retrieved in the same length of time.

Raster The pattern obtained when the electron beam sweeps across the screen vertically and horizontally without being modulated.

R-C circuit A circuit consisting of a combination of resistors and capacitors. The time constant of such a circuit is the product of the resistance and capacitance.

Read-only memory (ROM) A nonvolatile memory device containing permanent preprogrammed digital data. Data can be read out of this device on the data bus only when the proper location address is applied to the ROM address bus input and control lines. The ROM can be controlled either by a microprocessor or other device capable of generating the proper address. The ROM is preprogrammed at the time of manufacture.

Register A device used to store a certain number of digits within computer circuitry, often one word. Certain registers also may include provisions for shifting, circulating, or other options.

Reserved This term, when used in clauses defining the coded bit stream, indicates that the value may be used in the future for *Digital Television Standard* extensions. Unless otherwise specified within this standard, all reserved bits shall be set to 1.

Reset See *Clear.*

Resistor-capacitor-transistor logic (RCTL) Same as RTL except that capacitors are used to enhance switching speed.

Resistor-transistor logic (RTL) Logic is performed by resistors. The transistor produces an inverted output from any positive input.

Resolution A measure of the capability of a television system to reproduce detail. It is generated by the imaging device, the television, CRT, or the system bandwidth. Generally, an increase in bandwidth results in increased resolution. Resolution in a digital system often refers to the number of levels that the system can express. This is related to the number of bits in the digital word. A digital word can resolve as many levels as the number of bits can express. For example, 4 bits have a resolution of 2^4 (16) different levels, while 10 bits can express 2^{10} (1024) levels. Thus, in expressing levels within the same overall range, longer data words can resolve smaller changes in level, since the overall range is divided into smaller increments.

Resolution chart A pattern of black-and-white lines used to determine the resolution capabilities of equipment.

Retrace time Time during which the electron beam returns to the left side of the screen. Flyback time.

Retrace timing capacitor The capacitor in parallel with the horizontal output transistor that determines the rate at which the flyback current collapses.

Return trace Lines on the cathode-ray tube screen that are formed by the beam as it moves back to its starting position.

RGB (red, green, blue) The three primary colors of light used in color television systems. Some component video systems treat these three signals in parallel. Sometimes called *GBR*, due to the mechanical arrangement of the connectors in the SMPTE interface standard.

Rise time A measure of the time required for a circuit to change its output from a low to a high level.

Safety capacitor Name given to the retrace timing capacitor because of the critical role it plays in controlling the pulse amplitude, which, in turn, determines the high voltage.

Sampling The process of repeatedly measuring the instantaneous levels of an analog signal at regular intervals in order to produce a digital representation of the signal.

Sandcastle Signal used in many luma/chroma circuits that is a combination of (1) horizontal flyback pulse, (2) horizontal blanking pulse, and (3) vertical blanking pulse. It gets its name from its appearance as a "sandcastle."

SAP Systems assurance program.

SAV (sign of active video) The timing reference signal (see *TRS*) that indicates that the active (picture) area of a video line is beginning and that the retrace process has been completed.

Sawtooth A voltage or current, the variation of which, with time, follows a saw-tooth configuration.

Sawtooth voltage A voltage that varies between two values at regular intervals. Since voltage drops faster than it rises, it gives a waveform pattern resembling the teeth of a saw. Used in television to help form the scanning raster.

Scan-derived supply A voltage source from a secondary winding on the flyback transformer that is powered by the normal scanning current in the flyback.

Scanning The process of breaking a picture into elements by means of a moving electron beam.

Scanning line A horizontal line, composed of elements varying in intensity, the width of which is equal to the diameter of the scanning electron beam.

Scanning raster See *Raster.*

Schmidt system An optical system adapted for television projection receivers in which the light from the image is collected by a concave mirror and directed through a correcting lens onto a screen.

Schmitt trigger A fast-acting pulse generator that produces a constant-amplitude pulse as long as the input exceeds a threshold dc value. Used as a pulse shaper, threshold detector, etc.

SCR See *System clock reference.*

Scrambling As related to digital video, the process of rearranging digital data in order to break up data that would produce long sequences without transitions in the digital state, such as normally happens when all the bits are set either to 1 or 0. The alteration of the characteristics of a video, audio, or coded data steam in order to prevent unauthorized reception of the information in a clear form. This alteration is a specified process under the control of a conditional access system.

SDTV See *Standard definition television.*

Second anode The positively charged electrode in the electron gun that accelerates the beam.

Serial digital Digital information that is transmitted in serial form. Often used to refer to serial digital-video signals of either composite or component format.

Serializer A device that converts bit-parallel digital signals to serial digital.

Serial operation Pertaining to the manipulation of information within logic circuitry so that digits of a word are transmitted one at a time along a single line. Although slower than parallel operation, the circuitry is considerably less complex.

Serrated vertical pulse A vertical pulse broken into shorter-duration pulses so that the horizontal oscillator does not fall out of synchronization during the vertical sync interval.

Set To restore a memory or storage device, counter, etc. to a "standard" state, usually the 1-state. Always the complement of clear.

Shading The process of correcting for distorted light distribution in the image by injecting a voltage into the signal.

Shift register An element in the digital family that uses flip-flops to perform a displacement or movement of a set of digits to one or more places, right or left. If the digits are those of a numerical expression, a shift may be the equivalent of multiplying the number by a power of the base.

Side bands The radiofrequencies on each side of the carrier produced by modulation.

Signal An electrical wave.

Slice A series of consecutive macroblocks.

SMPTE (Society of Motion Picture and Television Engineers) This organization of engineers establishes recommended practices and standards for the industry.

SMPTE 125M The document established by SMPTE that defines the standard for bit-parallel component digital video. It combines elements from CCIR 601 and CCIR 656.

SMPTE 244M The SMPTE standard that defines bit-parallel composite digital video.

SMPTE 259M The SMPTE standard that defines bit-serial transmission of both component and composite digital video.

Source stream A single, nonmultiplexed stream of samples before compression coding.

Splicing The concatenation performed on the system level or two different elementary streams. It is understood that the resulting stream level must conform totally to the *Digital Television Standard.*

Spot The point of light produced by the electron beam as it strikes the fluorescent screen.

Stagger tuning The tuning of amplifier stages to slightly different frequencies in order to obtain broadband response.

Standard definition television (SDTV) This term is used to signify a *digital* television system in which the quality is approximately equivalent to that of NTSC. This equivalent quality may be achieved from picture sources at the 4:2:2 level of ITU-R Recommendation 601 and subjected to processing as part of the bit-rate compression. The results should be such that when judged across a representative sample of program material, subjective equivalence with NTSC is achieved. Also called *standard digital television*. See also *Conventional definition television.*

Start codes 32-bit codes embedded in the coded bit stream that are unique. They are used for several purposes, including identifying some of the layers in the coding syntax. Start codes consist of a 24-bit prefix (0×000001) and an 8-bit stream_id.

Station selector The switch or tuning element in the receiver that is used to select the desired television signal.

STD See *System target decoder.*

STD input buffer A first-in, first-out buffer at the input of a system target decoder for storage of compressed data from elementary streams before decoding.

Still picture A coded still picture consists of a video sequence containing exactly one coded picture that is intracoded. This picture has an associated PTS, and the presentation time of succeeding pictures, if any, is later than that of the still picture by at least two picture periods.

Strobe A sampling pulse that is used to enable a register, flip-flop, counter, etc.

Subcarrier This is the carrier that modulates chroma information in a composite color television signal. In NTSC, its frequency is 3.579545 MHz. All other sync signals are phase-locked to the subcarrier.

Summing amplifier circuit A circuit consists of an amplifier (usually an op amp) with a multiple-wired OR input and a single combined output. The output signal is the analog sum of all inputs.

Sweep The uniform motion of the electron beam across the face of a cathode-ray tube.

Sync pulses Pulses transmitted as part of the video signal for the purpose of synchronizing the sweep circuits in the receiver with those in the transmitter.

Sync word A special bit pattern reserved for the synchronizing information in a digital video signal. These special words are used to signal the equipment as to the start and end of blanking times (in component digital video). In addition, they allow the deserializer to correctly locate the word boundaries in the serial data system.

Synchronization The process of maintaining the frequency of one signal in step with that of another.

Synchronizing generator An electronic generator that supplies synchronizing pulses to television studio and transmitter equipment.

Synchronous A transmission standard for digital bits to be "slaved" to an accurate clock at the transmitter and receiver.

System clock reference (SCR) A time stamp in the program stream from which decoder timing is derived.

System header The system header is a data structure that carries information summarizing the system characteristics of the *Digital Television Standard* multiplexed bit stream.

System target decoder (STD) A hypothetical reference model of a decoding process used to describe the semantics of the *Digital Television Standard* multiplexed bit stream.

Tearing An effect observed on the screen when the horizontal synchronization is unstable.

Television channel The group of frequencies allotted to a television station for the transmission of the sound and picture signals.

Test pattern A geometric pattern containing a group of lines and circles used for testing the performance of a receiver or transmitter.

Three-state device Any integrated circuit device in which two of the states are conventional binary (1–0) states and the third state is a high-impedance state. When in the third state, these devices present a high impedance to their respective output lines to reduce power drain and allow access to the common bus lines by other devices.

Timecode A sequential digital signal that is recorded along with the audio and video information on a videotape to indicate a time position for each video frame on the tape. The information typically gives a read-out in hours, minutes, seconds, and frames for each frame of the tape. A coding standard, developed by SMPTE, has been accepted worldwide. The timecode information can be recorded on an unused audio track or other longitudinal track of the videotape.

In the latter, the timecode is called *LTC (longitudinal time code)*. Another method records the timecode during the vertical blanking interval, which is called *VITC (vertical interval time code)*. In digital video recorders, the timecode is recorded as ancillary data during the digital vertical interval.

Time multiplex As applied to component digital video, this is the process of interleaving three parallel-bit streams so that they can be transmitted on one set of physical channels. Specifically, the 8 or 10 bits of the Cb signal are sent first, followed by the Y sample values for the same pixel, and then by the Cr sample. Thus all three sets of values are sent on one channel.

Time stamp A term that indicates the time of a specific action such as the arrival of a byte or the presentation of a presentation unit.

TOV Threshold of visibility.

Trailing edge The trailing edge of a pulse is that edge or transition that occurs last. The trailing edge of an HI clock pulse is the IH-to-LO transition.

Transistor-transistor logic (TTL) A modification of DTL that replaces the diode cluster with a multiple-emitter transistor.

Transport stream packet header The leading fields in a transport stream packet up to and including the continuity_counter field.

TRS (timing reference signal) The collective name for the synchronizing words in component digital video. There are two types: the SAV and the EAV.

TRS-ID (timing reference signal–identification) A special synchronizing word in composite digital video that allows line and field identification for synchronizing purposes and, more important, allows the deserializer to establish the word framing in the serial data stream.

Truncate The process of eliminating the LSBs of a digital word, for example, when converting from a 10-bit word to an 8-bit word. This process must be handled with care in equipment design to avoid introducing distortions into the signal. To round off a number.

Turnstile antenna One or more layers of crossed, horizontal half-wave antennas arranged vertically on a mast, resembling an old-fashioned turnstile. Used in television and other ultra-high frequency systems when a symmetrical radiation pattern is desired.

Universal asynchronous receiver transmitter (UART) A device that provides the data formatting and control to interface serial asynchronous data. Input serial data are converted to parallel data for transfer to the microprocessor or microcomputer via the data bus. Conversely, output parallel data, from the data bus, are converted to serial data in the output.

Variable bit rate Operation where the bit rate varies with time during the decoding of a compressed bit stream.

VBV See *Video buffering verifier.*

Vcc Supply voltage.

Vertical blanking The time after the downward beam movement (field) when the electron beam is turned off.

Vertical blanking pulse A pulse transmitted at the end of each field to cut off the cathode-ray beam while it is returning to the top of the picture for the start of the next field.

Vertical centering control The potentiometer in the vertical positioning circuit that raises or lowers the entire image on the screen.

Vertical frequency The rate at which the electron beam makes one scanning cycle from the top of the screen to the bottom of the screen and back.

Vertical hold control A potentiometer that varies the natural frequency of the vertical sweep oscillator so that it synchronizes with the applied sync pulses.

Vertical resolution A measure of the ability of a system to reproduce fine horizontal lines.

Vertical retrace The time during which the electron beam returns from the bottom of the CRT to the top.

Vertical scanning The motion of the electron beam in the vertical direction.

Vertical side-band transmission A type of transmission in which one side band is suppressed to limit the bandwidth required.

Video Latin, meaning "I see."

Video amplifier A wide-band amplifier for video frequencies. In a television receiver, this term generally refers to the amplifier located after the second detector, the frequency response of which extends from approximately 30 Hz to about 4 MHz.

Video buffering verifier (VBV) A hypothetical decoder that is conceptually connected to the output of an encoder. Its purpose is to provide a constraint on the variability of the data rate that an encoder can produce.

Video frequency The frequency of the signal voltage containing the picture information that arises from the television scanning process. In the current television system, these frequencies are limited to 4 MHz.

Video sequence A video sequence is represented by a sequence header, one or more groups of pictures, and an end_sequence code in the data stream.

Video transmitter The radio transmitter used for transmitting the picture signal.

View finder A term applied to an attachment to a television camera that enables the camera operator to observe the area covered by the camera.

Viewing screen The face of a cathode-ray tube on which the image is produced.

VSB Vestigial sideband.

Waveform The shape of an electrical signal represented in a graphic form. The representation usually shows increasing voltage or current on the *Y* axis and time on the *X* axis. This is most commonly shown on an oscilloscope.

Width control The control in the horizontal sweep circuit that varies the size of the picture in the horizontal direction.

Word The term *word* denotes an assemblage of bits considered as an entity. For example, a 16-bit address word contains 16 bits.

Word framing The process of separating the digital words from an asynchronous serial data stream. In order to correctly extract the digital words, the deserializer must determine correctly where the most significant bit of one word ends and the least significant bit of the next begins. This process is called *word framing.*

Y Abbreviation for luminance signal.

Y/C An input or output connector that provides the luminance (Y) signal and chroma (C) signal separately.

YCrCb A name commonly given to component digital video that conforms to SMPTE 125M. The name is derived from the abbreviations for the luminance and each of the color difference channels.

Yoke A coil of wire specially shaped to deflect the electron beam when a sawtooth current is applied. It consists of a set of vertical windings and a set of horizontal windings. Also, a set of coils placed over the neck of a cathode-ray tube that produces horizontal and vertical deflection of the electron beam when suitable currents are passed through them.

Index

Index

D

E

F

G

H

I

K

L

M

N

O

P

Q

R

S

T

About the Author

Robert L. Goodman is an electronics author, consultant, lecturer, and test instrument designer, with more than 40 years of experience in the electronics service field. He has written more than 50 technical books, published by McGraw-Hill, and more than 150 technical articles for trade magazines such as *Electronics Now*, *Electronics Servicing*, and *Electronics Technician/Dealer*.